PICTORIAL
ASTRONOMY

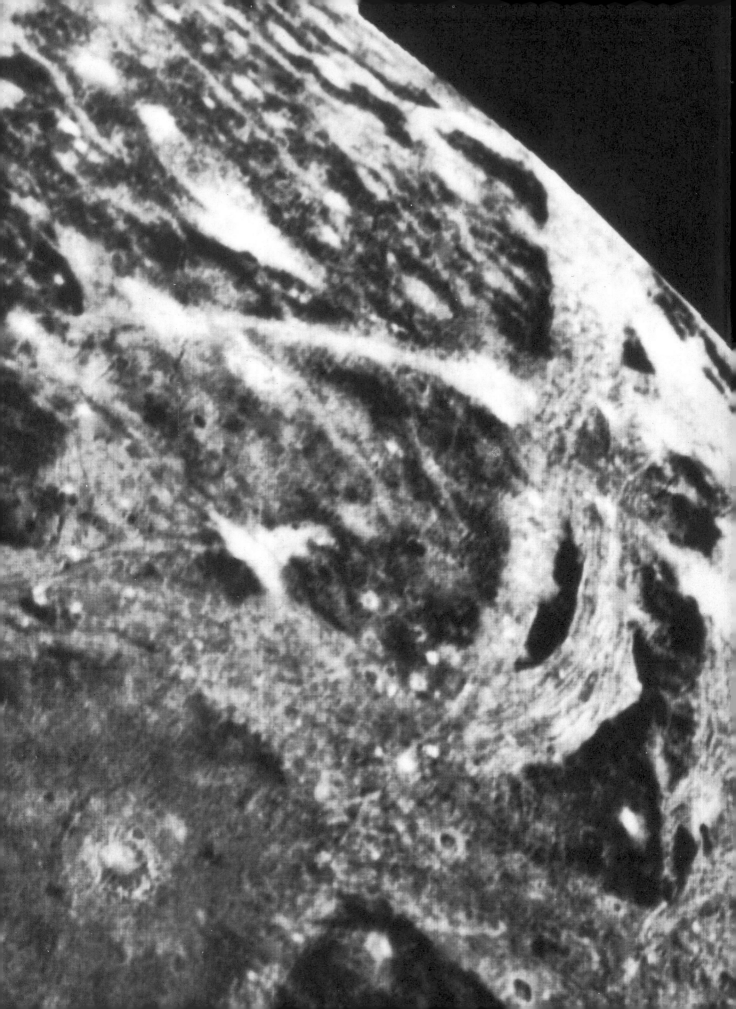

PICTORIAL
ASTRONOMY

Fifth Revised Edition

Dinsmore Alter
Clarence H. Cleminshaw
John G. Phillips

1817

HARPER & ROW, PUBLISHERS, New York

Cambridge, Philadelphia, San Francisco, London
Mexico City, São Paulo, Sydney

ACKNOWLEDGMENTS

Title page: Ganymede, Jupiter's largest satellite, photographed by Voyager 1 on March 5, 1979, from a range of 151,800 miles (253,000 kilometers). (NASA.)

Page viii: Solar corona photographed at the total eclipse of June 8, 1918, Green River, Wyoming. (Mount Wilson and Palomar Observatories.)

Page 71: Series of nine photographs of the moon taken at different phases. The images are inverted from that normally seen by the eye and are presented as they are seen in a telescope. (Lick Observatory.)

Page 76: Map of the moon based on composite photograph, from Lick Observatory photographs.

Page 96: Total eclipse of the sun photographed February 15, 1961, in Romania. (Ewing Galloway.)

Page 178: Ikeya-Seki Comet (2 November 1965). (Smithsonian Astrophysical Observatory.)

PICTORIAL ASTRONOMY (*Fifth Revised Edition*). Copyright © 1983 by Clarence H. Cleminshaw, John G. Phillips, and Helen Alter Johnson. Previous editions copyright © 1974, 1969 by Clarence H. Cleminshaw, John G. Phillips, and Ada Alter; 1963 by Dinsmore Alter, Clarence H. Cleminshaw, and John G. Phillips; 1956, 1952, and 1948 by Dinsmore Alter and Clarence H. Cleminshaw. All rights reserved. Printed in the United States of America. No part of this book may be used or reproduced in any manner whatsoever without written permission except in the case of brief quotations embodied in critical articles and reviews. For information address Harper & Row, Publishers, Inc., 10 East 53rd Street, New York, N.Y. 10022. Published simultaneously in Canada by Fitzhenry & Whiteside Limited, Toronto.

Library of Congress Cataloging in Publication Data

Alter, Dinsmore, 1888-1968.
  Pictorial astronomy.

  Includes index.
  1. Astronomy—Popular works. I. Cleminshaw,
Clarence H. (Clarence Higbee), 1902-   . II. Phillips
John G. (John Gardner), 1917-   . III. Title.
QB44.2.A45   1983   523   81-47878
ISBN 0-06-181019-3   AACR2

83 84 85 86 87 10 9 8 7 6 5 4 3 2 1

Contents

Introduction

Ever since the human race developed the ability to reason, it has speculated about the nature of the physical world in which we find ourselves. Every civilization has developed through its folklore a theory of its evolution and the role of mankind in this physical world. While astronomy may fairly be said to be the most ancient of the sciences, it is, at the same time, one of the most modern. The past few years have witnessed an almost explosive series of discoveries that have revolutionized and expanded our concept of the physical world. Pulsars, quasars, neutron stars, and black holes were unknown only a few years ago; today they are the subject of intensive scientific research. Satellites have visited all of the inner planets of the solar system. The Voyager flybys of Jupiter and Saturn have been of tremendous scientific importance and popular interest. Now that we have demonstrated our ability to survive in outer space, we can look forward in years to come to increasing exploration and exploitation of the resources and environment of space.

The fifth edition of *Pictorial Astronomy* has been extensively rewritten to include all of the most recent discoveries. As in previous editions, we have included a greater percentage of illustrations than is found in the standard textbooks. The general level of detail in the discussion is based on the authors' extensive experience in presenting public lectures, planetarium discussions, "cultural" university courses, and in their work with amateur astronomy societies. The tables of data that carry answers to most of the questions asked by laymen have been revised where necessary. An important feature is the glossary of terms with explanations that are more detailed than usual.

As in earlier editions, the general slant of the book is toward the lay reader or amateur astronomer. It does not pretend to be a text for students who are majoring in the physical sciences; however, numerous colleges and universities have adopted earlier editions for their "cultural" courses. Such use has been kept in mind while making this present revision.

THE AUTHORS

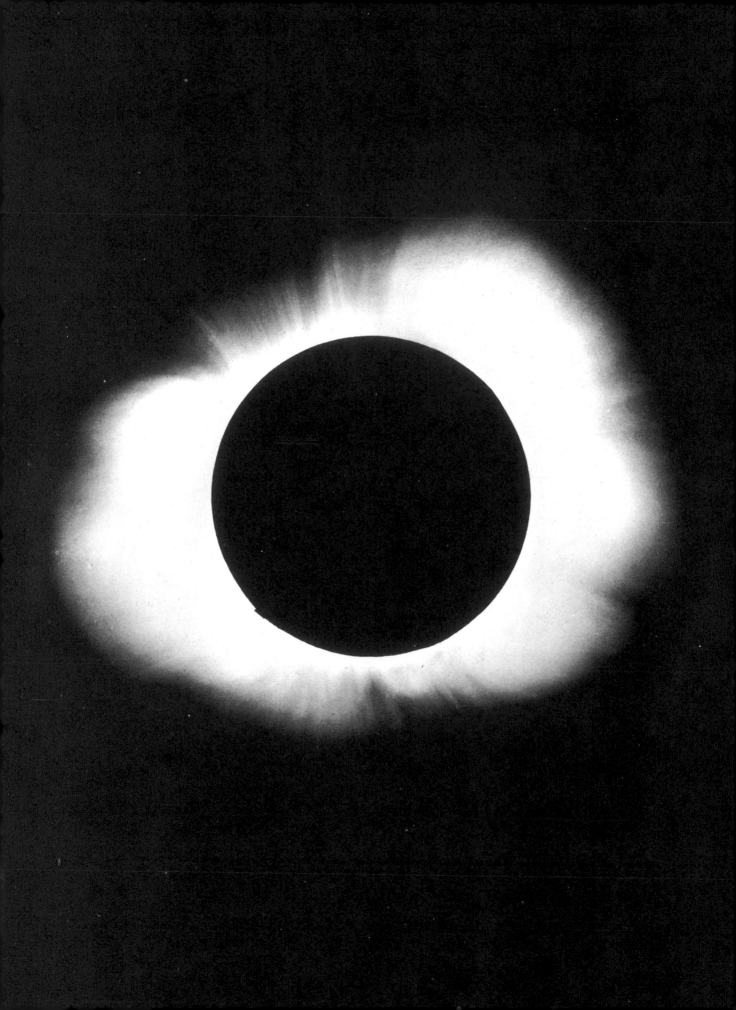

SECTION ONE

The Sun

CHAPTER **1**   **SOLAR PORTRAIT**

Until a few centuries ago, our sun was thought to be a unique body without parallel in all the heavens. To many ancient peoples it was a god. To the later Greeks, it was a huge, glowing rock. A few of them realized that it was even larger than the earth. Today we know that it is only an ordinary star; it is among neither the largest nor the smallest of our galaxy. Indeed, there is at least one star in the large Magellanic Cloud (a cloudlike patch of stars in the southern heavens) which gives roughly 400,000 times as much light as the sun does. On the other hand, it would require about 500,000 stars like the least luminous star known to equal our sun's radiation.

All that we can know about the sun comes from its effects on the planets, and from a study of the photosphere and the region above it. (The photosphere is the luminous envelope of gas surrounding the sun, and it is the "surface" we see.) The density of the gases at the photosphere is only a hundredth of that of our air, but the photosphere appears to us through our telescopes as if it were an opaque liquid sea. Two causes contribute to this opaqueness. One is that we are looking at the photosphere through hundreds of thousands of miles of still thinner gases. These are the sort of gases which produce the familiar blue haze around distant objects in mountainous areas on the earth. The second cause is more technical. The ions which are present in abundance at high temperatures interfere with light waves.

The photosphere is a very thin layer. Only a hundred miles above it, still a negligible distance in the sun, the densities of the gases have decreased to a hundred-thousandth of those in the photosphere—that is, to a ten-billionth the density of water. As we proceed from below the photosphere into the sun, we find that the gases become denser at an ever increasing rate, reaching about 160 times the density of water at its center. There is a similar steady increase in temperature, ranging from the photospheric temperature of 5,480°C to a central temperature over 15,000,000°C. The tenuous gases above the photosphere may be regarded as the sun's "atmosphere." The gases immediately above the photosphere, forming a layer about 300 miles thick, constitute the *reversing layer,* so called because it is responsible for most of the absorption lines that abound in the spectrum of the sun. Careful measurements of the locations of these absorptions in the spectrum make it possible to determine the chemical composition of the solar atmosphere. Above the reversing layer are found the *chromosphere* and *corona* (described later).

Astrophysicists have spent much time building theoretical models of stars that act externally like our sun. They start with the mass and mean density of the sun; its effective temperature and the amount of energy radiated each second; the densities of its outer parts; its rotation; studies of sunspots, prominences, and of calcium and hydrogen clouds; and all other possible observations of the external

1

SOLAR DATA

Diameter

times earth's	109.1
miles	864,600
kilometers	1,391,400

Mass

times earth's	333,500
tons	2.2×10^{27}
kilograms	1.99×10^{30}

Density

mean	1.41
in chromosphere 100 miles above photosphere (estimated)	10^{-9}
at photosphere (estimated)	10^{-7}

Gravitational acceleration at photosphere

times earth's	27.94
feet per sec.^{2}	900

Apparent angular diameter

mean	31′ 59″.3
maximum	32′ 30″
minimum	31′ 28″

Distance from earth (miles)

mean	93,000,000
maximum	94,500,000
minimum	91,500,000

Mean parallax 8″.798

(a recent discussion of the positions of Eros gives 8″.790.)

Effective temperatures

	F	C
average for disk	9,900°	5,480°
center of disk	10,400	5,780
in sunspots (approx.)	7,200	4,000
chromosphere (approx.)	90,000	50,000
outer corona (roughly)	10^6	10^6

Energy output

ergs per second	3.8×10^{33}
horsepower	5×10^{23}

Spectral class G2 (Gee Two)

Solar constant

calories per cm^2 per minute	1.97
ergs per cm^2 per second	1.37×10^6

Magnitude

Absolute	+4.69
Apparent	−26.91

Inclination of solar equator to ecliptic 7° 10′.5

Rotation period at equator (days, hours) 25^d 1^h

Date when closest January 4

Date when farthest July 5

(These dates vary due to leap year, and also progress about one day each 60 years, due to changes of the earth's orbit.)

Sunspot period

mean	11	years
longest observed interval	17.1	,,
shortest observed	7.3	,,

Strength observed in magnetic fields

times earth's field	100 (mean)
	200,000 (maximum)

Elements known on the earth but not yet found on the sun

Atomic Number	Name	Atomic Number	Name
43	Technetium	88	Radium
61	Promethium	89	Actinium
84	Polonium	91	Protoactinium
85	Astatine		All from 93 and higher,
86	Radon		produced artificially
87	Francium		and not known in nature.

parts of the sun. To these they add the laws of physics as determined both in the laboratory and by other means. Then they attempt to manufacture a theoretical ball of gas which represents the sun as we observe and measure it. Several such theoretical models have been constructed that would act, as observed from the earth, very much as does our sun. From these models various estimates for the rate at which temperature, pressure, and density increase as one approaches the center of the sun have been obtained. One model, which is quite well liked, places the temperature at the center of the sun as around 40,000,000° F. Another one places it far higher.

One of the most conspicuous features of the sun is its dark *sunspots*. But even when these are ab-sent, the solar disk is far from being a monotonous expanse. Both direct visual observation and photography reveal the surface as being completely mottled by small lighter and darker areas. These *granules*, once known as "rice grains," are not persistent phenomena but change in just a few minutes. The brighter parts of the granules are about 100° C hotter than their surroundings. The granules are probably hot bubbles of gas penetrating into the solar atmosphere from below the photosphere.

Another noteworthy feature of the sun is that the center of the disk is far brighter than the edge, or *limb*. The diagram shows clearly how absorption of radiation by the higher, cooler part of the sun darkens the limb.

2

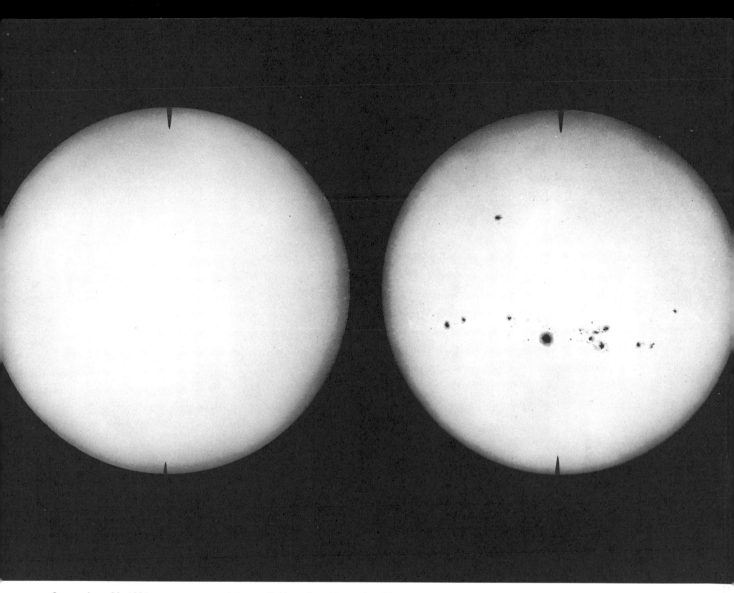

Sun on June 22, 1931, near sunspot minimum (left), and on November 30, 1929, near maximum (right). Black wedges marked on photographs show north and south poles of Sun. (Mount Wilson Observatory.)

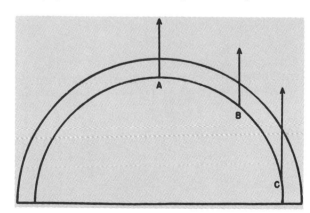

Darkening of solar limb.

Radiation from *C* toward the earth must pass through a longer path of upper gases before it escapes than does radiation from *A*. As a result, the limb appears much less bright than the center of the disk.

Radiation from C on the limb of the sun must travel a far longer path through the solar gases to escape to us than does that from A; therefore, less of it reaches us. It must be remembered, however, that no neat boundaries exist on the sun, like the lines of the diagram. Instead, there is a gradual change in densities above the photosphere.

Careful observation of the solar disk shows the existence of fairly large areas that are brighter than most of the remainder of the sun. These are called *faculae*. They generally occur in regions associated with sunspots before the latter appear, and persist after the sunspots disappear.

Faculae are not to be confused with the clouds of hydrogen and of calcium observed by the spectroheliograph, a specialized form of spectroscope

3

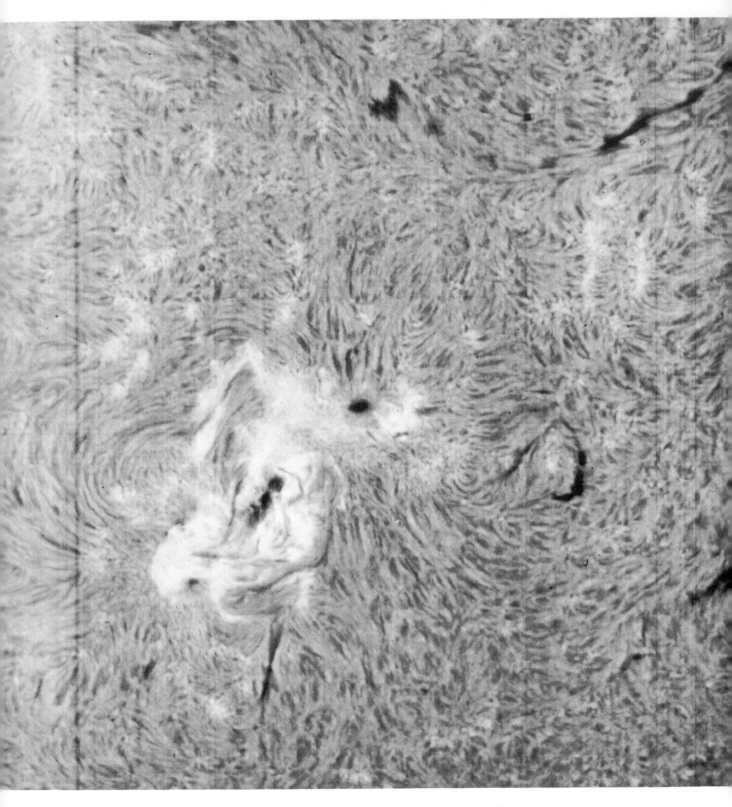

Hydrogen spectroheliogram showing association of hydrogen clouds with sunspot group. Black parts are dark only in comparison with surrounding areas. (Mount Wilson and Palomar Observatories.)

4

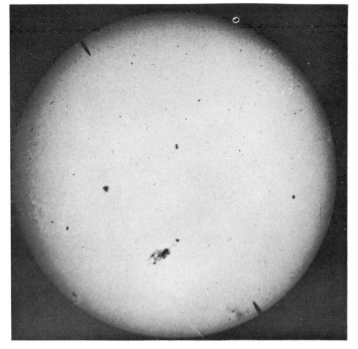 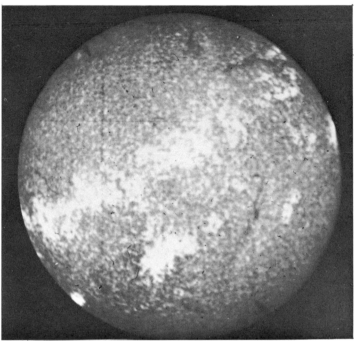

Identical views of sun as recorded in an ordinary photograph (left) and in a spectroheliogram using only light of calcium (right). (Mount Wilson Observatory.)

that permits us to use the light of only one common chemical element at a time. These clouds, named *flocculi* by the American astronomer G. E. Hale, are present constantly and are both bright and dark.

The bright clouds are brighter merely in the sense that an unusual concentration of one element exists in the area; they do not appear bright against the surrounding area until all other light is excluded. Some of them resemble the granules, others the faculae, in general appearance. The darker flocculi are clouds of gas much higher than the photophere and therefore much cooler. Such gases would screen out from the light coming up to them that radiation which they themselves chemically can emit.

Bright calcium flocculi are common in the neighborhood of sunspots, and hydrogen flocculi often float above such spots.

The phenomena of flares, prominences, and the corona will be discussed in the next chapter, followed by a chapter on sunspots. The process that is responsible for the tremendous energy given out by the sun is described in Chapter 45.

5

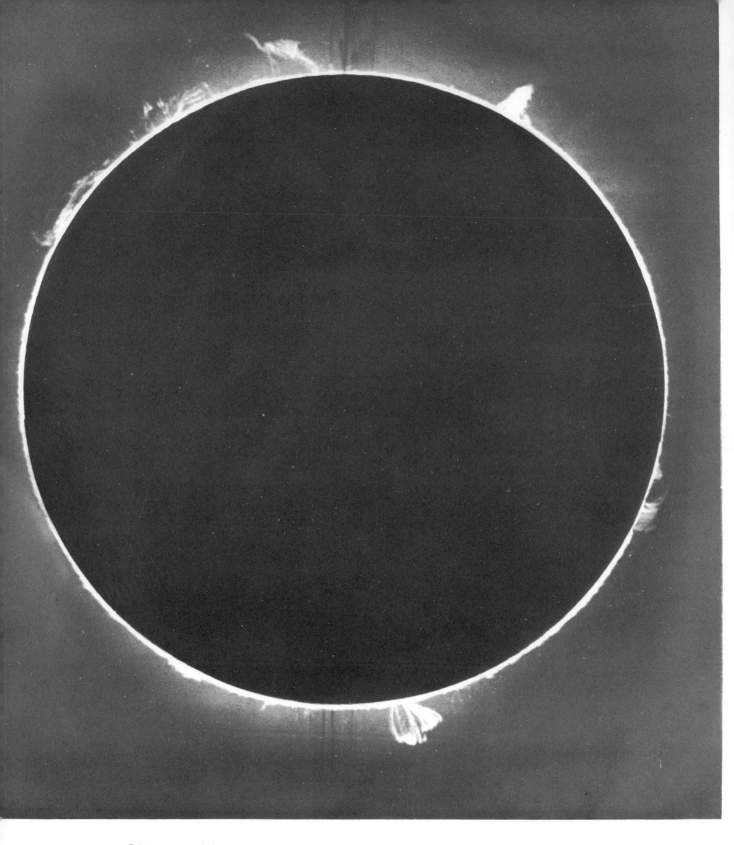

Calcium spectroheliogram of prominences, December 9, 1929. (Mount Wilson Observatory.)

6

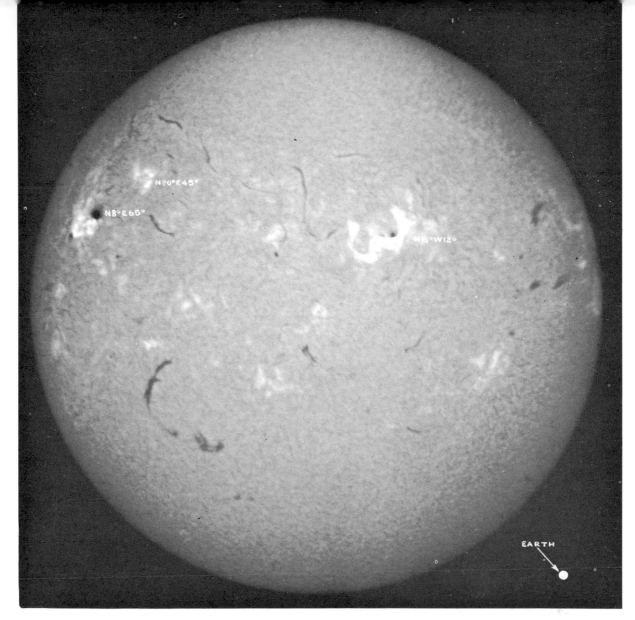

Three flares appear in this filtroheliogram showing disk of sun as photographed in helium light on June 18, 1959. Large flare in northwest (above right of center) is at maximum. Topmost flare in northeast is just starting, while the one slightly below and nearer limb is fading. (McMath-Hulbert Observatory of the University of Michigan.)

CHAPTER **2**   FLARES, PROMINENCES, AND THE CORONA

During a total eclipse the moon shuts out the intense light of the sun, and the yellow *corona* is revealed, sometimes reaching out as far as a solar diameter or more. Close to the black moon's edge a reddish ring can be seen. This colored layer of the sun's atmosphere is called the *chromosphere*. Rising from it are flamelike clouds of luminous gases called *flares* and *prominences*.

The accompanying photograph of the chromosphere and prominences suggests a total eclipse of

7

DISK PASSAGE OF A LARGE PROMINENCE

22 JAN. 59

24 JAN.

25 JAN.

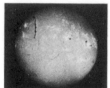
26 JAN.

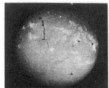
27 JAN.

28 JAN.

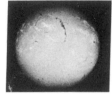
29 JAN.

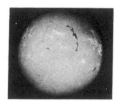
30 JAN.

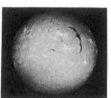
31 JAN.

2 FEB. 59

4 FEB.

ERUPTION AT WEST LIMB

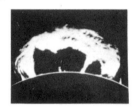

PROMINENCE ERUPTS 5 FEB. 59

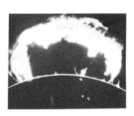

Hydrogen spectroheliograms. Rotation of sun carries prominences and a sunspot group across the disk. Prominences appear as dark streaks against photosphere, but bright when seen beyond limb, as in first and last pictures of upper sequence. At bottom is the prominence that was the principal dark streak on the disk, as it changes form on day after leaving disk. (Sacramento Peak Observatory, Geophysics Research Directorate AFRD.)

the sun, but it was not taken during an eclipse. The black disk is not the moon, but was produced artificially. The main part of the sun is hidden. All that appears is a thin ring of light with clouds extending out from it. The picture was made at the Mount Wilson Observatory by an instrument known as a spectroheliograph. Until the invention of this device none of the outer portions of the sun could be photographed except during a total eclipse of the sun.

Flares provide one of the sun's most dramatic displays. They may appear suddenly but infrequently in the neighborhood of a group of sunspots. Flares are much brighter than the background surface of the sun, especially when observed with a spectrohelioscope adjusted to use the light of hydrogen, which they emit. When flares are observed at the limb of the sun, their behavior is similar to that of prominences, although they usually have shorter lives. The bright hydrogen of the flare rises very rapidly, with velocities of the order of 400 miles per second. During this rise the flare's brightness increases dramatically. When the flare reaches a height of about 40,000 miles, it gradually fades away and disappears from sight about an hour after the first eruption. In addition to emitting hydrogen radiation, flares must also emit strong X-ray radiation. As a result, when a flare occurs on the solar meridian, we experience almost immediately a radio fade-out over the entire illuminated hemisphere of the earth. The X-ray radiation disrupts radio communications in a way similar to that of the magnetic storm, which is described in Chapter 4.

Prominences consist of great volumes of incandescent gases whose density is less than one millionth of that of our atmosphere at sea level. They are composed mainly of hydrogen, with some helium and ionized calcium. They assume a variety of forms, but can be arranged in six main types according to their shape and behavior. These are called active, eruptive, spot, tornado, quiescent, and coronal.

Active prominences are the most common. An active prominence consists of a thin sheet of gas standing on edge and connecting with the chromosphere in only a few points. Such prominences average 6,000 miles thick, 30,000 miles high, and 50,000 miles long, They generally become visible high up in the sun's atmosphere. Very little outward

Spectroheliograms of four unusual prominences are shown here and on the next two pages. (The last one is from McMath-Hulbert Observatory of the University of Michigan. The other three are from Sacramento Peak Observatory, Geophysics Research Directorate AFRD.)

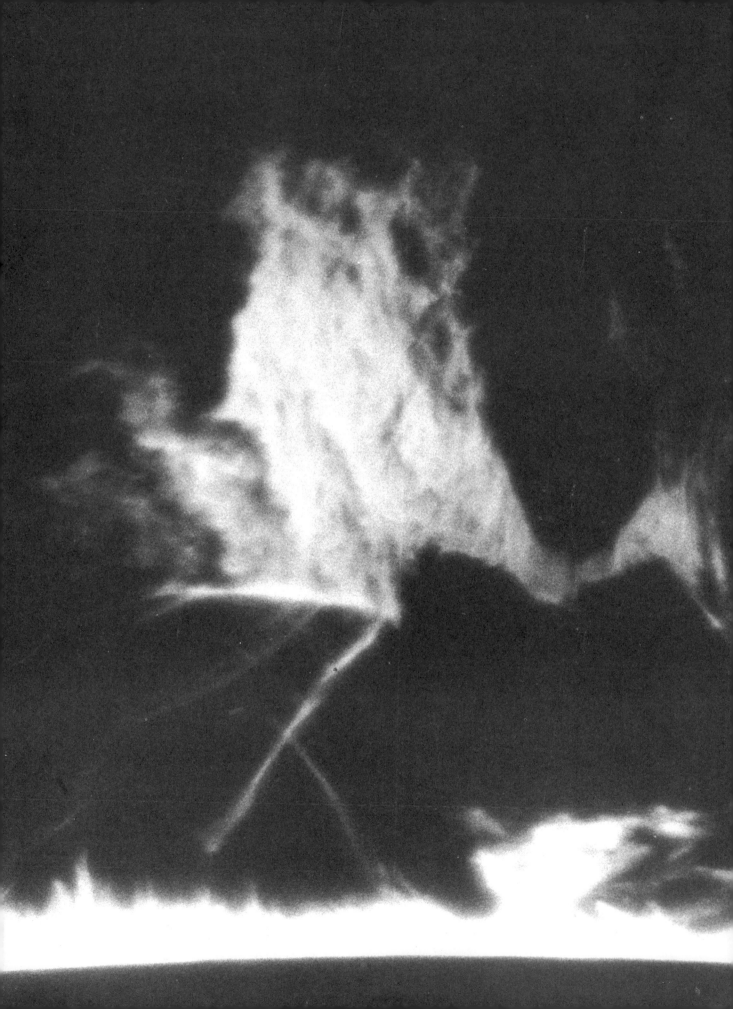

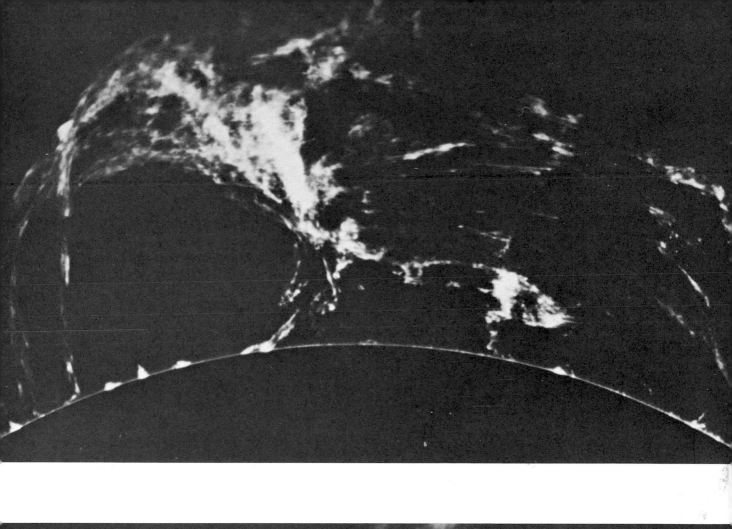

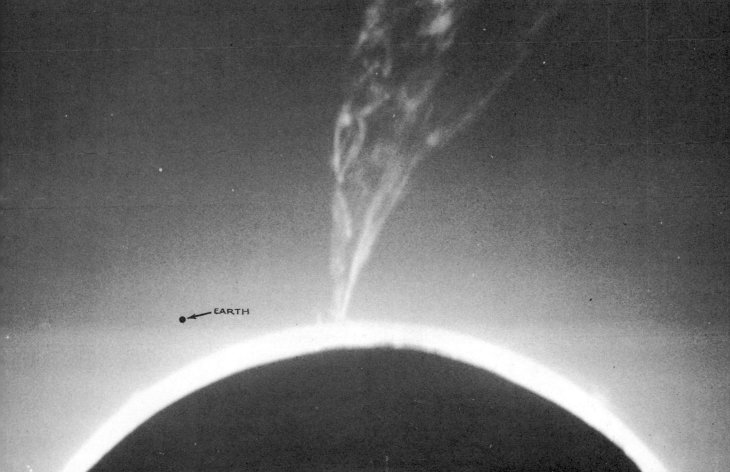

EARTH

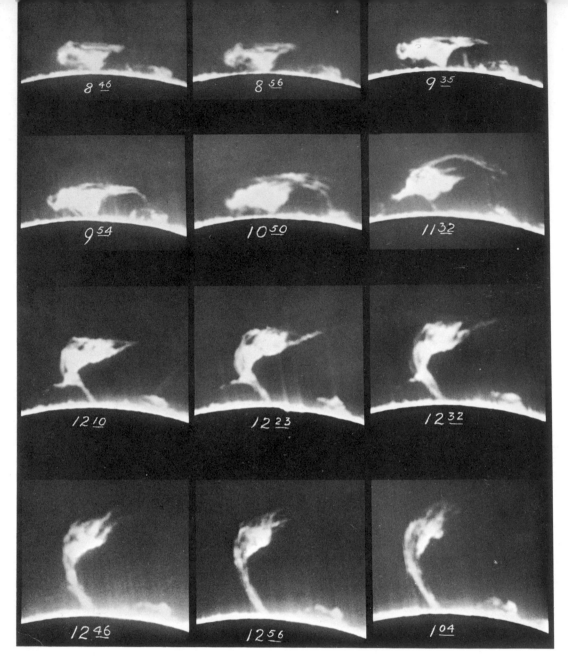

Spectroheliogram showing changes in an active prominence during 4 hours and 18 minutes on June 18, 1929. (Mount Wilson Observatory.)

motion is observed. Knots and streamers pour from the prominences into one or more invisible centers of attraction in the chromosphere. They act as if they were attracted by some electrical or magnetic force, and not by gravitation alone. The material making up the prominence moves along curved lines with a speed of about 50 miles a second, but sometimes of more than 100 miles a second. The speed may be constant for a few minutes, then suddenly increase and remain con-

stant for another few minutes until the next sudden increase occurs. Finally the prominences are torn apart and pulled into the sun.

The accompanying series of twelve photographs from the Mount Wilson Observatory shows the development of an eruptive prominence. The numbers indicate the times when the photographs were taken. An eruptive prominence, like the active type, moves with uniform velocity, which suddenly increases at intervals, but the direction is

12

outward from the sun rather than inward. The velocities range from a few miles per second to the record speed of 452 miles per second observed in 1937 at the McMath-Hulbert Observatory of the University of Michigan. Heights of 250,000 miles are commonly reached. This distance is equal to that from the earth to the moon. The record height of 1,056,000 miles for an eruptive prominence was observed at Mount Wilson in 1946. This is 191,000 miles greater than the sun's diameter.

In eruptive prominences each successive velocity is a multiple of those preceding. For example, the velocity may suddenly change from 40 miles a second to 80, and later to 120. This law does not always apply in active prominences, because they move in curved lines, while eruptive prominences move in straight lines. Probably some of the material in eruptive prominences is permanently detached from the sun, but the behavior of active prominences suggests that much of it is returned to the sun.

Active and eruptive prominences are found everywhere on the solar surface and are not necessarily associated with sunspots. But a third type, the sunspot prominence, occurs only over spots or in regions where spots are about to form or have just disappeared. Although sunspot prominences assume different forms, they often consist of broken filaments, in the shape of a fan, converging toward one spot. Sometimes an arch forms over a spot, and the motion is down both branches of the arch into the spot.

Several other types of prominences have been observed. The tornado prominence is named for its shape and motion. It has not been photographed very often and appears to be in very rapid rotation. Quiescent prominences change very slowly in appearance. They may last for several days and sometimes may reach a third of the way around the sun. Coronal prominences are long streams coming down from the region of the corona and lasting only a few minutes each. The forces that produce these different types of prominences are not yet understood.

A spectacular sight during the few moments of a total solar eclipse is the outermost envelope of the sun, the corona. It is brightest at its base and fades as it recedes from the main body of the sun. The outer parts have filaments that sometimes extend well over a million miles above the solar surface. Indeed, during the last century, Langley observed

an eclipse from the top of Pike's Peak and followed the filamentary structure to much greater heights. The shape of the corona shows a striking correlation with the sunspot cycle. If a solar eclipse is observed during a time of minimum sunspot activity, the filamentary streamers show their maximum length at low solar latitudes straddling the solar equator, and the streamers emanating from the solar poles are quite short. On the other hand, at times of great sunspot activity the streamers have more closely the same length in all solar latitudes.

Studies of the corona have shown that it can, for convenience, be divided into two parts. The *inner* corona, extending outward to about a third of the diameter of the sun, is pale yellow, while the fainter *outer* corona is made up of faint pearly white streamers. Part of the light of the corona is produced by the scattering or diffraction of light caused by dust and molecules around the sun, somewhat similar to the way in which high cirrus clouds in the earth's atmosphere produce a halo around the moon at night. Another part of the corona's light is produced by a number of spectral emission lines, known as coronal lines. For a long time after their discovery these coronal lines were quite a mystery because they could not be produced by the usual spectroscopic methods in the laboratory. It was even suggested at one time that the lines were produced by an unknown element named coronium.

In 1939, however, W. Grotrian and Bengt Edlen showed that the lines are produced by a solar halo of quite ordinary elements, such as iron, nickel, and calcium. The atoms of these elements, however, are in quite an extraordinary condition, each having lost up to half of their normal quota of "orbital" electrons. (See explanation of spectra on page 204). This condition is remarkable because such a high degree of ionization (loss of electrons) would require, under ordinary conditions of gas pressure, a temperature of about a million degrees. This is almost 200 times as hot as the underlying photosphere! However, because of its very low density, the actual heat content of the corona is low. It used to be thought that the high ionization was produced by shock waves ("sonic booms") progressing outward through the corona. Disturbances rising from the region of granulation in the photosphere eventually reach the velocity of sound in the ever-decreasing density of the chromosphere, and highly energetic shock waves appear.

13

Corona during eclipse. This type of corona is commonly found near epochs of sunspot minimum. Fine wisps emerge from polar regions, while strong streams parallel solar equator. (Photographed at Fryeburg, Me., Aug. 31, 1932, by Lick Observatory eclipse expedition.)

Typical corona observed near epoch of sunspot maximum. Corona is brighter at limb than when at sunspot minimum, but does not extend in long streamers. It differs little at equator and at poles. (Photographed at Aswan, Egypt, Aug. 30, 1905, by Lick Observatory expedition.)

However, recent results from an orbiting solar observatory (OSO-8) showed that the energy in the shock waves is very much too small, so that this mechanism had to be abandoned. As of 1981 no satisfactory alternative had been proposed.

The visible light contributed by the corona to the total light of the sun is negligible. The situation is quite different, however, at the very much longer and invisible wavelengths now studied by the radio astronomers. For example, at a wavelength of 17 meters, the sun has a diameter twice that observed by ordinary optical methods. The photosphere of the radio sun lies in the outer corona! Apparently it has no sharp boundary but fades gradually with increasing distance from the solar center. The outer corona tends to absorb radiation at radio wavelengths because of the abundance of the free electrons in the corona. As mentioned earlier, the atoms have lost most of their electrons because of the low gas density in the coronal filaments.

The advent of high-flying rockets, orbiting X-ray telescopes, and most notably, Skylab, has made it possible to observe the X-ray radiation from the sun from above the obscuring atmosphere of the earth. X-ray photographs of the sun show that the corona has a very complex structure, with large and frequent eruptions. Most of the brighter regions form loops connecting widely spaced areas on the solar surface. Dark "coronal holes" are found at the poles of the sun, and sometimes at lower latitudes. The coronal holes turn out to be regions where the corona is cooler and more quies-

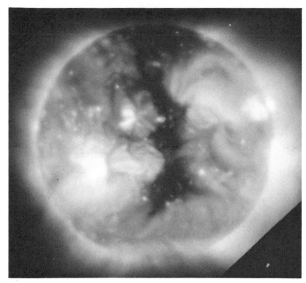

A soft X-ray photograph of the corona taken June 1, 1973, obtained by the Skylab S-054 experiment. The solar corona is the very thin outer portion of the Sun's atmosphere. Structures with temperatures higher than one million degrees can be observed in X rays. The loops, arches and other features seen in the photograph are produced by the interaction of the Sun's magnetic field and the ionized gas of the corona. (NASA.)

cent than in the bright regions. The looped structures of the bright regions suggest strongly that they are the result of complex and changing structures in the outer solar magnetic field. It is very likely that when theories are developed to explain behavior of the magnetic field of the sun, it will be found that many of the phenomena that we observe, such as prominences, flares, and sunspots, can also be explained, and that we will have found the source of the high energy of the corona.

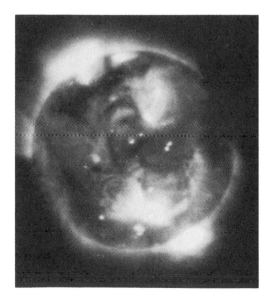

Large-scale structuring associated with a complex of activity. Within the structure are active regions (large, bright points) and small coronal holes (small, dark areas). (Illustration courtesy of G. Vaiana, American Science and Engineering, and Harvard College Observatory.)

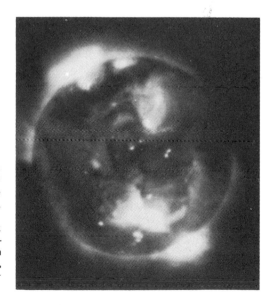

CHAPTER 3 SUNSPOTS: REFRIGERATORS OF THE SUN

Of the thousands of inventions of the twentieth century that have made life more comfortable and pleasant, few can compare with the electric refrigerator. Man can be justly proud of this great development in the production of cold. The invention, however, was not a new one. Thousands of millions of years ago, nature built the first refrigerators of the model that our generation has been using, and she has continued to build them ever since. Interestingly enough, her refrigerators are described by the physicist as being even more efficient than the man-made ones. Not only that, but nature's refrigerators are built on a fantastically greater scale, some being so large that many spheres the size of our earth could be stored within a single one. Man calls nature's refrigerators sunspots, and in general has not recognized that all he has done in his refrigerators has been to imitate rather imperfectly the processes carried on by these interesting solar storms.

Since we are comparing parts of the sun with refrigerators, it may be well to consider first the manner in which refrigerators work. They depend upon the fact that when a gas is compressed by pressure its temperature increases, and that when it is permitted to expand it cools. Often, as in the case of ammonia, the gas is liquefied when compressed in the refrigerator. The diagram shows the essential parts of that machine. The coil marked H contains a gas at low pressure. K is the chamber to be cooled. A and B are the piston and cylinder of a pump that draws the gas from H through the intake valve C, compresses it, and forces it through the valve D to the high pressure pipe E. From there it can escape slowly through the small valve G back to the low pressure side H. It expands and cools as it escapes.

If this were all that was done, the gas would increase in temperature as it entered E and decrease as it expanded into H. The heating and cooling would follow alternately, and little change would take place in the room K.

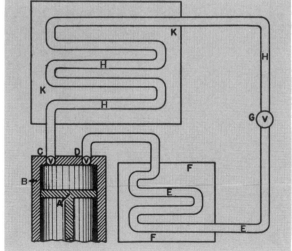

Essential parts of a refrigerator.

16

However, most of the tube E is contained in a jacket F through which some coolant runs. This coolant absorbs heat from the compressed gas and, therefore, when the gas again escapes into H it is cooler than when it left at the valve C. As a result the room K is cooled and drops in temperature till finally it is nearly as cold as the expanding gas entering from the valve G. Of course, modern refrigeration has added many parts to this simplest scheme, but, essentially, every refrigerator functions on the principles just described.

On the sun, the room K, to be cooled, is a small part of what to us looks like a surface for that body —the photosphere. Actually there is no surface, for all of the sun is gaseous. For some unknown reason the gases in a certain region just beneath this "surface" are pushed outward from the center of the sun and break through the photosphere. As

they do this they expand and cool just as do the expanding gases of our little refrigerators. The amount of cooling is almost incredible. At the "surface" the temperature is in the neighborhood of 10,300° Fahrenheit. Below this it is still hotter. But in the center of a sunspot (the room K of the refrigerator) the temperature has dropped to about 7,200°.

Because sunspots are cooler than the normal surface of the sun, they also are less bright. Thus they appear black in contrast with the much brighter background. If we could see a sunspot on a dark background, instead of on the sun's dazzling surface, it would appear brighter than almost all sources of light on the earth.

The sun does not have a satisfactory system of pump and valves; therefore, eventually, the uprush of expanding and cooling gas stops, and the refrig-

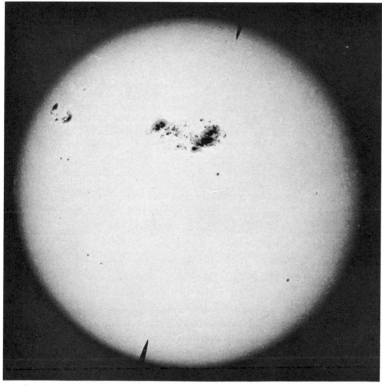

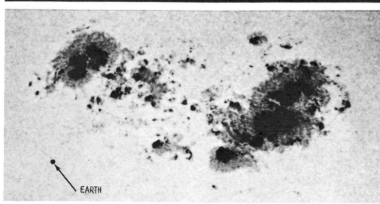

One of greatest of sunspot groups, photographed April 7, 1947. Spot on bottom photograph represents size of the earth in comparison. (Mount Wilson and Palomar Observatories.)

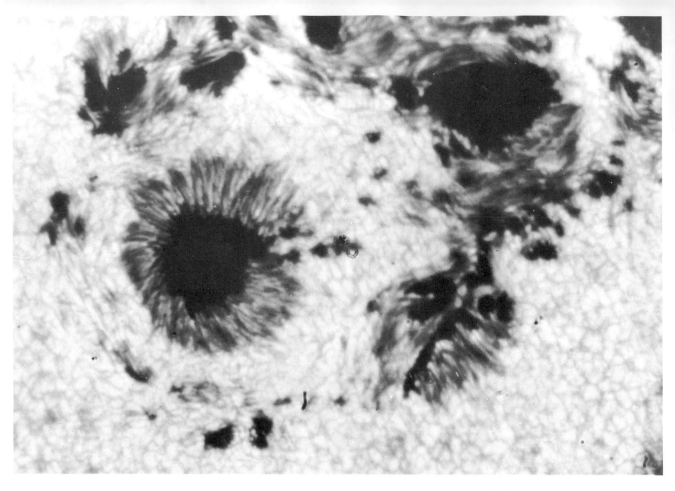

Direct photograph of sunspot group and surrounding area, made by a 12-inch telescope carried by balloon to altitude of 80,000 feet, where thin air has very little effect in distorting the image. (Project Stratoscope of Princeton University, sponsored by Office of Naval Research and National Science Foundation.)

erator (sunspot action) very rapidly disintegrates.

The record of a sunspot in 1840 and 1841 that persisted for 18 months is questioned today. The longest certain record is about five months. The average life is one or two weeks.

A spot appears first as a tiny, apparently black dot on the sun. This is the beginning outward trickle of the rising, expanding gases. If the gases, having broken through, are plentiful, they will enlarge the opening. The dark spot then grows, often to such a size that we, 93,000,000 miles away, can see it with the naked eye. The center of a spot is called the *umbra*, and its diameter varies from 500 miles in the smallest spots to 50,000 miles in the largest. The shaded portion around the umbra is the *penumbra*. Its diameter is usually double and sometimes triple that of the umbra. The boundary between these parts is very sharp, as is also the boundary between the penumbra and the bright photosphere. To the eye the umbra appears black,

although merely by contrast. Actually it gives approximately one-fourth the radiation that is received from the normal part of the sun. The part of the radiation that we see as light, however, is cut down more than this. Astronomers examining this light from the blackest part of a sunspot have found that the lowered temperature changes the quality of the light, making it quite different from that received from the rest of the sun.

In 1908 George E. Hale discovered that sunspots are very powerful magnets, with fields that may be as much as ten thousand times that of the earth, which is about half a gauss. Such strong fields must be produced by intense electric currents circulating around the spot, estimated to be as strong as ten million amperes in some cases. It used to be thought that sunspots were gigantic vortices, like tornadoes, but the motions of gases in spots seem to be inward toward the spot center at low elevations, and outward at higher elevations.

18

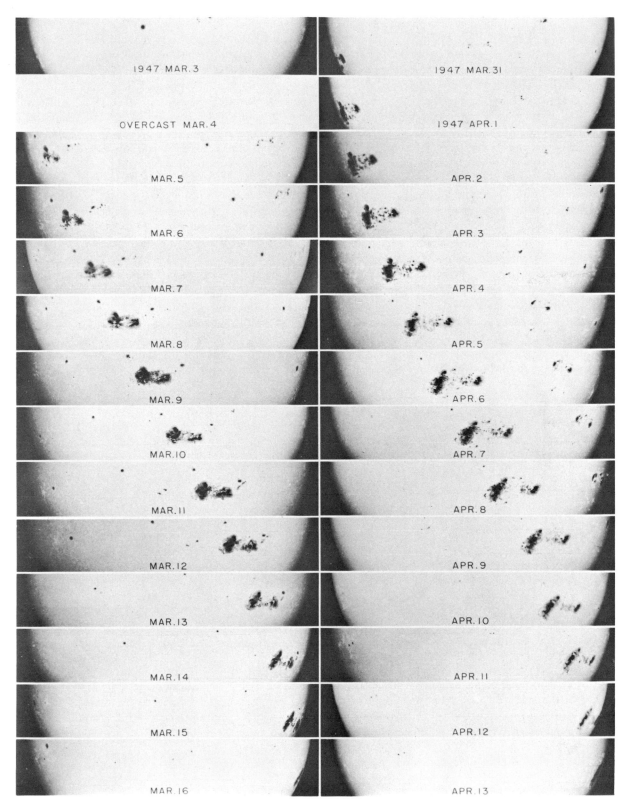

Day-by-day record of great sunspot group of 1947 during two passages on solar disk. (Mount Wilson and Palomar Observatories.)

The spot actually behaves like a pump; gas moves upward and along the lines of force of the strong field. The outward expansion cools the upper layers producing the lower temperature. A spot thus impedes the outward flow of solar radiation over a large area on the solar surface. This radiation has to get out somewhere, perhaps the reason bright faculae surround sunspots.

When gases are as highly heated as are those on the sun, the outer parts of the atoms making up the gases are broken up. The little pieces that come loose are electrons and have negative charges of electricity. The remaining central part of the atom is left with a positive charge. These charged particles take part in the sunspot whirlpool or vortex. A magnetic field always is produced wherever electrically charged particles move. The sunspot vortices often produce very strong magnetic fields. The direction of rotation determines the direction of the magnetic field. Since the two spots of a pair rotate in opposite directions, they have opposite magnetic fields. These fields affect the light which passes through them, and astronomers

disk and disappear from view at the western edge. If they last long enough, they will reappear at the eastern edge. Observation of the spots enables us to determine the sun's period of rotation. This is about 25 days at the equator, but increases with the distance from the equator. It is about 28 days at a latitude of 45° and about 35 days near the poles. Since few spots are found beyond 40°, another method for determining the rotational period has to be used for the higher latitudes.

In 1843 a German apothecary by the name of S. H. Schwabe announced that the number of spots varied in a period of about ten years. For eighteen years he had followed the hobby of looking at the sun with a small telescope on every clear day and counting the number of spots. In 1833 and 1843 he had found no spots on about half the days of observation. For several years around 1828 and 1938 he had seen spots every day he looked.

The world was not yet ready for Schwabe's findings. Only after he had worked for another dozen or so years did his results receive recognition. At about that time other observers showed

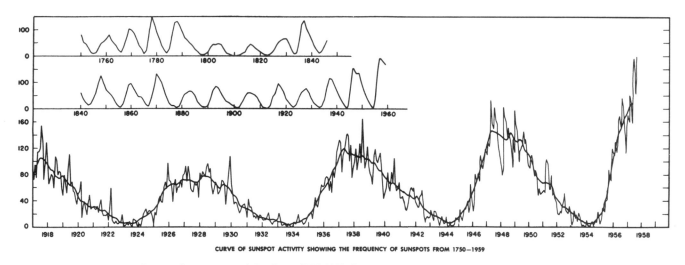

CURVE OF SUNSPOT ACTIVITY SHOWING THE FREQUENCY OF SUNSPOTS FROM 1750—1959

Curve of sunspot activity from 1750-1959. (Mount Wilson and Palomar Observatories.)

measure the light changes to get the strength and shape of the magnetic fields that are found there.

Sunspots generally appear in groups elongated in a direction approximately parallel to the sun's equator. Usually the spots at the eastern end of the group disappear first, leaving only a western spot. Thus single spots can be seen, but each is the remainder of an old group. The rotation of the sun on its axis causes the spots to move across the sun's

that the daily fluctuation of the magnetic compass needle increased and decreased in a ten-year period. Also, violent agitations of the magnetic needle, distinct from its regular daily fluctuation, showed a periodicity of ten years. The sunspots and magnetic disturbances were precisely in step, rising and falling together. Later the northern lights were also found to be correlated with sunspot activity.

In presenting a medal to amateur astronomer

20

Schwabe, the president of the Royal Astronomical Society said: "Twelve years hath Schwabe spent to satisfy himself—six more years to satisfy, and still thirteen more to convince, mankind. For thirty years never has the sun exhibited his disk above the horizon of Dessau without being confronted by Schwabe's imperturbable telescope, and that appears to have happened on an average of about 300 days per year. This is, I believe, an instance of devoted persistence unsurpassed in the annals of astronomy. The energy of one man has revealed a phenomenon that had eluded even the suspicion of astronomers for two hundred years."

Since that time the *average* length of the sunspot period has been found to be eleven years, but each of the last three periods has been only ten years. There are considerable fluctuations, the observed intervals ranging all the way from about seven to seventeen years. The irregularity is obvious from an inspection of successive cycles since 1750, shown in the accompanying diagram.

In the nineteenth century, following Schwabe's discovery, a Swiss astronomer named R. Wolf collected all the available observations of sunspots and assigned a number to each day as an index of sunspot activity. The Wolf Number for any day is equal to the total number of spots plus ten times the number of groups, the numbers being reduced to a uniform scale by a correction depending on the equipment of the observer. Even though the periodicity had not been noted until 1843, enough earlier observations were found to extend the curve back another century.

The lower part of the diagram shows the Wolf Numbers averaged by months. It is seen that the cycle does not progress in a regular way, but in pulsations which vary in length from five to as much as fifteen months.

Groups of spots are usually bipolar. In the cycle from 1913 to 1923 the polarity of the western spots in groups north of the sun's equator was like that of the north-seeking pole of a compass needle. The eastern spots in these groups were of opposite polarity. South of the equator these magnetic conditions were reversed. In the cycle ending in 1913 and in the one beginning in 1923 the magnetic relations were exactly opposite to those described above. Thus the complete cycle is really about 22 years in length.

If one were to ask an astronomer why we have sunspots, he might reply that there is a lack of thermodynamic equilibrium. Such a statement really tells us very little. We do not know why some years have very few of such upsets on the sun, whereas others have a great many. Probably a good many different causes combine. One tiny factor, quite probably, is the small tidal effect of the planets on the sun. This must be regarded as a very small part of a complete explanation. It does, however, help a little in understanding the variations in number from year to year.

In the years 1923, 1933, 1944, and 1954, we had very few spots. But after each of these years the number of sunspots increased till we got years of maximum, in 1928, 1937, 1947, and 1957. The maxima are not all equally strong, and the time intervals between maxima are not constant. As already mentioned, the average interval is about eleven years.

Many attempts have been made to connect sunspots with all sorts of phenomena on the earth. Compass needle fluctuations and the aurorae (northern and southern lights) vary with sunspot activity. Certain severe radio disturbances are also connected with sunspots. Connections of only theoretical interest have been found between sunspots and terrestrial temperature and rainfall. However, many published claims are quite fantastic. One London physician in 1920 connected sunspots to influenza epidemics, and they have been related to wheat prices and the stock market. It is unfortunate that authors have done so much to mislead the public with fantastic claims for the far-reaching effects of sunspots.

The curve of sunspot activity observed over the past two hundred years would lead one to believe that the eleven-year (or twenty-two-year) solar cycle is a permanent feature of solar behavior. However, such is probably not the case. A number of lines of evidence indicate that during the seventeenth-century reign of Louis XIV (the "Sun King") the sun was markedly free of disturbances. For instance, while records of sunspot sightings in Japan, Korea, and China go back to at least 28 B.C., and have been fairly systematically observed ever since the invention of the telescope about 1610, there were remarkably few reports of sunspots from 1645 to 1715. This has been called the "Maunder Minimum" after the nineteenth-century physicist who was one of the first to call attention to this anomaly. In addition, reports of the structure of the solar corona seen during solar eclipses occurring in this same interval made no mention of the intense coronal streamers associated with strong magnetic fields on the solar surface. There were also very few reports of aurorae, which, as we have mentioned above, have a frequency of occurrence that correlates well with solar activity. The most convincing evidence that this was a period of minimum solar activity is to be found in measures of the concentration of a heavy isotope of carbon in tree rings. This is because this particular isotope of carbon is produced in the earth's atmosphere by cosmic rays, which are enhanced in intensity when the sun is inactive. The carbon combines with oxygen in the atmosphere to form carbon dioxide, which is assimilated by trees. By studying the abundance of the heavy carbon isotope in tree rings it has been found that its concentration increased by 20 percent during this period. Interest-ingly, the tree rings also showed that there was an even earlier period of solar inactivity from 1460 to 1550.

It appears, therefore, that during the past 500 years the sun has undergone at least two protract-ed periods of changed magnetic activity, with probably a parallel change in the sun's energy out-put. John Eddy of the National Center for Atmospheric Research in Boulder, Colorado, has pointed out that the Maunder Minimum corresponded with the coldest period of the "Little Ice Age" that chilled northern Europe, and also with a period of protracted drought in the American Southwest that was also revealed by the study of tree rings. There is, of course, no reason to expect that similar protracted periods of solar inactivity will not take place in the future, and that they will have serious effects on the earth's climate.

CHAPTER **4** **THE NORTHERN AND SOUTHERN LIGHTS**

Perhaps the most beautiful sight in the heavens is the polar aurora, or aurora polaris. Aurora is a collective term that refers both to the northern lights, also called the aurora borealis, and the southern lights, or aurora australis. The aurora's form changes constantly and its coloring and intensity vary rapidly. The fainter displays are usually white with a greenish or yellowish tinge. With increasing brilliance, red, pink, violet, green, and yellow become prominent. The first appearance is sometimes a quiescent arch. This may suddenly be followed by many streamers, like searchlights, but changing and flickering. Sometimes the display begins with nearly vertical curtains of light, the folds of which change in form.

The altitudes of the aurora are measured by at least two observers in telephone communication who take photographs at the same instant from different places. The photographs shown were made by Stormer and Birkeland at stations in Norway nearly twenty miles apart. The apparent shift of the aurora with respect to the planet Venus is shown plainly. Measurement of this angular shift, known as parallax, makes it possible to compute the height of the aurora. About 60 miles is the lowest altitude. The record height for the top of an auroral streamer is about 680 miles.

For many years scientists have known that the occurrence of an aurora is often accompanied by a "magnetic storm." The magnetic field of the earth is disturbed, as is shown by the behavior of a compass needle. This also interferes with radio and telegraph communication.

The number of auroral displays and magnetic disturbances varies from year to year, increasing from a minimum to a maximum and back to a minimum again in an average period of about eleven years. The number of spots on the sun varies in a similar manner. It is believed that the sun ejects tremendous numbers of electrical particles at times of intense sunspot activity. These particles augment the ever present "solar wind" of charged particles spiraling out from the sun and embracing the entire planetary system. Thus the solar wind can be considered to be "gusty." Its particles take about a day to make the trip from the sun to the vicinity of the earth, traveling at supersonic speeds. When the solar wind hits the magnetic field surrounding the earth it produces a shock front about 40,000 miles above the earth's surface and a long tail extending

Photographs of aurora. The photograph at the top is an example of a rayed arc type. (V. P. Hessler—Geophysical Institute at University of Alaska.)

out from the dark side of the earth. The particles in the solar wind cannot penetrate the shock front, but the intense bursts or "gusts" of particles from spots or active solar regions apparently affect the front in such a way that accelerations are produced in particles in the earth's magnetosphere below the shock front. The actual nature of the interaction is not clear, but it must be present, because the phenomena of aurorae, magnetic storms, and the like do definitely correlate with increased solar activity. These accelerated particles then come into the earth's atmosphere along the lines of force of the earth's mag-

observed not far from the center of the sun. The Department of Terrestrial Magnetism of the Carnegie Institution of Washington sent out a warning to radio operators to watch for disturbed conditions about September 18. Violent fluctuations in the direction and intensity of the earth's magnetism began late in the evening of the 17th. They reached a maximum of activity during the next afternoon. Vibrations in electrical power transformers were noted, and a variation of one per cent in voltage was recorded. Extensive interference with long-distance radio, telegraphic, and tele-

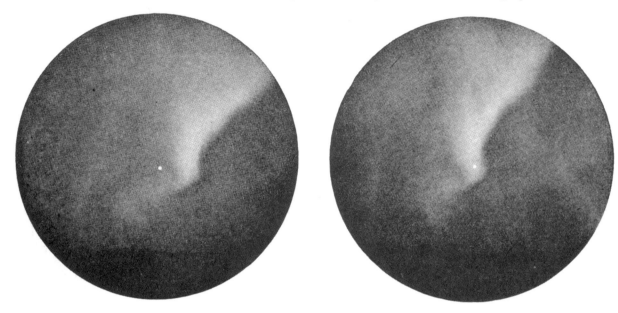

The aurora and the planet Venus photographed at the same instant from two places about twenty miles apart. The apparent shift of the aurora makes it possible to compute its height.

netic field and produce a glow in the atmosphere by a process similar to that producing the glow in a neon sign. These lines of force are approximately parallel to the earth's surface in the regions of the equator. Thus few of the particles get into the atmosphere near the equator, with the result that there are few auroral displays there. In northeastern United States the lines are within 20° of perpendicular, so that auroral displays are not uncommon in the northern part of the country. The area of maximum auroral intensity is a circular belt around each of the magnetic poles, where the particles can follow the almost perpendicular lines of force into the atmosphere.

One of the greatest displays seen in this country in recent times occurred on September 18, 1941. Several days earlier a large group of sunspots was

phonic communications was experienced.

People in the eastern part of the country who watched the sun set on the evening of September 18 noticed a strangely persistent glow in the northwestern sky. Soon after dark the sky was filled with rays, apparently converging to a point about 20° south of the zenith. The convergence was only apparent. The particles from the sun were traveling along the magnetic lines of force, which are tilted about 20° southward at Washington. Auroral rays are actually parallel, but they appear to converge toward what is called the magnetic zenith, just as railroad tracks appear to converge in the distance.

At the height of the display, voltage fluctuations of $2\frac{1}{2}$ percent were recorded in Baltimore. One operator reported that the vibrations in an electric power transformer built up in intensity with the increase in the brilliance of the aurora. The rapid changes in the earth's magnetism that occur during such a magnetic storm induce electric currents in the earth in much the same way in which electric currents are generated commercially. Power lines, a part of the conducting material of the earth, are overloaded by the earth's magnetic currents and produce the effects described.

The interference with radio communication can also be explained. Radio waves travel in straight lines, but can go around the curved surface of the earth by being reflected from electrified layers in the atmosphere. When these layers are bombarded by large streams of electrical particles from the sun, they become disturbed and do not reflect radio waves as readily. So radio communication may be interfered with during a magnetic storm, which, of course, has nothing to do with the ordinary thunderstorm.

The electrified layers of air from which radio waves are reflected form what is known as the *ionosphere*. This begins at a height of about 60 miles and extends beyond this point up for more than 100 miles. It has been already mentioned that auroras are most common at a height of about 60 miles. This is the region where the long-enduring trains left by some meteors are observed. These trains are not well understood, but may be in the nature of an electrical afterglow resulting from the passage of certain meteors.

The great magnetic storm of 1941 occurred about four years after a maximum in the sunspot cycle. The accepted explanation is that at this stage of the cycle the spots are located near the equator of the sun. Streams of particles from spots at the center of the sun's disk are more likely to hit the earth than particles from spots at higher latitudes on the sun. These particles produce electric currents that must amount to millions of amperes. Thus when one admires the beauty of the aurora display, he is actually looking at the path of a very intense electric current.

SECTION TWO

The Earth

CHAPTER **5**   **TERRESTRIAL PORTRAIT**

The earth is one of the smaller of the nine planets, having a mean diameter of 7,918 miles. It is almost a perfect sphere, but the equatorial diameter is 27 miles longer than the polar diameter. Thus it is an oblate spheroid, meaning that it is like a sphere, but flattened at the poles. Oblateness is the fractional degree of flattening and is expressed by putting the difference between the greatest diameter and the least (7,927 − 7,900 = 27) over the greatest diameter. The result is 1/297. Recent observations of the motions of artificial satellites prove definitely that the surface of the earth departs very slightly from the oblate spheroid.

The earth is so nearly spherical that its surface and volume can be computed with sufficient accuracy by the formulas for a perfect sphere, if the mean diameter is used. The mean diameter is not equal to half the sum of the polar and equatorial diameters. The three axes of symmetry drawn through the earth's center at right angles to each other are the polar diameter and two equatorial diameters. Therefore, the mean diameter is equal to one-third the sum of the polar diameter and the two equatorial diameters. The result is 7,918 miles.

The average density of the earth of 5.5 grams per cubic centimeter is much greater than the average density of rocks found at the earth's surface, which is close to 3 grams per cubic centimeter. The inference is that the density of the material of the earth increases with depth, reaching a value greater than 5.5 grams per cubic centimeter at the earth's center. This is confirmed by the behavior of earthquake waves as they travel through the interior of the earth. The speed of propagation of earthquake waves increases in the denser, deeper layers of the earth, so that measurements of the time required for a passage of the waves through the interior lead to knowledge of the density distribution. Of greatest interest is the discovery of a terrestrial core about 2,000 miles in radius, which has a density between 10 and 12 grams per cubic centimeter and which is probably composed of liquid iron. The innermost part of the core, extending somewhat less than 1,000 miles from the center of the earth, may not be liquid.

It may be of interest to describe one method for finding the mass of the earth. The first successful determination was made in 1774 by N. Maskelyne, the Astronomer Royal of England. Two stations were set up in Scotland on the same meridian on either side of an abruptly rising hogback mountain of very regular contour. The distance between the stations and the dimensions of the mountain were accurately measured. The two stations were separated by a distance of a little over 4,000 feet or 41″ of geographical latitude. This is the angle that the plumb lines at the two stations would have made with each other if there were no mountain there. Then the astronomical or "true" angle between the stations was recorded by means of

Satellite photographs of large areas enable geographers to make more accurate maps while providing an improved understanding of geological features and their relationships. Dominating this Gemini 11 photo of Libya is the circular Murzuch sand sea. The rugged Tassili n'Ajjer mountains are in the foreground; the Mediterranean Gulf of Sirte is visible at top. Some geographic and geologic features of Libya were virtually unknown and unexplored until revealed from space by such satellite photographs as this. (National Aeronautics and Space Administration.)

TERRESTRIAL DATA

Polar diameter	7,900 mi.	12,714 km.
Equatorial diameter	7,927 mi.	12,757 km.
Mean diameter	7,918 mi.	12,742 km.

Oblateness (fractional degree
   of flattening)   1/297

Polar circumference	24,860 mi.	40,009 km.
Equatorial circumference	24,902 mi.	40,076 km.
1° of latitude at poles	69.4 mi.	111.7 km.
1° of latitude at equator	68.7 mi.	110.6 km.

Land area 57,000,000 sq. mi.
          149,000,000 sq. km. (29% of total)
Water area 140,000,000 sq. mi.
          361,000,000 sq. km. (71% of total)
Total area 197,000,000 sq. mi.
          510,000,000 sq. km.

Volume 26×10^{10} cu. mi. 11×10^{11} cu. km.
                            $(1.1 \times 10^{27}$ cu. cm.)
Mass 6.6×10^{21} tons 6×10^{27} grams
Density   (grams per cu. cm., water=1) 5.5
Nautical mile (minute of arc along a great
    circle)   6,080 feet
                           1.15 statute miles
                           1.85 kilometers
Statute mile   5,280 feet
                           0.87 nautical mile
                           1.61 kilometers
Kilometer (originally intended to be 1/10,000
    the distance from equator to pole)   3,281 feet
                           0.54 nautical mile
                           0.62 statute mile

Length of one degree of longitude at different latitudes:

Latitude	Statute Miles	Nautical Miles	Kilometers
0°	69	60	111
10°	68	59	109
20°	65	56	104
30°	60	52	96
40°	53	46	85
50°	44	38	71
60°	35	30	56
70°	24	20	38
80°	12	10	19

Greatest height above sea level   29,140 feet
   (Mt. Everest)
Greatest depth in ocean   35,600 feet
   (Challenger Deep in the Mariana
   Trench off Guam)
Dip of horizon in minutes of arc equals approximately
   the square root of observer's height in feet above sea
   level. (Example: Dip=6 minutes of arc for height of
   36 feet.)
Distance of sea horizon in statute miles equals approximately 1.3 times the square root of observer's height
   in feet above sea level. (Example: Distance = 130 miles
   for height of 10,000 feet.)
Curvature of earth in feet equals about 2/3 of the
   square of distance in miles. (Example: Curvature=6
   feet at a distance of 3 miles.)

Atmospheric pressure at sea level equals 15 pounds per
   square inch, supporting a column of mercury 30
   inches (76 centimeters) high.
Atmospheric pressure diminishes about one-half for
   each 3.5 miles (5.6 kilometers). (Examples: Pressure
   at 7 miles=$7\frac{1}{2}$ inches; pressure at 10.5 miles=$3\frac{3}{4}$
   inches.)
Greatest height of aurora=680 miles (1,100 km.)
Atmospheric refraction=35 minutes of arc at horizon,
   5 minutes of arc at altitude of 10°, 1 minute of arc at
   45°, 0 at zenith.
Temperature of free air (away from mountains) decreases about 1° Fahrenheit for each 300 feet of
   ascent up to about 7 miles, where it is —67° F.
Temperature in deep mines and borings increases downward at an average rate of about 1° Fahrenheit per
   60 feet.

Albedo (reflecting power) of earth	0.29
Velocity of escape	7 mi./sec. (11 km./sec.)
Period of earth's rotation	$23^{h}\ 56^{m}\ 4^{s}.09$
Inclination of equator to orbit	23° 27′
Orbital velocity	18.5 miles (29.8 km.) per second
Eccentricity of orbit	0.01674

Four kinds of year (see note below):

Tropical (ordinary) year	$365^{d}.24220$
	($365^{d}\ 5^{h}\ 48^{m}\ 46^{s}$ or 31,556,926 seconds)
Sidereal year	$365^{d}.25636$ ($365^{d}\ 6^{h}\ 9^{m}\ 10^{s}$)
Anomalistic year	$365^{d}.25964$ ($365^{d}\ 6^{h}\ 13^{m}\ 53^{s}$)
Eclipse year	$346^{d}.62$ ($346^{d}\ 14^{h}\ 53^{m}$)
Precessional cycle	25,800 years

Equinoxes move westward at rate of 50″.3 per year

NOTE: The year is the period of the earth's revolution around the sun, or of the sun's apparent motion
around the celestial sphere. There are several kinds of
year depending on different points of reference. The
true period of the earth's revolution is called the sidereal year, which is the interval of time in which the sun
apparently completes a revolution with reference to
the stars.

   The tropical year is the interval between two successive returns of the sun to the vernal equinox. It is the
ordinary year of the seasons on which the calendar is
based. Because of the precessional motion of the earth,
the vernal equinox moves westward and goes to meet
the eastward moving sun, making the tropical year 20
minutes shorter than the sidereal year.

   The anomalistic year is seldom used. It is the interval
of time between two successive passages of perihelion,
the point of the earth's orbit nearest to the sun. Since
the perihelion is moving eastward about 11″ a year, this
kind of year is nearly five minutes longer than the
sidereal year.

   The eclipse year is the interval between successive
arrivals of the sun at the same node, which is one of the
two intersections of the moon's path with the sun's apparent path on the celestial sphere. Since the moon's
nodes are moving westward about 19° a year, the
eclipse year is about 19 days shorter than the three
other kinds of year.

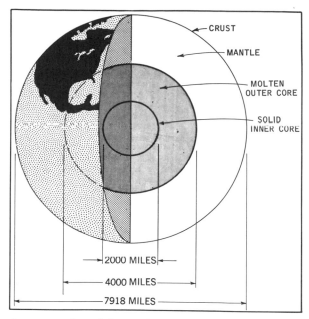

A cross section of the earth.

(Adapted from *Planets, Stars, and Galaxies*, Inglis, John Wiley and Sons.)

star sighting as 53″. This means that the deviation between 41″ and 53″ was caused by the lateral gravitational attraction between the mountain and the plumb line. The apparent zenith as pointed out by the plumb line was shifted by 6″ at each station, so that the astronomical difference of latitude was increased by 12″ over the geographical. The angle

of 6″ is so small that if the plumb line were 35 inches long, the weight at the lower end of it would be drawn over only about 1/1000 of an inch.

In round numbers the pull of the earth was about 35,000 times the pull of the mountain. To find the relative masses, we must also allow for the difference in distances. Calling the distance to the center of the earth 4,000 miles and the distance from one station to the center of the mountain 2/5 mile, we get a ratio of distances of 10,000. Since the force of gravitation varies inversely as the square of the distance, we square 10,000 and get 100,000,000. Multiplying this last figure by 35,000, we find that the ratio of the earth's mass to that of the mountain was about 3,500,000,000,000.

By means of deep borings into the mountain and measurements of its size, its approximate mass was determined. Multiplying the mass of the mountain by the ratio of the earth's mass to it, we get the mass of the earth. Other methods for finding the earth's mass are more accurate, but there is not the space here to describe them.

All miners know that the temperature of the earth increases as they descend from the surface. In general, the increase is one degree Fahrenheit for each 60 feet in increased depth. The central temperature must be several thousands of degrees. It is interesting that in 1862 Baron Kelvin calculated the age of the earth on the assumption that is was initially a completely molten globe; he found

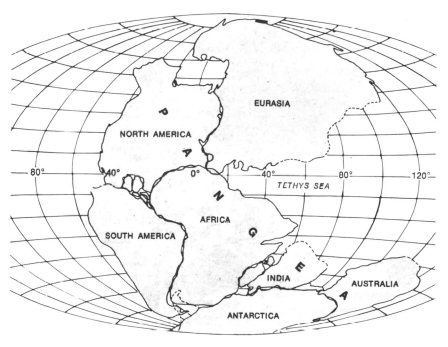

This reconstruction of continental positions at the beginning of the Mesozoic era, about 240 million years ago, shows the continents grouped into the single land mass known as Pangea. (After Dietz and Holden 1970.)

29

that it would have taken 100 million years to cool to its present surface temperature. This has proved to be a gross underestimation. We now know, from the dating of surface rocks via their radioactivity, that the earth is some 45 times as old as Baron Kelvin said it is. Back in 1862 they did not know that 80 percent of the heat found in deep mines comes from the radioactive decay of uranium, thorium, and potassium found in the crust itself, so that the earth is not cooling off nearly as fast as was estimated.

Anyone who has ever experienced an earthquake is acutely aware of the fact that the crust of the earth is far from being static. This crust is only a few miles thick, and is, in a sense, "floating" on the underlying rocks of the mantle, which are actually solid, but yet hot enough to be somewhat plastic to permit the crust to move. The crust is not continuous, but is actually broken up into a number of enormous sections called "continental plates" that are in motion relative to each other, either grinding past each other, moving apart, or colliding, forcing the edge of one to override that of the other. As one consequence, the Atlantic Ocean is gradually getting wider, with an up-welling of deeper material along a north-to-south mid-Atlantic ridge forcing the western and eastern Atlantic plates apart. Another example is the Indian subcontinent, which is moving northward, pushing up the Himalaya mountains. This movement of the continental plates is, of course, quite slow, running between one and five inches per year, but yet fast enough to change completely the familiar geographic arrangement of the continents in a few million years. Geologic and other evidence show that 250 million years ago the Americas and the Eurafrica land masses were locked together to form an enormous supercontinent called Pangea, which has subsequently broken up. Pangea, in turn, was the result of earlier movements that have been traced back to 550 million years ago, in the Late Cambrian. Back then, there was a vast supercontinent called Gondwana that was composed of what eventually became South America, Florida, Africa, Australia, India, Tibet, Iran, Saudi Arabia, Turkey, and Southern Europe. Another, somewhat smaller continent called Laurentia included most of northern North America, Greenland, and Scotland. Initially, these and a couple of other smaller plates were located along the equator, but by 430 million years ago Gondwana had drifted southward and was centered over the south pole. This was followed by a subsequent drift to the north, accompanied by a coalescence of most of the other continental fragments into it to form Pangea.

CHAPTER **6** **THE SKY AS SEEN FROM DIFFERENT LATITUDES**

As we travel northward or southward, the sky changes in appearance. The accompanying diagrams show which portion of the sky appears from various parts of the earth. In the first figure the plane of the earth's equator has been extended out into space. Like any other plane, it cuts all space into segments. Half of the sky is north of the equator and half is south. During a short interval, such as the lifetime of a man, the axis about which the earth is whirling keeps very nearly stationary, and, therefore, the equator also must remain very nearly on the same plane.

Since the distance to the stars is so great and since the earth's axis moves very little during the lifetime of a man, the stars appear to stay in the same position. An arrow pointing almost along the axis of the earth is in this picture directed toward Polaris, the north star, but if our picture were to be extended to include that star, it would be necessary, using this tiny scale, to carry it north 5,700,000 miles beyond our printed page. Polaris is not the nearest star. The nearest one is south of the equator of the earth—indeed, so far south that it has not been visible for many centuries from the latitude of Los Angeles. But even for this star we would have to extend the arrow, which is pointed toward it, by 93,000 miles.

These numbers illustrate the distances of the stars from us and from each other, on a scale in which the earth is a little ball two inches in diameter. Because of the inconceivable magnitude of these distances, we find that the stars, despite their velocities, do not change position rapidly as viewed by us. In other words, if a star is north of the present

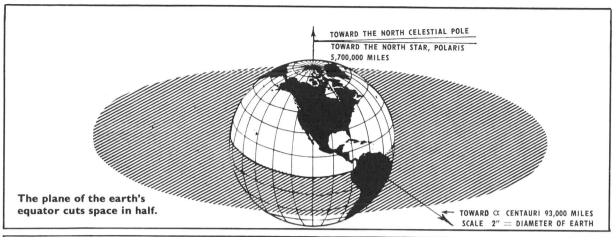

TOWARD THE NORTH CELESTIAL POLE

TOWARD THE NORTH STAR, POLARIS
5,700,000 MILES

The plane of the earth's equator cuts space in half.

← TOWARD α CENTAURI 93,000 MILES
SCALE 2" = DIAMETER OF EARTH

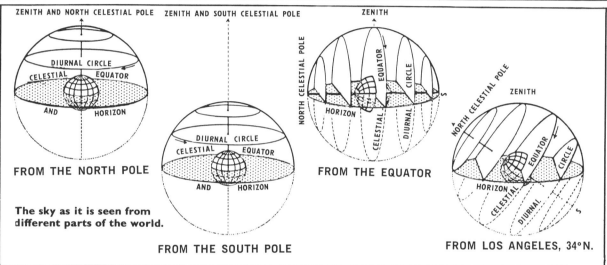

ZENITH AND NORTH CELESTIAL POLE   ZENITH AND SOUTH CELESTIAL POLE   ZENITH

DIURNAL CIRCLE

CELESTIAL   EQUATOR

AND   HORIZON

FROM THE NORTH POLE

DIURNAL CIRCLE

CELESTIAL   EQUATOR

AND   HORIZON

The sky as it is seen from different parts of the world.

FROM THE SOUTH POLE

ZENITH

NORTH CELESTIAL POLE

EQUATOR

DIURNAL CIRCLE

HORIZON

CELESTIAL

FROM THE EQUATOR

NORTH CELESTIAL POLE

ZENITH

EQUATOR

CIRCLE

HORIZON

CELESTIAL

DIURNAL

FROM LOS ANGELES, 34°N.

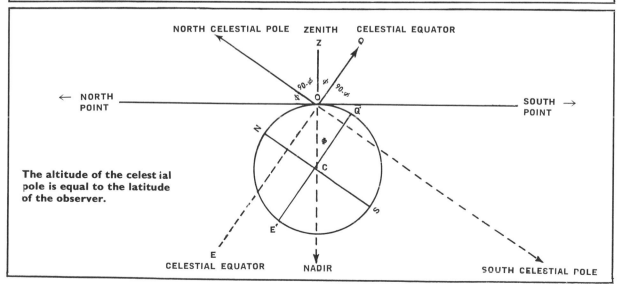

NORTH CELESTIAL POLE   ZENITH   CELESTIAL EQUATOR

Z

Q

90-φ   φ   90-φ

← NORTH POINT

O

Q'

SOUTH POINT →

N

Φ

C

The altitude of the celestial pole is equal to the latitude of the observer.

S

E'

E

CELESTIAL EQUATOR

NADIR

SOUTH CELESTIAL POLE

plane of the equator, its angle of elevation from this plane will change very little over a century.

In order to catalogue and chart the positions of the stars, we picture the earth as surrounded by a blue sphere of infinite radius—the *celestial sphere*. If the plane of the equator be carried out through this imaginary sphere, we obtain a circle which we name the *celestial equator*. From this circle it is customary to measure positions of stars for our star catalogues. The angular distance of a star north or south of the celestial equator is known as the star's *declination*. The point toward which the axis of the earth is pointed in the north is called the *north celestial pole*, and the opposite point of the sky, the *south celestial pole*. The whirling of the earth causes all objects of the sky to appear to revolve around these two points in daily circles and, therefore, to move parallel to the celestial equator.

The next four figures show the sky as viewed from four different latitudes—the north pole, the south pole, the equator, and the latitude of Los Angeles. Regardless of where an observer is actually standing, he has the illusion of being on top of the globe. Because of this instinctive feeling, the observer is represented for each of these places as being at the uppermost point of the earth. His horizon has appeared not to be changed, although in his traveling on the surface of the earth it is the horizon which actually has been tipped. Instead of the true motion, the whole sphere of the stars has appeared to him to shift.

In observing the stars there are two points at which the earth's axis will coincide with the zenith —the north pole and the south pole. To find the zenith of the sky at any given point, we use a plumb line. The direction in which a body falls is called the *plumb line*, or *vertical*. If we sight upward along an imaginary plumb line, it seems to cut the sky in the point we call the *zenith*. If a stone is dropped at the north pole, it will fall toward the earth along the axis of the earth. Therefore, the plumb line of an observer at the north pole is the earth's axis, which is directed toward the north celestial pole. For this observer the north celestial pole and the zenith coincide. He, therefore, will find his horizon coinciding with the celestial equator, and every star that is north of the equator must continually be above his horizon. The stars do not appear to rise or set for the observer at either pole of the earth. Due to the fact that the distances of the stars from the equator are not changing, any given star

must remain at all times the same distance above his horizon. It will, therefore, have a daily, or diurnal, path parallel to the horizon. This, of course, does not hold for the sun and the moon and the planets, which are millions of times closer to us than the stars, and whose distances from the celestial equator change continually.

The sun is north of the equator for half of the year. During that time it is above the horizon for an observer at the north pole. When on June 21st it gets $23\frac{1}{2}°$ north of the equator, it is above his horizon by that same distance.

When the observer has traveled halfway around the world from the north pole, to the south pole, he still feels he is on the top of the earth. Once more his plumb line coincides with the axis of the earth, but now the up direction is toward the south pole of the sky, and the north pole is at his *nadir*. The apparent direction of the daily motion of the stars is exactly opposite to what it was at the other pole. This last statement may be illustrated by spinning a globe on its axis. Observe the direction in which the globe is spinning and then turn it upside down while still in motion. You will now see that it is spinning in the opposite direction.

Looking at the fourth figure we see that the observer on the earth's equator is standing in the plane that traces out the celestial equator. If this observer drops a stone, it will fall toward the earth in this plane. His plumb line, therefore, must point upward toward the celestial equator, which also must pass through the zenith. Since the poles of the sky are 90° from the celestial equator, and the horizon is 90° from the zenith, they must lie on the horizon, and the north star must be very close to the north point of the horizon. The celestial equator here rises vertically instead of coinciding with the horizon as it does at the poles, and every star making its daily motion parallel to the equator must come up vertically from the horizon. Each star in the sky must be above the horizon half the time. At the north pole it was found that those stars north of the celestial equator were always visible and those stars south of the equator were never visible. The south pole of the earth reverses the conditions of the north pole.

At the equator there are no stars always visible and no stars always hidden. For latitudes between the poles and equator, the situation is a trifle more complicated. *No matter where a person is on the surface of the earth, the altitude of the celestial pole is equal to his latitude.* In other words, the number of degrees between the horizon and the axis of the earth is equal to the number of degrees between the observer's plumb line and the plane of the equator. The proof of this statement is not at all difficult. It requires only some simple high school geometry, which will be given at the end of this chapter. This little law may be illustrated thus: If one stands at the equator of the earth, the celestial pole is on his horizon. If he travels north 5°, the north pole appears to climb 5° above the horizon. When he gets 20° north, he finds the north celestial pole 20° above the horizon. When he is 70° north, the north celestial pole is 70° high. And when he reaches the pole of the earth, the celestial pole is at his zenith. If he goes 5° south from the equator of the earth, the south celestial pole is 5° above the horizon, and the north pole is 5° below it. And similar changes occur as he continues on toward the south pole of the earth. Those stars whose distances from the pole are less than the latitude of the observer can never set because the radii of their daily circles about the pole must be less than the distance from the pole to the horizon.

Astronomers call this region, within which the stars never set, the *region of perpetual apparition.* Around the opposite pole must be a region within which the stars cannot rise. This region is called the *region of perpetual occultation.* The fifth diagram shows these regions for the latitude of Los Angeles. The star Alpha Centauri is far enough south to be in the region of perpetual occultation. Within this same region lie the Southern Cross and some of the most beautiful of the constellations. For this latitude of 34° north, the region of perpetual occultation includes all stars within 34° of the south pole, or to express it differently, all stars more than 56° south of the equator. Within the region of perpetual apparition lie the Little Dipper and parts of the Big Dipper, of Draco, and of Cassiopeia. For observers far enough from the equator, the sun in summer will be in the region of perpetual apparition, and in the winter in that of perpetual occultation. As a result, such regions have the midnight sun of summer and the 24-hour night of winter.

There is some twilight until the sun is 18° below the horizon. An examination of the diagrams of page 37 shows that the sun goes "straight down" at the earth's equator and that, therefore, the twilight is very short. As we move away from the equator we find longer twilight, for the sun goes down at an angle. At the earth's poles twilight lasts several weeks. Indeed at the poles, except for effects of clouds, we have the minimum annual total of complete darkness!

PROOF OF THE LAW OF POLAR ALTITUDE

The latitude of the observer is always equal to the altitude, or height, of the celestial pole above the horizon. The rigid proof of this relationship is so simple, requiring only the most elementary high school geometry, that it seems well to give it here.

In the sixth diagram the observer is at the point O. His plumb line is the line ZOC, where C is the point at which the plumb line intersects the plane of the equator, near, but not necessarily at, the exact center of the earth. E'Q' traces the plane of the earth's equator, and EQ, through the point O, is parallel to it. If EQ were extended to infinity, it would eventually meet the extension of the plane of the earth's equator. Thus one can say that the line EQ points to the celestial equator, even though the celestial equator is actually found by extending the plane of the earth's equator. In a similar way, one can say that the line from the north to the south celestial pole passes through O. Now, by *definition,* the latitude of the observer is equal to the angle between his plumb line and the plane of the equator. This is the angle ZCQ'. It is customary in all languages to designate this angle by the letter ϕ. ZOQ is the angle the observer sees between the zenith and the celestial equator. This angle and his latitude are the exterior-interior angles formed by a line cutting parallel lines. They, therefore, are equal. Since it is 90° from the pole to the equator and also 90° from the zenith to any point on the horizon, it follows at once that the angle, north celestial pole-observer-north point, also equals the latitude. This last angle is the altitude of the celestial pole, the point around which the stars appear to make their daily circles.

This law explains the positions of the sky as shown in the earlier diagram and described in the main part of this chapter.

CHAPTER 7 EXPLORING THE EARTH'S ATMOSPHERE

When we look at the sky on a clear night, we may not realize that we are viewing the stars from the bottom of a vast, seething ocean of air. The use of such expressions as "vanished into thin air," and "as light as air" shows how unsubstantial we regard our atmosphere. Yet it takes but twelve cubic feet of air to weigh a pound. The planetarium chamber of the Griffith Observatory, which seats about 500 people, contains six tons of air.

We can determine exactly how much atmosphere there is without ever leaving the bottom of our ocean. The ordinary domestic barometer is weighing it for us all the time. When the barometer needle points to 30, there is as much weight over our heads as there is in a layer of mercury 30 inches thick. This is the same amount as there would be in a layer of bituminous coal about 25 feet thick, since the density of mercury is about ten times that of the coal. It is somewhat surprising that we can see through such substantial atmosphere. The pressure of the air amounts to nearly 15 pounds on every square inch of our bodies, so that a man of ordinary stature is exposed to a total pressure of about 14 tons. But since the air can penetrate the porous tissues and cavities of the body, the pressure is exerted equally in all directions, and under normal conditions we are not conscious of it.

Man has climbed mountains in search of facts about the atmosphere. When he reaches a height of $3\frac{1}{2}$ miles, he finds that his barometer reads only 15 inches instead of the usual 30-inch pressure existing at sea level. This means that half of the total substance of the atmosphere is within $3\frac{1}{2}$ miles above sea level. But it does not mean that the atmosphere ends at a height of 7 miles. The extreme height of the atmosphere must be more than 600 miles above the earth's surface. Indeed, it could be considered to extend to an altitude of 10,000 miles, depending upon what is considered to be atmosphere. Nevertheless, the bulk of the atmosphere is within a matter of miles from the earth's surface. The air is most dense in its lower layers, where there is a great deal of air above it pressing it down.

Mount Everest, the highest mountain in the world, has an elevation of $5\frac{1}{2}$ miles. Since the atmospheric pressure at the top gives a barometer reading of $9\frac{1}{2}$ inches, only one-third of the air remains above it. Airplanes and manned balloons have attained much higher altitudes. When Major David G. Simons, an Air Force doctor, reached a height of nearly 19 miles (100,000 feet) in a sealed gondola attached to a helium-filled plastic balloon in 1957, he was above 99 per cent of the atmosphere. The sky appeared, in his words, as "a dark, purplish black, a color I have not seen before and can't really describe." Since there was practically no air to interfere, the stars did not

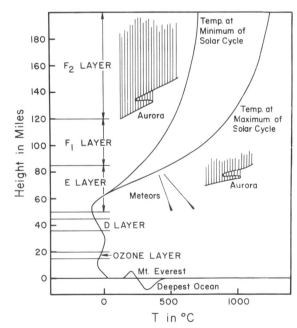

Temperature distribution in the earth's atmosphere.

Also shown are locations of the ozone layer, the *D, E,* and *F* ionized layers, and the mean heights of meteors and aurorae. (Adapted from *Elementary Astronomy,* Struve, Oxford University Press.)

34

twinkle, but shone steadily. (An even higher altitude was reached in a balloon in May, 1961, when Lieutenant Commander Victor G. Prather and Commander Malcolm Ross ascended to almost 22 miles above the earth's surface and recently the X-15 rocket plane reached an altitude above 41 miles.) It is not merely the greater brightness of the sun which hides the stars from us in the daytime. It is because the air scatters the sunlight and sends it to us from all parts of the sky, that the fainter light of the stars does not show. An observer on the moon could see the stars plainly in full daylight, because the moon has no atmosphere.

One recent high-altitude balloon flight secured a photograph of the sun (see page 17). Another such flight gave a spectrogram showing that the atmosphere of Venus does contain water vapor. The results were possible only because there is a small amount of atmosphere above the balloons.

The scattering of sunlight by air is not the same for all colors. The sunlight may be thought of as made up of waves of different length, the short waves being blue, the long waves red, and the other colors in between. The atmosphere scatters a larger proportion of the blue, short-wave light than of the red, long-wave light. This takes from the sun the bluish-white glare it would otherwise have. It is the predominance of this scattered blue light that makes the sky blue. When an observer sees the sun near the horizon, the sunlight comes to him through a greater thickness of air than when he sees the sun higher in the sky. The sunset colors are orange and red, because they are of the longer wave lengths of light that are absorbed the least and penetrate through the air between the observer and the setting sun. The reddening of the sun was observed in a very striking way in 1883, when the volcano Krakatoa, on an island west of Java, erupted and threw up great clouds of volcanic dust, which were carried around the world. For several months, the sunrises and sunsets were of an indescribable magnificence.

In addition to the manned balloon, the unmanned balloon and rockets have played an important part in exploring the atmosphere to more than a hundred miles in altitude. Artificial satellites have revealed the presence of radiation belts several thousands of miles above the surface. The records obtained by balloons and rockets with self-recording thermometers show that there is a decrease of about 1° Fahrenheit for every 300 feet rise in elevation up to an altitude of about seven miles. This marks the beginning of the stratosphere, where the temperature changes very little with altitude. The base of the stratosphere varies with weather conditions, with the season, and with the latitude. Its height is about eleven miles over the equator, about seven miles over the temperate zones, and about four miles over the poles.

From the fact that twilight lasts until the sun is about 18° below the horizon, it is known that the air must be at least about 50 miles high. Twilight is due to reflection from air on which the sun is still shining after the sun has set at the horizon. If there were no air around the earth, the setting of the sun would abruptly change day to night, as though by the snap of a switch.

Above 50 miles, the air is not dense enough to reflect the sunlight appreciably, but it is still dense enough to make meteors visible. These streaks of light, which are commonly known as "shooting stars," are caused by small swiftly moving celestial bodies, most of which are probably no larger than grains of sand. Ordinarily these particles are invisible, but those entering the earth's atmosphere are heated enough by friction to become visible at an average height of about 75 miles. Occasionally a bright meteor leaves a persistent train that may be observed for some time. An interesting example is a brilliant meteor that passed over the northern coast of France on February 22, 1909, at 7:33 p.m. The luminous train left in its wake remained visible for more than two hours, during which time the train drifted over England in a northwest direction at a speed of 120 miles an hour. The exact nature of a meteor train is not known, but it is probably an electrical phenomenon similar to the afterglow produced by an electrical discharge in a vacuum tube. Meteor trains are formed in a rather thin layer of the atmosphere at an average height of about 55 miles, though the visible paths of the meteors that produce them may extend far above and below this height.

The region in which meteors are seen marks the lower limit of what is called the *ionosphere*, the atmospheric layer from which radio waves are reflected. In the early days, it was a puzzle to explain how wireless waves, which were supposed to travel in a straight line like light, could go around the curved surface of the earth. Now it is known that the radio waves are reflected back to the earth from several electrified layers of the atmosphere, and in this way are "bounced" around the earth's curved surface. One of the electrified layers is the Kennelly-Heaviside layer, at an average height of about 65 miles; another is the Appleton layer, at an average height of about 150 miles.

In addition to balloon, rocket, and airplane flights, another method of studying the atmosphere is by using sound as an explorer. Scientists found that a great explosion can be heard within a circle having an average radius of about 60 miles. Outside this area is a ring of about 125 miles in breadth in which the sound cannot be heard. Beyond this zone of silence is a zone in which the sound is again audible. It was suggested that the sound waves are reflected back from a layer of air having a higher temperature than that of the stratosphere. Evidence from a study of meteors also indicates the existence of such a hot layer.

Early in 1947 announcement was made of the direct measurement of temperatures up to about 75 miles. As had been suspected, a hot layer was found between 30 and 40 miles high. Thermometers in V-2 rockets launched at White Sands, New Mexico, recorded a temperature of 170° Fahrenheit in this layer. Measurements showed that from 40 to 50 miles, the air grows colder again. Between 50 and 75 miles is another torrid zone even hotter than the lower one. This is due partly to the presence of dust particles, which absorb and hold the heat.

The first increase in temperature is produced by a layer of ozone, which has the property of absorbing and holding more heat from the sun than the air below it. Ozone is a heavy variety of oxygen, having three atoms to the molecule instead of the usual two. It absorbs most of the ultraviolet radiation coming from the sun.

The Explorer artificial satellites and Pioneer lunar probes launched from Cape Canaveral during 1958 showed that the atmosphere of the earth extends enormously farther from the surface than had been previously thought. The word "atmosphere" includes all gaseous or dusty material in the neighborhood of the earth, under the influence of the earth's magnetic or gravitational fields. For example, cosmic ray detectors carried aloft by these missiles revealed the surprising fact that the earth is surrounded by a radiation belt of diminishing intensity. They have become known as the "Van Allen radiation belts" in honor of physicist James A. Van Allen, the director of the research team making the discovery. The first layer begins about one thousand miles above the earth's surface and extends to between three and four thousand miles from the surface. The second layer is located at an altitude of about ten thousand miles. At least the first layer has the form of a belt around the earth, centered over the equator. Current theories indicate that the layers consist of charged particles—electrons and protons—trapped by the earth's magnetic field. It has been suggested that the charged particles are of solar origin, but this question is still controversial. More recent flights have shown that the outer belt is variable in strength and depth, occasionally even merging with the inner belt. Undoubtedly it is strongly influenced by changes in solar activity.

The reason that human beings are so well suited to life in our atmosphere is that we are the offspring of millions of generations of ancestors, whose organs have been gradually adapting themselves to their environment. Our air contains about 78 percent nitrogen, 21 percent oxygen, nearly 1 percent argon, .03 percent carbon dioxide, and traces of several other gases, but the very minute quantity of ozone is as important to us as any of the rest.

Our atmosphere performs certain functions favorable to the development of living beings. In the first place, the atmosphere acts as an efficient blanket that smooths out fluctuations of temperature. If there were no atmosphere, the daily range of surface temperatures would be similar to that measured on the surface of the moon. When the sun is at the lunar zenith, the surface temperature reaches the boiling point of water. Soon after sunset, the temperature has dropped to −60° Fahrenheit, while at lunar midnight it has plummeted to −240°. Conceivably living beings could adapt to such extremes of temperature by burrowing into the ground, but it is probable that such conditions would retard the extensive evolutionary changes we have experienced here on the earth.

Another benefit provided by the atmosphere is its ability to shield us from most of the meteoritic material continually raining down at us from outer space. Most meteoroids never penetrate closer than 30 or 40 miles above the surface. As they pass through the tenuous outer atmosphere, frictional heating melts the outer surface of the meteoroid, the melted material is sloughed off, and soon the entire object is consumed. When one realizes that meteoroids come in at velocities up to more than 44 miles per second, it is obvious that this atmospheric shield provides important protection. Without that shield our daily life would become even more risky than it is already!

At elevations above 30 miles from the earth's surface, the density of the atmosphere is so low that the ordinary concept of temperature is meaningless. The high temperatures recorded in the ionized layers of the ionosphere are caused simply by the charged particles attaining high velocities when the atoms in these layers are ionized by absorption of sunlight. If a slowly moving artificial satellite at these elevations were shielded from the direct rays of the sun, its temperature would be very low, since it could absorb little heat from the thin surrounding atmosphere.

The average person finds it paradoxical that the air near the earth grows colder as one ascends and approaches the sun. One reason is that the lower strata of the air are not directly heated by the sun's radiation, which passes through the air largely unabsorbed. The earth's surface absorbs the sun's heat and re-radiates it in long-wave radiation, which the air is able to absorb. In this way the air nearest the earth is heated the most. Water vapor in the air has much to do with the absorption of long-wave radiation from the earth and the heating of the lower atmosphere.

Thus the atmosphere acts like the glass in a greenhouse, letting a great deal of the solar radiation pass through, but preventing much of the heat radiated by the earth from escaping on out into space.

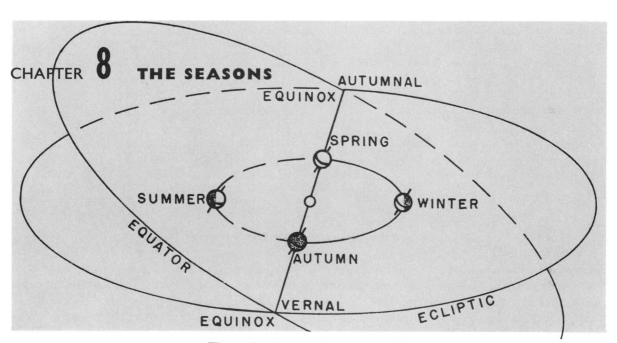

CHAPTER 8 THE SEASONS

The earth at four positions in its orbit.

The above diagram shows the earth at four positions in its orbit. It shows the earth at the beginning of winter, spring, summer, and autumn. The sun is in the center. As the earth moves around the sun, it appears to us that the sun moves in the large circle marked "ecliptic" lying in the same plane. The circle marked "equator" is the projection of the plane of the earth's equator against the sky. The diagram shows how the southern hemisphere is more illuminated than the northern during our fall and winter and how this is reversed during the spring and summer. It also shows that the part of the ecliptic against which we see the sun during spring and summer is north of the equator of the sky and that the other half is south of the equator. The axis of the earth is shown, about which it whirls daily like a top. That the North Pole is in continuous sunlight during the spring and summer is obvious from this diagram, as well as the fact that the South Pole enjoys this advantage during

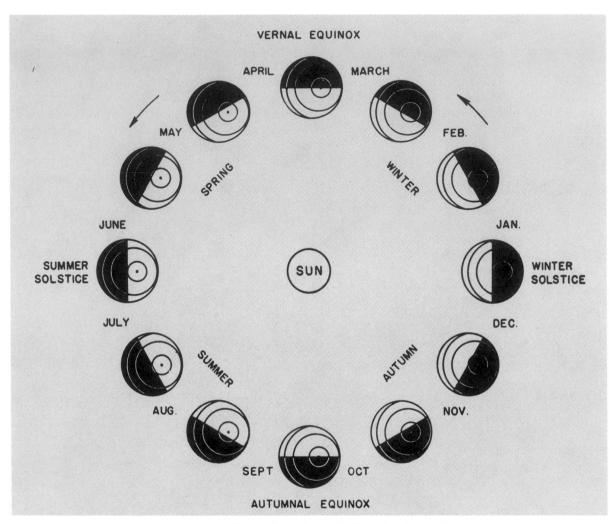

Positions of the earth on about the 21st of each month.

our fall and winter also is obvious in the diagram.

There are two reasons why we have seasonal fluctuations in the length and warmth of the day. These cyclic changes are due to the fact that the axis upon which the earth rotates, rather than being perpendicular to the earth's orbit around the sun, deviates $23\frac{1}{2}°$ toward its plane. This causes variations both in the length of the day and the intensity of the sun's heat.

Let us first examine why the days become longer as the summer season approaches. The second diagram shows the earth at monthly intervals, as viewed from a direction perpendicular to the plane of its orbit. In each of the 12 positions, the North Pole is shown at the center of two circles. The smaller one is the Arctic Circle and the other is the Tropic of Cancer. One-half of the equator appears tilted up at the left of each position. The sun illuminates one-half of the earth all the time, and the North Pole is in the sunlit half during the six months from about March 21 to September 23. The sun remains above the horizon at the North Pole and below the horizon at the South Pole 24 hours a day during that time. On June 21 the sunlight reaches $23\frac{1}{2}°$ beyond the North Pole, so that the sun can be seen for 24 hours that day at all places within that distance of the North Pole. The boundary line is the Arctic Circle, which has a latitude of $66\frac{1}{2}°$ North. On December 22 all of that circle is in the dark half of the earth, the sun remaining below the horizon all day. (The latitude would not be exactly $66\frac{1}{2}°$, because the air bends the light from the sun and raises its upper edge about half a degree above the horizon.)

38

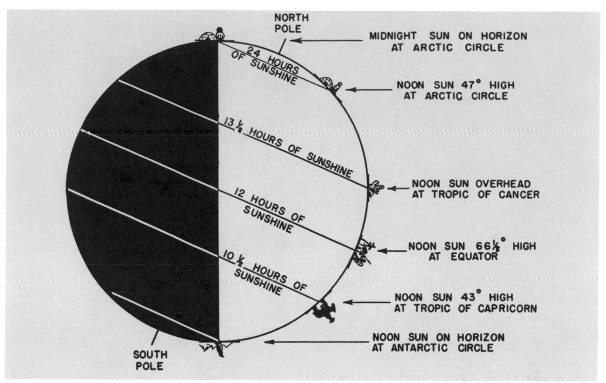

NORTH
POLE

MIDNIGHT SUN ON HORIZON
AT ARCTIC CIRCLE

24 HOURS
OF SUNSHINE

NOON SUN 47° HIGH
AT ARCTIC CIRCLE

13 ½ HOURS OF SUNSHINE

NOON SUN OVERHEAD
AT TROPIC OF CANCER

12 HOURS OF
SUNSHINE

NOON SUN 66½° HIGH
AT EQUATOR

10 ½ HOURS OF
SUNSHINE

NOON SUN 43° HIGH
AT TROPIC OF CAPRICORN

NOON SUN ON HORIZON
AT ANTARCTIC CIRCLE

SOUTH
POLE

The earth on June 21.

The earth on December 22.

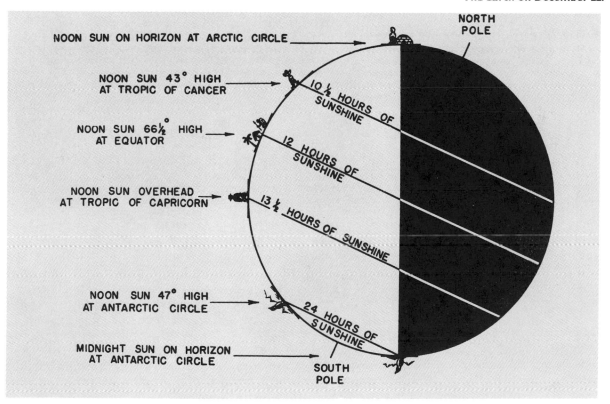

NORTH
POLE

NOON SUN ON HORIZON AT ARCTIC CIRCLE

NOON SUN 43° HIGH
AT TROPIC OF CANCER

10 ½ HOURS OF
SUNSHINE

NOON SUN 66½° HIGH
AT EQUATOR

12 HOURS OF
SUNSHINE

NOON SUN OVERHEAD
AT TROPIC OF CAPRICORN

13 ½ HOURS OF SUNSHINE

NOON SUN 47° HIGH
AT ANTARCTIC CIRCLE

24 HOURS OF
SUNSHINE

MIDNIGHT SUN ON HORIZON
AT ANTARCTIC CIRCLE

SOUTH
POLE

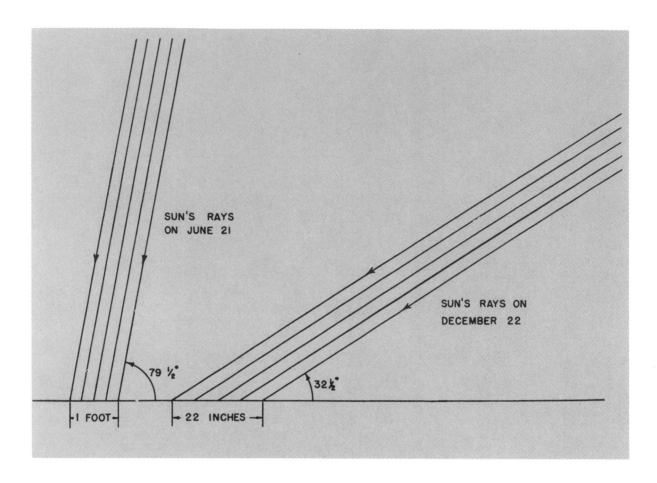

Earth's surface at Los Angeles—34° north latitude.

This diagram demonstrates the fact that not only the number of hours of sunlight, but also the altitude of the sun produce the heat of summer. The noonday altitude of the sun, for the days marked, is shown. At the lower altitude the same amount of energy from the sun is spread over nearly twice the area and, therefore, it takes nearly twice as long to warm the surface by a given amount. At the poles the sun is never more than $23\frac{1}{2}°$ high. However, when highest, the steady 24 hours of sunshine more than over-balance the fewer hours when it is higher at the equator.

The preceding diagrams show the earth at the solstices as viewed from a point in the plane of the earth's orbit. On June 21 an observer at the Arctic Circle sees the sun just above his northern horizon at midnight and 47° above his southern horizon at noon. He has 24 hours of sunshine and is just on the boundary of the Land of the Midnight Sun. In the center of that area is the North Pole, where the sun's altitude remains at about $23\frac{1}{2}°$ during the 24 hours of June 21. If one moves southward $66\frac{1}{2}°$ from the North Pole, to a latitude of $23\frac{1}{2}°$ North, the sun's noonday altitude increases by $66\frac{1}{2}°$ and appears overhead on that date. This circle of latitude is the Tropic of Cancer. The word "tropic" is derived from a word meaning "turn," referring to the turning of the sun southward after June 21. Cancer is the Latin word for "crab," and the sun

originally appeared in the constellation of that name when it reached the summer solstice. Two thousand years of precession, the slow conical motion of the earth's axis, have caused the summer solstice to move from Cancer into the adjoining constellation, Gemini.

If the noon sun on June 21 is directly above an observer at the latitude of $23\frac{1}{2}°$ North, it is $10\frac{1}{2}°$ south of the zenith of a person at a latitude of 34° North, such as at Los Angeles. The zenith distance of the noon sun on that date is equal to the difference of latitude between the observer and the Tropic of Cancer. As another example, Philadelphia, at a latitude of 40° North, is $16\frac{1}{2}°$ north of the Tropic of Cancer. Therefore, the noon sun on June 21 is $16\frac{1}{2}°$ south of the zenith, or $73\frac{1}{2}°$ above the southern horizon.

Everyone south of the Tropic of Cancer faces north to see the noonday sun on that date. It is $23\frac{1}{2}°$ north of the zenith at the equator, and $47°$ north of the zenith at the Tropic of Capricorn, which has a latitude of $23\frac{1}{2}°$ South. The Antarctic Circle is $66\frac{1}{2}°$ south of the equator and $90°$ south of the Tropic of Cancer. Therefore, the sun there is on the horizon at noon and below the horizon the rest of the day. Within about $23\frac{1}{2}°$ of the South Pole the sun does not rise on June 21.

The hours of sunshine on June 21 vary from 24 in the Arctic region to none in the Antarctic region. At the equator there are 12 hours of sunshine, not only on June 21, but on every day of the year. As one moves northward from the equator, the hours of sunshine on June 21 increase slowly at first and then more rapidly as the Arctic Circle is approached. As one moves southward from the equator, the hours of sunshine on June 21 decrease slowly at first and then more rapidly as the Antarctic Circle is approached.

With the sun high in the sky and above the horizon for more than 12 hours each day, the northern hemisphere of the earth has summer at this time. With the sun low in the sky and above the horizon for less than 12 hours each day, the southern hemisphere has winter.

On December 22 the conditions in the two hemispheres are reversed. The sun's rays reach $23\frac{1}{2}°$ beyond the South Pole and fall $23\frac{1}{2}°$ short of

the North Pole. An observer at the Tropic of Capricorn finds the noon sun overhead. The name of that circle is derived from the fact that the sun originally appeared in the constellation of Capricornus when it was farthest south at the point called the winter solstice. This is called the *winter* solstice, because the sun reaches that point at the beginning of winter in the northern hemisphere, but of summer at this time in the southern hemisphere.

For every place on the earth the altitude of the noon sun changes by $47°$ from June 21 to December 22. For instance, in Los Angeles the noon sun is $79\frac{1}{2}°$ high on June 21 and only $32\frac{1}{2}°$ high on December 22. The effect of this change is shown in the diagram of the two beams of sunlight of equal width. The December beam strikes the earth's surface so obliquely that it covers nearly twice as much ground as the June beam. With the same amount of heat in each beam, it is obvious that the ground would be heated more rapidly in June. Also, in December the sun's rays have to penetrate a greater thickness of air before they reach the ground, and so they are subject to more absorption and scattering.

Thus we see that the seasons are caused by the changes in the sun's altitude and in the daily duration of sunlight. These, in turn, are a result of the tipping of the earth's axis away from the perpendicular to the earth's orbit around the sun.

CHAPTER **9** **THE SUN'S DAILY PATH ACROSS THE SKY**

The changing position of the sun in the sky from hour to hour and from day to day is of interest especially to architects, who want to know when and how far the sun will shine in the windows of different sides of a building. The accompanying diagrams show the sun's daily path across the sky at the latitudes of $34°$ and $42°$ North. From these two diagrams, one can determine the approximate position of the sun with respect to the horizon of any place in the United States at any hour of the year.

The sun's position with reference to the horizon is expressed by altitude and azimuth. *Altitude* is

the angular distance above the horizon measured perpendicularly to the horizon. Its maximum is $90°$ at the zenith. Altitude is marked on the diagrams at intervals of $10°$ along the vertical line in the center. That line represents the celestial meridian, which the sun crosses at noon. The altitude of the sun for any time of the day can be read from the concentric circles $2°$ apart on the diagram.

Azimuth is the angular sweep measured along the horizon in a clockwise direction. Astronomers measure it from the south point, navigators from the north point. In our diagrams, which follow the navigators' rule, north is $0°$, east is $90°$, south is

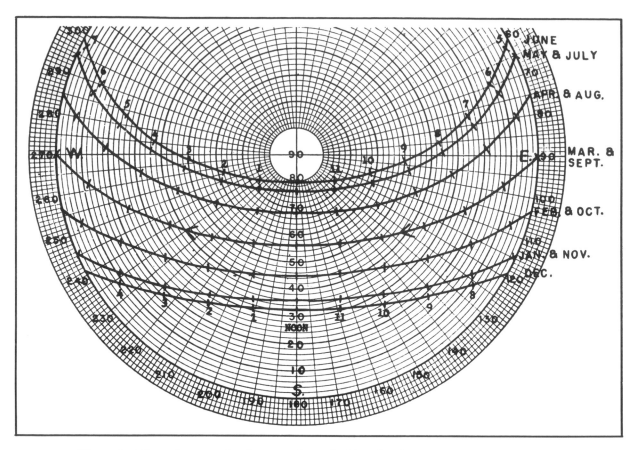

The sun's daily path across the sky on or about the 21st of each month at a latitude of 34° N.

180°, and west is 270°. Each degree of azimuth is shown in the circular band around the outside of the diagram, and numbers from 60 to 300 indicate the azimuth at intervals of 10°. Lines radiating from the center mark the azimuth at intervals of 5°.

The sun's daily path across the sky on or about the 21st day of each month is indicated by means of seven curved lines. The upper one is for June, and the lower one is for December. Each of the other five is for two months. For instance, the path on March 21 is the same as on September 23.

Each path is divided into hours. Numbers along the upper and lower paths show the hours that would be indicated by a sundial. This is known as local apparent sun time. Standard time will differ from this, depending on the observer's longitude and latitude. (See discussion of time, pages 46-49.) However, this is not important for our purpose, which is to show the general course of the sun across the sky and not its exact position at any particular instant of time.

It is interesting to see how much the sunrise and sunset points move during the year. The azimuths of their extreme positions for a latitude of 34° N. are as follows:

	Sunrise	Sunset
June 21	61°	299°
Dec. 21	119°	241°
Difference	58°	58°

In other words, on December 21 the sun rises 29° south of east and sets 29° south of west. The arc of the horizon between the east point and the sunrise point is called *amplitude*. On June 21 it is $23\frac{1}{2}°$ at the equator and increases to 90° at the Arctic Circle, where the sun is up for 24 hours on that day. At a latitude of 42° N. the amplitude on that day is $32\frac{1}{2}°$.

On March 21 and September 23 the sun is on the celestial equator, which intersects the celestial meridian at a distance from the zenith equal to the

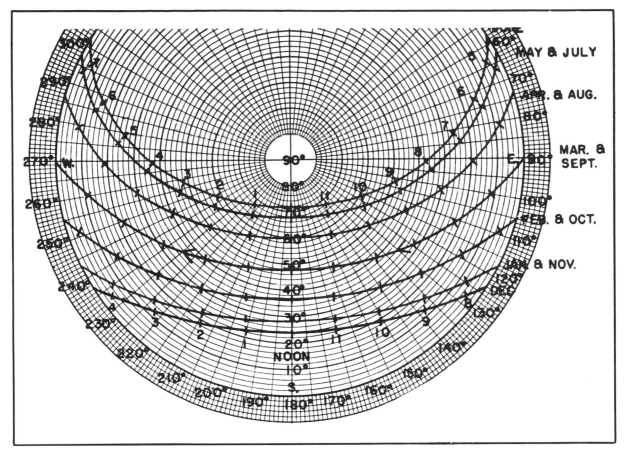

The sun's daily path across the sky on or about the 21st of each month at a latitude of 42° N.

latitude. At a latitude of 34° N. on these dates the noon sun is 34° south of the zenith. Its altitude is 56°, which is found by subtracting 34° from 90°.

On June 21, at a latitude of 34° N., the sun is $23\frac{1}{2}°$ north of the celestial equator, and its maximum altitude at noon is $79\frac{1}{2}°$, which is the sum of 56° and $23\frac{1}{2}°$. On December 21 the sun is $23\frac{1}{2}°$ south of the celestial equator and its noon altitude is $32\frac{1}{2}°$, which is the difference between 56° and $23\frac{1}{2}°$. The sun's meridian altitude varies 47°. This range is twice $23\frac{1}{2}°$ and is the same for all latitudes.

The approximate duration of sunlight can be estimated from the diagram for a latitude of 34° N. The upper curve for June 21 shows that the sun rises at about 4:45 A.M. and sets at about 7:15 P.M., giving a duration of $14\frac{1}{2}$ hours. On May 21 and July 21 the sun is up for 14 hours. The duration of sunlight in March and September, when the sun is on the celestial equator, is 12 hours. There are nearly 10 hours of sunshine on the shortest day.

CHAPTER **10** WHAT MAKES OUR CLIMATE?

Often it is said that all of our weather comes from the sun. In a very real sense this is true, for if the earth were carried far from the sun, our atmosphere would freeze, and a perpetual uniform condition soon would exist over the earth's whole surface. Of all the celestial bodies, only the sun has

43

any noticeable effect on the climate of the earth.

In another sense, however, the sun contributes not more than half of our climate. Conditions on the earth are almost as fundamental to climate as is the radiation received from the sun. Climate is a complex subject. As an aid to understanding it, let us start with an extremely simplified case and add the appropriate modifying effects, one by one.

The first aspect of climate that we shall examine is the heat that a body can absorb and emit. Consider a minute particle in space exposed to the sun's radiation. Calculations will tell almost perfectly the conditions attained by it. Following such a study, a smooth, dry planet of normal size, but without atmosphere, can be considered. (Our moon is a good example of an airless body.) In this case also, simple calculations will predict accurately what will take place. These two simplified cases will form the main subject matter of this chapter. After such a study has been completed, the effects of atmospheres of varying densities can be contemplated. These must then be modified by considerations of the gases contained in them, by their moisture and cloud content, etc. Finally oceans, continents, mountain ranges, and a host of other things must add their contributions. Any detailed study would fill several large books. All that can be done here is to give the barest outline of the field.

Let us consider the case of a particle moving around the sun at a distance equal to the earth's. The solar radiation is to be taken as identical with that actually striking the upper atmosphere of the earth. If the particle is what the physicists call a *black body* (*i.e.*, a body that does not reflect), it absorbs all the radiation reaching it and sends its own radiation as a substitute into space. The latter radiation depends upon the temperature to which the particle (called the radiator) is heated. The amount and quality of the radiation change very rapidly with increase of temperature. The amount of energy radiated is proportional to the fourth power of the absolute temperature (*i.e.*, as the temperature multiplied by itself four times). Absolute temperature is measured from 273° Centigrade below the freezing point of water under standard conditions. In other words, 0° absolute temperature equals −273° Centigrade. Consider three absolute temperatures of $136\frac{1}{2}°$, 273°, and $409\frac{1}{2}°$. The middle one is that of freezing water and is twice the lowest temperature. The third is three times the lowest one. If the ratio of three temperatures on an absolute scale is 1:2:3, then the density of radiation emitted is in the ratio $1^4:2^4:3^4$, or 1:16:81.

At absolute zero, which is approximately the temperature that the earth would reach eventually if the sun were not shining on it, a black body would not radiate at all. But as soon as it absorbed solar radiation and got warmer, it would begin to send out its own radiation. Such radiation, of course, would be of wave lengths far longer than the eye can perceive. As this radiation increases, it becomes more and more nearly equal to the amount of radiation received. Finally the black body reaches such a temperature that it radiates exactly as much energy as it absorbs from the sun. It then has reached a *steady state*, and the temperature remains unchanged as long as the incoming solar radiation is constant.

At the distance of the earth from the sun, the temperature that would produce this condition in a "black" particle is 277° absolute, or 4° above the standard freezing point of water.

It is easy to compute the temperature of such a black body if placed at the distance of Mars, $1\frac{1}{2}$ times the earth's distance from the sun. It would receive 4/9 as much radiation as at our distance and, therefore, radiate only 4/9 as much energy when the steady state was reached. The particle would automatically assume the lower temperature that required it to radiate this much. An extremely simple calculation by the fourth power law gives this temperature as 226° absolute, or 51 Centigrade degrees below the corresponding temperature at the earth's distance. Conversely, at the distance of Venus, the temperature would be higher than the earth's, since Venus is closer to the sun.

Another factor affecting climate is the extent to which the earth reflects and absorbs heat. Suppose that the particle were a perfect reflector instead of absorber. It would absorb no heat from the sun, because no heat would get through its surface. If initially cold, it would continue in such a condition. Its temperature would not change whether it was close to or far from the sun!

All actual bodies have qualities between these extremes. No body absorbs all radiation and none reflects all of it. Some parts of the earth, such as snow, reflect most of the sun's light; others, like black soil, absorb most of it. On the average, the surface of the earth absorbs 68% and reflects 32%. The calculated temperature for an average terrestrial particle existing alone is 252°, as against the 277° of a black body.

Now let us see what actually happens to an airless body. The calculation of this temperature assumes that the particle can radiate freely. But a particle on a planet cannot radiate downward very much, due to the other particles there. It will lose some heat downward by *conduction*, but for most of the materials on the earth conduction can account for only a small fraction of the energy received from the sun. Actually, the particle can radiate only about half as much as does an independent particle. As a result, if the sun were to shine continually on one side of an airless earth, the parts directly beneath would get far hotter than the preceding calculation has shown. On the actual earth there is an additional factor. The earth's atmosphere acts somewhat like a greenhouse, permitting the solar radiation to enter rather freely but blanketing down the long-wave outgoing energy from the earth. We must now consider a large airless planet as a whole. The sun can shine only on half of it at one time. The other half is absorbing no heat, but continues to radiate and therefore to cool. If there is no air present the cooling is rapid. This condition is partially approximated on Mars where the air has a density far less than our own. Even at the equator of Mars the daily temperature variation is greater than is the annual change for most parts of the earth. The phenomenon can be observed on our highest mountains.

The moon furnishes a particularly striking example of this condition. During the total lunar eclipse of June, 1927, Nicholson and Pettit measured the surface temperature of the rocks and dust of the moon. It was +70° Centigrade as the eclipse started, −60° at the beginning of totality, and −120° at the end. This change of 190° C, which occurred in less than three hours, is far greater than the temperature difference between the hottest and coldest spots on the earth.

Assuming that the moon is a perfect absorber and that there is no conduction to the rock beneath, its maximum temperature is calculated to be about 389° absolute, or 116° Centigrade, which is quite nearly the measured maximum value of 101° C. The moon reflects only 7% of the light that falls on it, and therefore this "black body" temperature is not far from the actual one. Allowing for the 7% reflection, the calculated value is 111°. There are several possible explanations for the small discrepancy between the measured and the actual temperatures.

Climate is also determined by the intensity of the light falling on the earth. The last part of this discussion has concerned an area on an airless planet, with the sun at the zenith. Such an area is the one marked C on the diagram. The other areas at A and B will not get as hot even though they are at the same distance from the sun.

The planet Mercury provides a good example with which to study this. Mercury is a practically airless planet with a very long period of rotation of about 60 days. It must, therefore, have its highest temperature near the place where the sun is at the zenith. This ranges from 285° C to 400° C, depending on where Mercury is in its eccentric orbit. From this place the temperature drops in all directions along the surface. The side away from the sun may be below the freezing point of water. The moon,

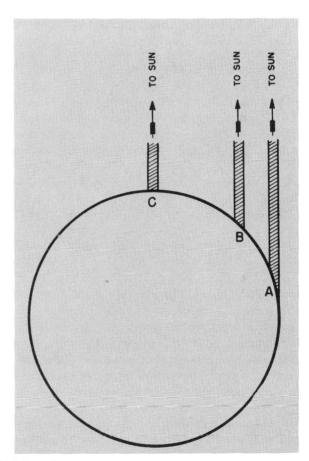

Effect of solar altitude on temperature.

Each of the three "bundles" of radiation has the same amount of energy and, therefore, the same heating effect. From A the sun is seen low in the sky. From C the sun is at the zenith. The "bundle" striking A is spread over a far greater area than that at C and gives less heat per square foot. As a consequence A is not heated to nearly as high a temperature as C. B is intermediate.

which also rotates very slowly, shows a similarly great range in temperature.

The mean temperature at any spot on the airless planet depends on a number of factors. The simplest case is a planet whose equator lies in the plane of its orbit around the sun. This case would apply to the earth if the $23\frac{1}{2}°$ "obliquity of the ecliptic" could be made zero and the air drained away. Under such conditions each point on the earth would have sunshine 12 hours. The altitude of the sun would vary each day between zero, at the instant of sunrise, and 90° minus the latitude of the place, at the instant the sun crossed the meridian each noon. Obviously the most energy would be received at the equator, where the sun passes through the zenith at noon. The least energy would be received at the poles. Thus the mean temperature would decrease steadily as one left the equator, going either north or south. Mathematically, the law of temperatures is not difficult. It can be solved by the use of sophomore college calculus.

The introduction of an obliquity (*i.e.*, an angle between the planes of the equator and the orbit) complicates the problem somewhat, but still leaves it solvable by the same general mathematical procedure as before. Seasons are introduced and some rather surprising results occur. The latitude of hottest weather travels north and south during the year from and toward the equator. If the obliquity is large enough, the hottest places during the respective summers are at the poles. On our earth the North Pole in June and the South Pole in December receive more heat per 24 hours than any other places on the earth. If it were not for the water on the earth, the poles would take their turns as the hottest spots of all.

The amount of energy received from the sun is called the *insolation*. The density of air and the amount of water vapor in it decrease the effect of insolation in producing differences of climate from one latitude to another. If our air were several times denser than it is and if our sky were perpetually covered by heavy clouds, there would be very little difference of temperature from day to night and comparatively little between the equator and the poles. The seasonal effects also would be much reduced.

Geographical features play an enormous part in the study of climate. They provide most of the complications that have made a complete understanding impossible. If it were not for them, most of the problems involved in our winds and our expanding and contracting air masses could be solved with fair accuracy.

CHAPTER **11** THE TIME OF DAY

The rotating earth is the master clock for all of humanity. The direction in which it turns we call east. We keep track of this rotation by observing the apparent westward motion of the heavens which it causes. One rotation is called a day, but there are several different kinds of day, depending on what celestial object is used as a time reckoner.

Suppose that the middle of a total eclipse of the sun occurs at noon and the star Regulus appears just above the eclipsed sun. The moon, the sun, and the star are all on the celestial meridian, the imaginary circle running north and south through the zenith. We point a telescope at them and leave it fixed in position. The turning of the earth quickly carries them out of the field of view. The next day we find that the three objects have separated. Assuming that we could see a star so close to the sun, we would note that the sun is no longer directly below Regulus. The sun has moved about one degree to the east and appears in the still-fixed telescope about four minutes after Regulus. The moon has moved about 13 degrees southeast of the star. By pointing our telescope a little farther south, we can see the moon cross the meridian nearly an hour after the star.

The interval between two successive passages of a star across the middle of the field of view of our fixed telescope is called a sidereal day. It is the period in which the earth completes one rotation. The corresponding interval for the sun is a solar day, which is about four minutes longer than the sidereal day. The corresponding interval for the moon is a lunar day, which averages about 50 minutes longer than a solar day. If the moon rises at sunset today, it will rise about an hour after sunset tomorrow. It rises later each day, because it is moving rapidly eastward among the stars. That is due to its revolution around the earth, which is completed in a little less than a month. The sun also appears to move eastward with respect to the stars, but that is due to the earth's revolution around the sun once a year.

Since our ordinary affairs are regulated by the sun and not by the moon or stars, we use the solar day and solar time. However, the sun is not a reliable timekeeper. Its apparent eastward motion with respect to the stars is not uniform. There are two reasons for this. In January the earth is 3,000,000 miles nearer to the sun than in July, and it travels a little faster. This makes the sun seem to travel faster in winter than in summer. The other reason for inaccuracies in the solar day is that the earth's equator is inclined to its orbit. The sun's apparent annual motion on the celestial sphere is parallel to the equator when it is at the solstices at the beginning of summer and of winter. At those times it makes more rapid progress to the east than when its motion is at an angle to the equator as in March and September. Only eastward motion affects the length of the day.

Because of the irregularity of the earth's orbit around the sun, an imaginary sun has been invented. It is a smooth-running timekeeper, its day being the same length throughout the year— the average of all the apparent solar days. This average sun is called the "mean" sun, and the time it keeps is "mean" solar time, which we use in everyday life. The difference between apparent solar time and mean solar time is called the equation of time. The maximum discrepancy between mean solar time and apparent solar time is a little over 16 minutes. The extent of the discrepancy between the two is often marked on the globe of the earth with a figure "8," called the *analemma*. By means of this figure one can tell at what latitude on the earth an observer would have to stand throughout the year to see the sun at the zenith at noon.

When the sun is exactly south, it is on the celestial meridian, and a sundial reads noon. The time is also apparent noon for all places on the corresponding terrestrial meridian from pole to pole. At places to the east the sun is past their meridians,

and the time is P.M. This is the abbreviation for the Latin words "post meridiem," meaning "after the meridian." The sun has not yet reached the meridian of a place to the west, where the time is A.M. This is the abbreviation for the Latin words "ante meridiem," meaning "before the meridian."

It is obvious that local time varies in an east and west direction. Formerly each community kept its own local solar time. Boston clocks were about 16 minutes faster than Philadelphia clocks, which were about 8 minutes faster than Washington clocks. As means of travel and communication became more rapid, the many different kinds of time caused much inconvenience. In 1883 the United States was divided into four standard time belts running north and south. In the following year an international conference met in Washington to establish a uniform system of time and longitude for the world. A majority of the nations represented favored the measuring of longitude from the meridian of Greenwich, England, where the Royal Observatory was located.

Since the earth turns through 360° in 24 hours, it turns through 15° in one hour. Standard meridians are marked off around the world 15° (or one hour) apart, beginning with the meridian of Greenwich. Each standard meridian is in the middle of a zone, which averages 15° in width. The boundaries between zones may be irregular, for reasons of convenience.

The standard time of any zone is the local mean solar time of the standard meridian in the middle of that zone. Eastern Standard Time is the local time of the meridian 75° west of Greenwich. All places in that zone use the same time, which is five hours behind Greenwich Time. Central Standard Time is six hours behind, Mountain is seven, and Pacific is eight.

Thus we see another reason why the sundial does not agree with the clock. In addition to the correction for the sun's irregular motion, we must allow for the difference between local time and standard time. The longitude of Los Angeles is about 118° 15′ West. It is 1° 45′ east of the 120th meridian. Since 1° equals four minutes of time, the local time of Los Angeles is seven minutes ahead of Pacific Standard Time, which it uses. The longitude of San Francisco is about 122° 30′ West, which is 2° 30′ west of the 120th meridian. So its local time is ten minutes behind Standard Time. The difference of local time between two places is equal to

47

their difference of longitude. San Francisco is 17 minutes behind Los Angeles in local time. Its longitude is 17 minutes, or $4\frac{1}{4}°$, to the west.

Daylight Saving Time is one hour ahead of ordinary standard time. How it saves electricity may be illustrated by an example. On February 8

the line there is always a different day, the later day being on the west side. The need of a date line can be more easily seen if we imagine an airplane flying westward along the Arctic Circle. If its speed is a little over 400 miles an hour, it travels westward as fast as the earth is turning eastward at that

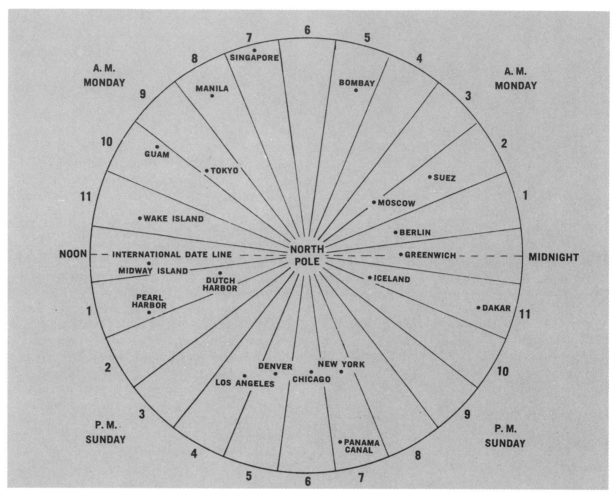

Standard time zones from equator to North Pole.

Irregularities in boundaries are not shown. Direction of earth's rotation is counterclockwise. Midnight and International Date Line are always the limits of a day.

the sun sets in Los Angeles at 5:30 P.M., Pacific Standard Time. By six o'clock people begin turning on their lights. Under Daylight Saving Time, the time of sunset is 6.30 P.M., and lights are turned on at seven o'clock, saving one hour's use of electricity.

Halfway around the world from the meridian of Greenwich is the International Date Line. It runs north and south through the Pacific Ocean along or near the 180th meridian. On opposite sides of

latitude. The sun will remain practically fixed in the sky with respect to the flier's horizon. If he starts at noon, the time will be noon for him all the way around the world. Without a date line his day would not have changed, and yet 24 hours would have elapsed. It is obvious that some line must be established where a traveler's day will change.

There are several irregularities in the line to avoid dividing contiguous territory in Siberia, the Aleutian Islands, and South Pacific Islands.

Queer things can happen at the date line. On one of Admiral Byrd's trips to the Antarctic regions, the ship crossed the line going westward on Christmas Eve. When the men woke up the next day, they found that they had missed Christmas: the date was December 26. If one crosses the line going eastward after having finished a Thanksgiving dinner, he should be entitled to another such dinner the next day. One of the favorite conundrums of Professor Young of Princeton was: What is the greatest possible number of Sundays in February? The answer is ten, for the crew of a vessel making weekly sailings from Siberia to Alaska in a leap year and leaving Siberia on Sunday, February 1. This could have been done in 1948. The next time February will begin on Sunday in leap year in our present calendar will be in 1976.

Steamship companies would save money by running all their world tours to the west, because an eastward journey around the world consumes six more meals than a westward one. As one goes eastward, he sets his watch ahead each time he enters another time zone and eats each meal a little sooner than a person who stays at home. When he gets back he finds that he has eaten three more meals than the rest of us. A traveler to the west sets his watch back and misses out on three meals. The difference between the two journeys is six meals. So those who wish to reduce weight should take a world cruise to the west.

At the North and South Poles there is no time of day. The rotation of the earth on its axis causes the sun and the stars to move parallel to the horizon. The sun rises and sets only once a year. When it is noon in the United States, the sun is in the south. From the North Pole the sun is in the south 24 hours in the day. One might say that it is noon all day. However, the pole is only a point, and a person occupies more than a point. If it is noon at his nose, it will be 6 P.M. at his left ear, midnight at the back of his head and 6 A.M. at his right ear. He will have all the time in the world!

CHAPTER **12** OUR CALENDAR

The year is the period of revolution of the earth around the sun and is one of the three natural units for measurement of time. The other two are the day, defined by one rotation of the earth on its axis, and the month, marked by one revolution of the moon around the earth. No two of these units possess a common denominator, so that it is impossible to divide the year into an integral number of months or days without obtaining an unending decimal as a remainder. This is the cause of all the difficulties that the ancients encountered in their efforts to fit the lunar month of about $29\frac{1}{2}$ days evenly into the year of about $365\frac{1}{4}$ days. Twelve lunar months form a period of about 354 days, which is about $11\frac{1}{4}$ days short of the year. Attempts were made to keep the sun and the moon in step with each other by putting thirteen months in some years. The original Roman calendar was of this sort, but since there were no regular rules for the extra months, the calendar became a source of political intrigue and was much abused. The Julian calendar made the sun the standard of measurement of the year, and the moon was disregarded entirely.

The month is a convenient period of time and has been retained in the calendar, denoting an arbitrary number of days approximating the twelfth part of a year. Seven months have 31 days, four have 30, and one veers between 28 and 29. The details as to how the months came to have their present lengths are somewhat in dispute among historians. The Roman belief that there was luck in odd numbers probably was an important factor. February is so short because it was the twelfth month under the old calendar. When days were needed to make other months lucky, it seemed logical to take them from the end of the year. The fifth and sixth months of the old calendar, Quintilis and Sextilis, were renamed July and August, after Julius Caesar and his successor, Augustus Caesar, respectively.

December is the twelfth month, but its name tells us that at one time it was the tenth, from the Latin "decem." Likewise September, October, and

November were originally the seventh, eighth, and ninth months, respectively. The reason for the present discrepancy between name and number is that Julius Caesar changed the date of the beginning of the year from March to January. The old Roman year began with the first month of spring, during which the sun crossed the equator on its northward journey. Caesar evidently intended to start the year with the shortest day, when the sun was at the winter solstice about December 22. However, he delayed the beginning of the year out of regard for the customs and superstitions of the people, who had always begun the year with a new moon. The new moon nearest to the winter solstice fell on January 1, and so the Julian calendar was started on that date in the year 45 B.C. Thus we see that our New Year's Day is not related to any astronomical fact or logical reason.

The month of January was named in honor of Janus, the Roman God of gates and doors, hence, of all beginnings. (Compare "janitor," one meaning of which is "doorkeeper.") Janus is represented with two faces, looking in opposite directions. One looks behind, reflecting upon the past, and the other looks forward to the future. How appropriate to have the first month of the year named after this god of beginnings. Yet this is just an accident, for January was the eleventh month in the old Roman calendar and was made the first because of Caesar's decision to start the year with the first new moon after the winter solstice. The month was originally dedicated to Janus because it was the first month after the winter solstice, when the farmers in southern Italy began their work in the fields again.

The week is a unit of time made by man. Its length has varied from five to ten days. The number seven has always seemed to man to be a symbol of perfection. A good reason for applying it to the week was that the phases of the moon (new moon, first quarter, full moon, last quarter) occur at intervals of about seven days. There were also seven celestial bodies, known to the ancients, which wandered among the stars. They were the sun, the moon, and five planets. It is conceivable that if the ancients had known of the other three planets, later discovered with the telescope, we might now have a ten-day week. The days of the week received names from the seven wanderers.

The following table lists the celestial bodies and the English and French names for the days.

Celestial Body	English	French
Sun	Sunday	Dimanche
Moon	Monday	Lundi
Mars	Tuesday	Mardi
Mercury	Wednesday	Mercredi
Jupiter	Thursday	Jeudi
Venus	Friday	Vendredi
Saturn	Saturday	Samedi

Tuesday comes from Tiw, the Saxon god corresponding to Mars. Similarly Wednesday is from Woden (Mercury), Thursday from Thor (Jupiter), and Friday from Friga (Venus).

The arrangement of the days in the week is of astrological origin, being based on the superstition of a supposed relationship between the planets and terrestrial affairs. The seven objects were arranged in order of their supposed distance from the earth. Starting with the most remote, they were Saturn, Jupiter, Mars, the sun, Venus, Mercury, and the moon. Their distances were not known, but it was assumed that the faster an object appeared to move in the sky, the closer it was to the earth. This is not true in all cases. Mercury moves faster than Venus, but is farther from the earth. It was believed that each hour of each day was ruled by one of the seven objects, in the order named. Each day was named after the object ruling the first hour of that day. In addition to the first hour on Saturday, Saturn ruled also the 8th, 15th, and 22nd hours. The 23rd and 24th hours were ruled by Jupiter and Mars, respectively. This then brought the sun to the first hour of the next day, which was called Sunday. The sun also ruled the 8th, 15th, and 22nd hours of Sunday. The 23rd and 24th hours were ruled by Venus and Mercury, respectively. This brought the moon to the first hour of the next day, which was called Monday. The other days were named in a similar manner.

In the Julian calendar, the year was assumed to be $365\frac{1}{4}$ days long. The ordinary year drops the quarter day, becoming 365 days, or 52 weeks and one day. Every fourth year consists of 366 days, or 52 weeks and two days, the extra day making up for the quarter days that were dropped. The extra day is added at the end of February. It is called leap year, because this extra day causes any date after February in that year to *leap* over a day in the week and to fall two days later in the week than it did in the previous year.

The length of the year is not exactly $365\frac{1}{4}$ days, but is 11 minutes and 14 seconds shorter. This might seem like too small a quantity to worry about, but it constantly accumulates, amounting to a whole day in 128 years and to three days in 384 years. This accumulation began to cause trouble in the Julian calendar by the year 1582. At that time the date of the vernal equinox, which is used in determining the date of Easter, had slid back to March 11. If this had continued, Easter and other religious festivals connected with it would eventually have come near Christmas. So in 1582, Pope Gregory XIII, with the aid of an astronomer named Clavius, made a slight change in the calendar. In the Julian calendar, there were too many leap years. During each 384 years, three days were added that were not needed. To make a simple rule, the Pope ordained that three leap years in every four centuries should thereafter be ordinary years. The leap years chosen were the century years not evenly divisible by 400. Thus, in the new Gregorian calendar, named after the Pope, 1600 and 2000 are leap years, but 1700, 1800, and 1900 are ordinary years. This rule makes the average length of the Gregorian year only 26 seconds too long. More than 3000 years will pass before this calendar will be in error by one day. This is the calendar we use today, and it is interesting to note that all the observations of the heavenly bodies on which it depends were made with the naked eye before the invention of the telescope.

In order to restore the date of the vernal equinox to March 21, where it was in A.D. 325, when the rule for Easter was fixed by the Council of Nicaea, Pope Gregory decreed that the day after October 4, 1582, should be called October 15, thus dropping ten days. The new calendar went into effect at once in those countries that acknowledged allegiance to the Pope, but the Eastern Church and most Protestant countries did not adopt it until much later. The change was not made in Great Britain until 1752, when the difference between the two calendar systems had become eleven days, since 1700 was not a leap year under the new rule. Parliament decreed that the day following September 2, 1752, should be called September 14 instead of September 3. Although care was taken to guard against any possibility of injustice in the collection of rents and other payments, riots broke out and the cry was raised, "Give us back our eleven days!" It was even worse than that because of a change made in the date of the beginning of the year. Prior to 1752 the official date when a new year began was March 25. This was changed to January 1 for 1752, so that the year 1751 in England had no January or February, and it also lost the first 24 days of March. The ladies objected when they found themselves older, not only by eleven days, but by three months. The workmen rebelled at what they thought was the loss of a quarter of a year's pay.

To avoid confusion in the dates of events occurring in January, February, and March around the period of change, it was customary to give the years according to both styles. For example, January 15, 1753, according to the old style, was written January 15, 1753-54.

Although for ordinary purposes January 1 is New Year's Day in England, those who pay income taxes there know that the financial year of the British Exchequer begins on April 6. This is a survival from the time when the year began on March 25. In order that there would be no change in the length of the financial year of 1752, the calendar date of the beginning of the following financial year was put forward eleven days to April 5. For some reason the old style was retained for a while. Since 1800 was not a leap year in the Gregorian calendar, April 6 took the place of April 5 during the nineteenth century. After that the financial year was determined by the Gregorian rule.

History books tell us that George Washington was born on February 22, 1732. However, if there was a calendar on the wall in the Washington home on the day of George's birth, it read February 11, 1731. The English colonies were still using the Julian Calendar, with the year beginning on March 25. When George was twenty years old, the dropping of eleven days from the calendar made his birthday fall on February 22. Putting New Year's Day back about three months moved his birth into 1732.

Our calendar today can be improved by several simple changes. Since the year of 365 days consists of one day more than 52 weeks, the calendars of all years are not alike. A particular date progresses, year by year, through the days of the week. For example, in 1946 Christmas fell on Wednesday, in 1947 on Thursday, and in 1948 on Saturday, the latter because a date advances two days in a leap year. Also, Thanksgiving is irregular, falling on the fourth Thursday in November, but on an indefinite day of the month. By withdrawing one day from an

THE WORLD CALENDAR

JANUARY
S	M	T	W	T	F	S
1 (1)	2 (2)	3 (3)	4 (4)	5 (5)	6 (6)	7 (7)
8 (8)	9 (9)	10 (10)	11 (11)	12 (12)	13 (13)	14 (14)
15 (15)	16 (16)	17 (17)	18 (18)	19 (19)	20 (20)	21 (21)
22 (22)	23 (23)	24 (24)	25 (25)	26 (26)	27 (27)	28 (28)
29 (29)	30 (30)	31 (31)				

FEBRUARY
S	M	T	W	T	F	S
			1 (1)	2 (2)	3 (3)	4 (4)
5 (5)	6 (6)	7 (7)	8 (8)	9 (9)	10 (10)	11 (11)
12 (12)	13 (13)	14 (14)	15 (15)	16 (16)	17 (17)	18 (18)
19 (19)	20 (20)	21 (21)	22 (22)	23 (23)	24 (24)	25 (25)
26 (26)	27 (27)	28 (28)	29 (1)	30 (2)		

MARCH
S	M	T	W	T	F	S
					1 (3)	2 (4)
3 (5)	4 (6)	5 (7)	6 (8)	7 (9)	8 (10)	9 (11)
10 (12)	11 (13)	12 (14)	13 (15)	14 (16)	15 (17)	16 (18)
17 (19)	18 (20)	19 (21)	20 (22)	21 (23)	22 (24)	23 (25)
24 (26)	25 (27)	26 (28)	27 (29)	28 (30)	29 (31)	30 (1)

APRIL
S	M	T	W	T	F	S
1 (2)	2 (3)	3 (4)	4 (5)	5 (6)	6 (7)	7 (8)
8 (9)	9 (10)	10 (11)	11 (12)	12 (13)	13 (14)	14 (15)
15 (16)	16 (17)	17 (18)	18 (19)	19 (20)	20 (21)	21 (22)
22 (23)	23 (24)	24 (25)	25 (26)	26 (27)	27 (28)	28 (29)
29 (30)	30 (1)	31 (2)				

MAY
S	M	T	W	T	F	S
			1 (3)	2 (4)	3 (5)	4 (6)
5 (7)	6 (8)	7 (9)	8 (10)	9 (11)	10 (12)	11 (13)
12 (14)	13 (15)	14 (16)	15 (17)	16 (18)	17 (19)	18 (20)
19 (21)	20 (22)	21 (23)	22 (24)	23 (25)	24 (26)	25 (27)
26 (28)	27 (29)	28 (30)	29 (31)	30 (1)		

JUNE
S	M	T	W	T	F	S
					1 (2)	2 (3)
3 (4)	4 (5)	5 (6)	6 (7)	7 (8)	8 (9)	9 (10)
10 (11)	11 (12)	12 (13)	13 (14)	14 (15)	15 (16)	16 (17)
17 (18)	18 (19)	19 (20)	20 (21)	21 (22)	22 (23)	23 (24)
24 (25)	25 (26)	26 (27)	27 (28)	28 (29)	29 (30)	30 (1)

JULY
S	M	T	W	T	F	S
1 (2)	2 (3)	3 (4)	4 (5)	5 (6)	6 (7)	7 (8)
8 (9)	9 (10)	10 (11)	11 (12)	12 (13)	13 (14)	14 (15)
15 (16)	16 (17)	17 (18)	18 (19)	19 (20)	20 (21)	21 (22)
22 (23)	23 (24)	24 (25)	25 (26)	26 (27)	27 (28)	28 (29)
29 (30)	30 (31)	31 (1)				

AUGUST
S	M	T	W	T	F	S
			1 (2)	2 (3)	3 (4)	4 (5)
5 (6)	6 (7)	7 (8)	8 (9)	9 (10)	10 (11)	11 (12)
12 (13)	13 (14)	14 (15)	15 (16)	16 (17)	17 (18)	18 (19)
19 (20)	20 (21)	21 (22)	22 (23)	23 (24)	24 (25)	25 (26)
26 (27)	27 (28)	28 (29)	29 (30)	30 (31)		

SEPTEMBER
S	M	T	W	T	F	S
					1 (1)	2 (2)
3 (3)	4 (4)	5 (5)	6 (6)	7 (7)	8 (8)	9 (9)
10 (10)	11 (11)	12 (12)	13 (13)	14 (14)	15 (15)	16 (16)
17 (17)	18 (18)	19 (19)	20 (20)	21 (21)	22 (22)	23 (23)
24 (24)	25 (25)	26 (26)	27 (27)	28 (28)	29 (29)	30 (30)

OCTOBER
S	M	T	W	T	F	S
1 (1)	2 (2)	3 (3)	4 (4)	5 (5)	6 (6)	7 (7)
8 (8)	9 (9)	10 (10)	11 (11)	12 (12)	13 (13)	14 (14)
15 (15)	16 (16)	17 (17)	18 (18)	19 (19)	20 (20)	21 (21)
22 (22)	23 (23)	24 (24)	25 (25)	26 (26)	27 (27)	28 (28)
29 (29)	30 (30)	31 (31)				

NOVEMBER
S	M	T	W	T	F	S
			1 (1)	2 (2)	3 (3)	4 (4)
5 (5)	6 (6)	7 (7)	8 (8)	9 (9)	10 (10)	11 (11)
12 (12)	13 (13)	14 (14)	15 (15)	16 (16)	17 (17)	18 (18)
19 (19)	20 (20)	21 (21)	22 (22)	23 (23)	24 (24)	25 (25)
26 (26)	27 (27)	28 (28)	29 (29)	30 (30)		

DECEMBER
S	M	T	W	T	F	S
					1 (1)	2 (2)
3 (3)	4 (4)	5 (5)	6 (6)	7 (7)	8 (8)	9 (9)
10 (10)	11 (11)	12 (12)	13 (13)	14 (14)	15 (15)	16 (16)
17 (17)	18 (18)	19 (19)	20 (20)	21 (21)	22 (22)	23 (23)
24 (24)	25 (25)	26 (26)	27 (27)	28 (28)	29 (29)	30 (30)	Y (31)

The world calendar.

The larger numerals are the dates in the proposed world calendar. The smaller numerals are the corresponding dates in the calendar now in use. Year-end day, December Y, follows December 30 every year. Leap-year day, June L, follows June 30 in leap years.

ordinary year and two days from a leap year, the drifting week would be avoided. The year would consist of 364 days, or exactly 52 weeks. Every year would be like every other year. One's birthday would always fall on the same day of the week. How many of us could say offhand on what day of the week we were born? The 365th day would not be a part of any week. It would be placed at the end of the year, and would be called "Year-End Day." The extra day in leap year, "Leap-Year Day," would be inserted at the end of June.

Complications arise in the present calendar from the inequalities in the lengths of the months, ranging from 28 to 31 days. Some months have four Sundays and others have five. The number of working days varies from 24 to 27. This difference of three days is $12\frac{1}{2}\%$ of 24 working days. It is obviously difficult to compare commercial transactions of one month with those of other months, when the irregularities may run as high as $12\frac{1}{2}\%$. Also the half years are unequal. In an ordinary year, the first half contains 181 days and the second half 184 days. The quarters vary from 90 to 92 days.

Two plans have been seriously considered for reforming the calendar. Both involve the use of the year of exactly 52 weeks, with Year-End Day and Leap-Year Day as extra days. In the Thirteen-Month Calendar, each month contains 28 days, or exactly four full weeks, and so every month is like every other month. It would thus be very easy to find what day of the week any date falls on. The proposed plan would have every month begin on Sunday, and so the 1st, 8th, 15th, and 22nd would always be Sundays. And the 13th would always fall on Friday. Friday the 13th thirteen times a year! This alone might set many people against the adoption of such a calendar. But there are other, more serious objections. A year of 13 months cannot be divided, on a monthly basis, into any parts other than 13. The adoption of such a calendar would eliminate the quarter and the half year, which are so useful in our present calendar.

Another practical disadvantage of the 13-month plan is the great disturbance of dates. If the additional month, to be called Sol, were inserted in the middle of the year, 94 days would have to be removed from the other months. As a result, at least one-fourth of the people would find their birthdays switched into a different month.

The World Calendar seems to be the most workable plan. It could be adopted with a minimum of disturbance to our present system of dating. It is a perpetual, symmetrical, 12-month calendar. The quarters are identical, each one containing three months, 13 weeks, 91 days. The first month of each quarter contains 31 days and begins on Sunday. Each of the other eight months has 30 days. The second month of each quarter begins on Wednesday, and the third month of each quarter begins on Friday. Since the 31-day months have five Sundays and the 30-day months have four Sundays, every month has 26 working days, thus facilitating statistical comparisons.

This calendar has an advantage over any other proposed revision. It enables us to keep the dates we have, without material change. People born in January, February, September, October, November, and December would find their birthdays occurring on exactly the same dates in both calendars. Only eight days would have to be transferred from one month to another, as compared with 94 days in the 13-month calendar.

This calendar is sponsored by the World Calendar Association, New York City, publishers of the "Journal of Calendar Reform." The benefits that would result from the establishment of such a calendar are so numerous that it is hoped the efforts to secure its adoption will soon be met with success.

CHAPTER **13** THE PRECESSION OF THE EQUINOXES

The equinoxes are the two intersections of the sun's apparent annual path (the ecliptic) with the celestial equator. The sun reaches the vernal equinox about March 21, as it crosses the equator from the south side to the north side. In the second century B.C., the Greek astronomer Hipparchus found that the distances of certain stars from the vernal equinox had changed. For example, when

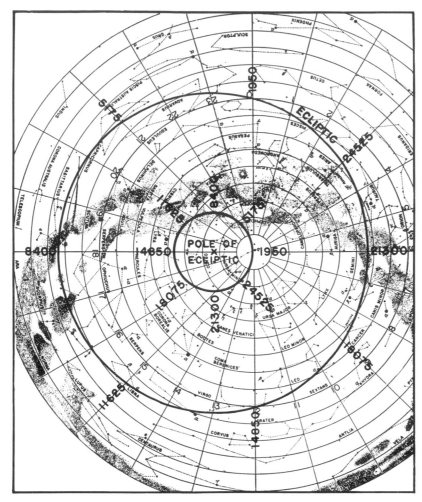

The path of the celestial pole and the ecliptic among the stars.

The small circle is the path of the north celestial pole among the stars. The numbers are the yearly dates, beginning in 1950, when it occupies the various positions, the circuit being completed in 25,800 years. The similar numbers around the ecliptic show the position of the sun among the stars at the instant spring begins at each of these epochs. These dates may be changed by 25,800 years either way, to secure a repetition.

he compared his observations of Spica with those made by earlier astronomers, he found that its eastward distance from the vernal equinox had increased by about two degrees in 150 years. This would give an annual rate of nearly one minute of arc. There was no change in the distance of the star from the ecliptic.

Other stars give similar results. The agreement between these motions was enough to justify Hipparchus in concluding that the change could be accounted for, not as a motion of individual stars, but as a westward change in the positions of the equinoxes.

Copernicus showed that the westward movement of the equinoxes could be explained by a wobbling motion of the earth like that of a spinning top, the axis of which is not quite vertical. Suppose

the top is immersed in water—up to its "equator" —and continues to spin. The surface of the water represents the plane of the ecliptic. As the top spins, with its axis slightly tipped, its "equator" will be partly under and partly above the water, cutting the water's surface at two opposite points. The wobbling of the top will cause these two points to move around along the surface of the water. Similarly the earth's precessional motion causes the equinoxes to move westward along the ecliptic, completing a circuit in about 26,000 years.

Thus the prolongation of each end of the earth's axis into the sky will trace out a circle among the stars. The center of the circle is called the pole of the ecliptic. A line from the sun perpendicular to the plane of the ecliptic intersects the sky in the pole of the ecliptic. The radius of the circle which is formed is $23\frac{1}{2}°$, since that is the inclination of the earth's axis away from the perpendicular to the ecliptic.

Polaris has been the pole star for only a few hundred years. At the time of Copernicus and Columbus, the earth's axis pointed $3\frac{1}{2}°$ away from Polaris. Now it points only $1°$ away, and by the year 2100 the minimum distance of $\frac{1}{2}°$ will be reached. After that the north celestial pole will move away from Polaris, and other stars will take their turns as the pole star. The brightest will be Vega, which will be $5°$ from the pole in A.D. 14,000.

The earliest pole star of which we have historical knowledge is Thuban, which is also known as Alpha Draconis. It is located about halfway between Mizar, the middle star in the handle of the Big Dipper, and the two stars at the end of the bowl of the Little Dipper. But it was the pole star when the Great Pyramid of Gizeh was built in Egypt and gives a clue to the date of construction. The pyramid has an inclined gallery directed to a point about $3\frac{1}{2}°$ below the pole. An observer at the bottom of this narrow passage could see Thuban as it crossed the meridian $3\frac{1}{2}°$ below the pole each day. Thuban passed at this distance about 3500 B.C. and again about 2100 B.C. Historical considerations decide against the latter date, and so the first date may be accepted as within a century or two of the time when the Great Pyramid was built. The pole was closest to Thuban about 2800 B.C., but then it was too near the pole for observation from the pyramid's gallery.

Copernicus' realization that the motion of the earth's axis is conical was a beautiful generalization whereby many facts were grouped into a single phenomenon. He did not explain why the earth wobbled like a top. That remained to be done by Newton. We cannot go into the mathematical details, but perhaps we can get the idea by referring again to a spinning top, the axis of which is tilted away from the vertical. Maintaining the same angle of tilt, the axis describes the conical movement called precession. Any force tending to change the direction of the axis of a whirling body produces a motion at right angles to its own direction. Thus the force of gravity which would tend to draw the axis of the top downward, actually causes the conical motion in a plane $90°$ from the vertical. We use precession when without touching the handle bars we lean to one side to ride a bicycle around the corner. Each wheel is a top: a gyroscope.

Just as the axis of the top leans away from the vertical, so the earth's axis is inclined $23\frac{1}{2}°$ from the perpendicular to the plane of its orbit around the sun. Because of centrifugal force, the rotating earth is slightly bulged at the equator, its diameter being 27 miles longer through the equator than from pole to pole. The gravitational pulls of the moon and of the sun on the earth's equatorial bulge tend to change the direction of the earth's axis. Again, as in the case of the top, the result is a conical motion of the axis.

The top takes only a few seconds to complete its reeling movement, but the period for the earth is nearly 26,000 years. The axis of the top moves in the direction in which the top rotates, since the force of gravity tends to tip the top farther over. The axis of the earth moves in the direction opposite to that of the earth's rotation, because the pulls of the moon and of the sun on the earth's equatorial bulge tend to straighten up the earth's axis.

It should be emphasized that while the north celestial pole moves among the stars, precession does not cause the North Pole of the earth to change its position on the earth's surface. Some people have tried to explain the ice ages (which are still unexplained) by assuming that the earth's poles once were hundreds of miles from their present positions. That is not true. A place having a latitude of $34°$, such as Los Angeles, is always $56°$, or about 3,864 miles, from the North Pole. The distance from any point on a spinning top to the end of its axis remains constant. Precession causes the whole top to wobble, but does not change the position of the axis within the top. The same holds true for the spinning earth.

Ordinarily we see the stars only at night, when the part of the earth on which we live is turned away from the sun. As the earth carries us around the sun, we look out at night in different directions at different times of the year, seeing such typical constellations as Orion in winter, Leo in spring, Scorpius in summer, and Pegasus in autumn. These seasons refer to the northern hemisphere, and are reversed in the southern hemisphere. The reason that we see Orion in the night sky in winter is that the northern hemisphere is tilted away from the sun when the night side of the earth is facing Orion. The noonday sun is lower then, and there are fewer hours of sunshine than when the earth is on the other side of the sun and its North Pole is tilted toward the sun.

However, after about 13,000 years, whenever

the earth is between the sun and Orion, the northern hemisphere will be tilted toward the sun. Thus Orion will become a summer constellation. The other constellations will change their seasonal times of appearance in a similar manner. At midnight in early March of the twentieth century, Leo is on the meridian high in the south and Orion is about to set below the western horizon. After about 6,500 years, an observer looking at the midnight sky in March will see Orion high in the south. After 13,000 years, its place will be taken by Pegasus, and after 19,500 years, Scorpius will be a spring constellation. At the end of the earth's precessional period Leo will be back in the spring.

During a precessional cycle, the number of trips the earth makes around the sun is one less than the number of years that elapse. The year of the seasons, called the tropical year, is about 20 minutes shorter than the sidereal year, which is the time required by the earth to make a complete revolution around the sun from a given direction in space to the same direction again. The tropical year is the time included between two successive passages of the vernal equinox by the sun.

Because of precession, the equinox moves slowly toward the west to meet the sun, which does not have to make quite a whole revolution of the sky to complete the year of our calendar.

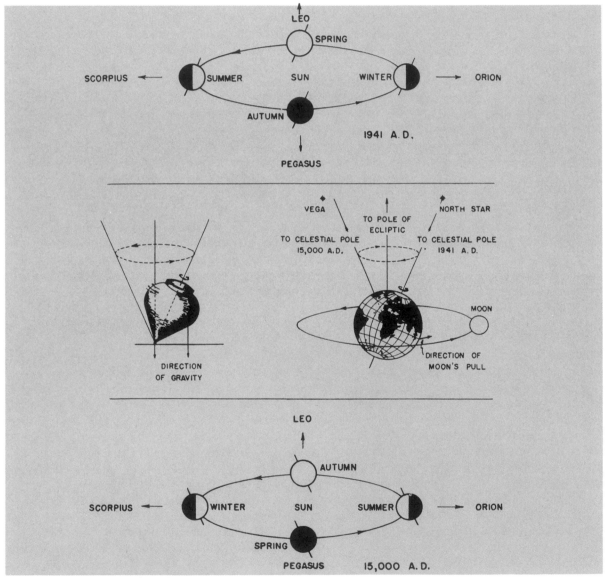

Effect of precessional motion of the earth.
The precessional motion of the earth will make Vega the pole star and Orion a summer constellation in A.D. 15,000.

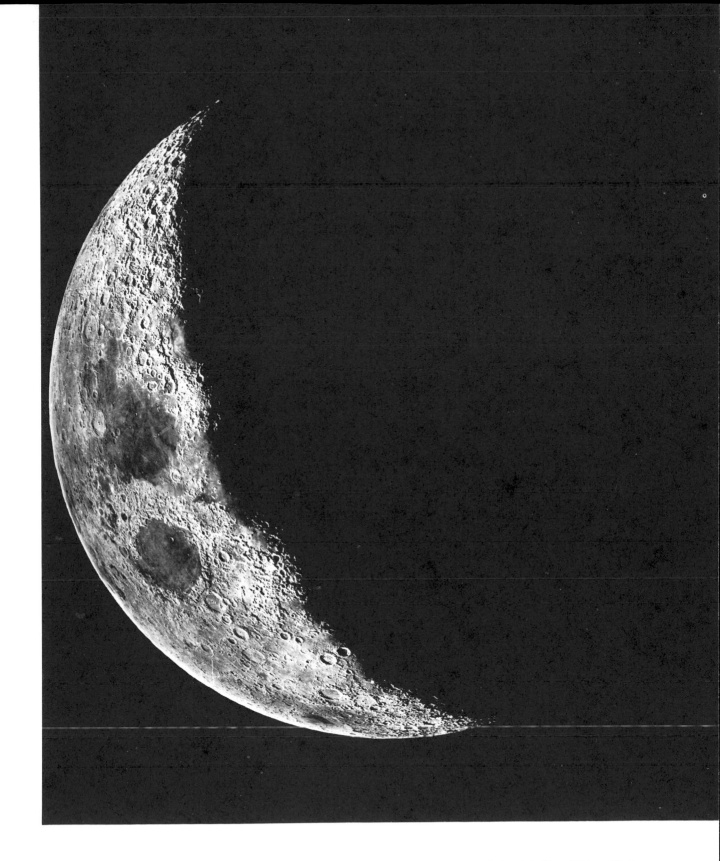

Crescent moon, 4½ days after new moon. Largest dark area is Mare Foecunditatis. The one just below it is Mare Crisium. Maria Nectaris and Tranquillitatis are just beginning to show at sunrise (right side). South is up, and west to the left. Under terrestrial conventions the sun rises in west, despite the fact that earth and moon rotate in same direction. (Lick Observatory.)

57

CHAPTER 14 SATELLITES EXPLORE THE EARTH

When the first artificial satellites were put into orbit around the earth by the U.S.S.R. and the U.S.A., they were justifiably hailed as great triumphs of our civilization's advancing technology. They were equally important, however, as heralds of a new era in the history of astronomy. The feeble "beep-beep" transmitted from the first Sputnik signaled dazzling possibilities, as sweeping as those opened up when Galileo turned his first crude telescope to the stars, or when radio telescopes were developed following the close of World War II.

Our ocean of atmosphere has always presented a major problem to the ground-based astronomer. Its turbulence produces the beautiful but endlessly frustrating twinkling of the stars and the tantalizing blurring of planetary images. The ozone molecules in the upper atmosphere completely block all short-wave radiations in the ultraviolet range, so that one can only guess about much of the radiation emitted by hot stars or the solar chromosphere. Similarly, in the infrared range, various other gases in the atmosphere (notably water vapor) chop out further sections of the spectrum, leaving only discouragingly small "windows" accessible to those studying the red stars.

Now all has been changed. We have seen an almost frantic rush to exploit the new technology, supported by enormous programs in this country and the U.S.S.R. Efforts of the past ten years have resulted in enormous advances, which we shall discuss in this and the next two chapters. In this chapter we shall describe what orbiting satellites have discovered about the earth itself, leaving for the later chapters the results of lunar and planetary explorations.

When a space vehicle is put into orbit around the earth, the size and shape of the orbit are dependent upon its speed and the gravitational attraction exerted on it by the earth. If the earth were a perfect sphere with a symmetrical distribution of material around its center, then the orbit of the space vehicle would be an ellipse in a stationary orbital plane relative to the stars. However, the earth possesses an equatorial bulge. Its diameter at the equator is some 26 miles greater than its diameter from pole to pole. We say that it is an oblate spheroid. As a consequence, the orbital plane of a space vehicle will rotate. This is similar to the gradual precession of the orbital plane of the moon. If the space vehicle is moving close to the earth's surface, observations of the rotation of its orbit can be used to check the degree of the equatorial bulge and to search for small deviations from a perfectly symmetrical oblate spheroid shape.

Observations of the orbits of several space vehicles show that there is a slight difference between the plotted equatorial bulge and that demonstrated by the orbits. In particular, there are significant deviations from perfect symmetry that must be in consequence of variations in the gravitational attraction of the material deep in the earth's interior. For instance, the orbits reveal that the

Earth Resources Technology Satellite-I composite photographs of San Francisco area. Notice clouds hugging the coast (lower left), San Francisco Bay (lower center), man-made structures jutting into the bay, and the two lakes along the San Andreas fault (upper left). (NASA.)

surface is too high, forming a bump, in the western Pacific just north of Australia. It is almost 200 feet high relative to a perfectly symmetrical oblate surface. Similar bumps about 100 feet high are centered in the Scandinavian peninsula and in the south Atlantic between the southernmost tips of Africa and South America. Corresponding depressions of about 60 feet are found in Antarctica and in the eastern Pacific, while a depression of almost 200 feet is located just to the west of India in the Indian Ocean.

The variations in the density of the earth's inner material may be caused by rising and falling currents of material flowing in its interior. A rising column would carry hotter material toward the surface. This hotter material would have a slightly lower density than the corresponding cooler material in a descending column. Thus, the bump in the sea level in the western Pacific would be an indication of a descending column at that point.

Today earth scientists are also proposing the existence of rising and descending columns in the upper mantle of the earth, associated with their currently well-established theory of continental drift. The descending column in the West-

ern Pacific may be associated with the system of deep oceanic trenches in that area, while the Eastern Pacific possesses a system of oceanic ridges, which may be the reason for the depression from true spherical symmetry in that area.

The Tiros program of weather satellites, the first of which was launched in 1960, is making possible the first study of the earth's atmosphere on a global basis. Television cameras transmit photographs of cloud cover that permit an almost continuous scrutiny of an area of weather activity. The warnings they provide of the advent of tropical hurricanes are well known. Of equal importance is the way the cloud patterns reveal the general circulation of the atmosphere all over the world. Before the satellites' photographs were available, a global study of atmospheric circulation and disturbance was impossible because great gaps existed in the requisite information. The gathering of oceanic data was dependent on ships, which left large sections of the globe very poorly covered. The polar regions and deserts also were excluded from reliable monitoring procedures.

Geophysicists and meteorologists hope that the Tiros cloud-cover photographs will aid the study

59

of the energy balance between the earth and its atmosphere. The temperature of the earth's surface is strongly influenced by the efficiency of the overlying atmosphere as a heat blanket. Clouds strongly affect the transmission of solar radiation to the surface, and also the reverse transmission of infrared radiation from the surface.

The atmospheric drag on satellites at altitudes over 100 miles above the surface provides information on the density of the upper atmosphere. It has been found, for instance, that at times of increased solar activity the upper atmosphere becomes significantly hotter and expands upward, resulting in a much greater density. Increases by a factor of 100 times have been reported. The resultant increased drag on satellites at such times is closely correlated with other phenomena produced by solar disturbances, such as magnetic storms.

The observed drag in the upper atmosphere also provides valuable clues to its probable chemical composition. It has been known for a long time that the composition of the atmosphere above 600 miles from the surface must be quite different from the familiar oxygen-nitrogen mixture that we breathe. At high altitudes it is composed largely of atoms and ions of the lighter elements. Data provided by the large Echo I spherical orbiting reflector showed that the upper atmosphere above 700 miles from the surface was probably composed mainly of helium and oxygen atoms. This has been subsequently confirmed by rocket observations of the number of electrons in the atmosphere between 300 and 1,300 miles high. Eventually, at very high altitudes, the helium must be replaced by hydrogen as a major constituent.

Geologists and environmentalists are also reaping profits from the orbiting satellite program. For instance, early in 1971 the Nimbus IV satellite photographed the entire area of Alaska and adjacent regions of Canada on a rare, cloudless day. At a height of 600 miles, geologists could recognize features on the photograph as small as 2 miles across. They could trace all of the major topographic features—mountain ranges, high plateaus, and lowlands—but to their delight they could also detect a system of lines interpreted as fault lines, many not yet recognized from surface surveying. It raises the possibility that such studies could enhance our knowledge of the geology of remote areas

on the earth's surface. In addition, photographs by the Earth Resources Technology Satellite of the earth in different colors through filters make it possible to recognize areas under cultivation, forests, etc., and even to trace areas where vegetation is suffering blight, drought, and other phenomena. It is probable that future generations will rely heavily on such information in many ways; for instance, in the prediction of crop yields, in prospecting for mineral resources, in mapping, in setting national and international environmental policies, in monitoring pollution standards, and even in census taking.

As we saw in Chapter 7, one of the earliest scientific triumphs of the space program was the discovery in 1958 of the Van Allen belts, intense zones of charged particles extending upward from an altitude of about 1,000 miles above the surface. The initial discoveries indicated the presence of two broad belts of particles circling the earth and oriented symmetrically to the magnetic equator. The inner belt was centered about 2,000 miles above the surface and was 3,000 miles thick, while the outer, larger belt lay at 10,000 miles altitude and was about 4,000 miles thick. From their orientation it is apparent that the belts are composed of particles trapped in the earth's magnetic field. The particles in the belts have been identified as mostly electrons, though there are some protons also, particularly at lower levels.

Subsequent observations have shown that the belts frequently change significantly in position and intensity. In fact, they occasionally merge. When the belts were first discovered, it was thought likely that they were the source of the charged particles that produce aurorae, when agitated by solar disturbances. However, the number of particles involved in an auroral display is so large that the belts would be depleted in about an hour. It is therefore apparent that the majority of the particles in aurorae come directly from space into the earth's atmosphere, along lines of magnetic force.

Since auroral phenomena and magnetic storms are known to be intimately associated with solar disturbances, outbursts of solar radiation and "gusts" in the "solar wind" of charged particles from the sun must be responsible for the intensified flow of particles into the earth's atmosphere. However, the connection between the solar wind and such geomagnetic phenomena is quite complex.

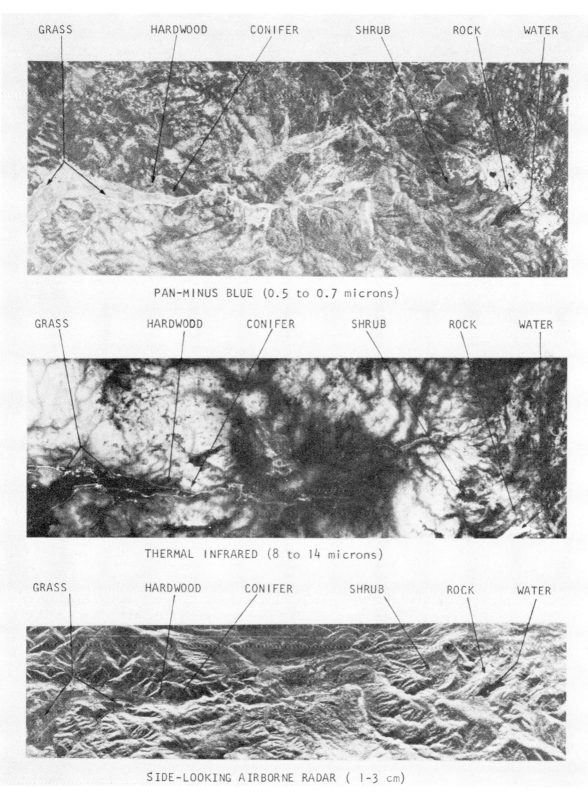

GRASS   HARDWOOD   CONIFER   SHRUB   ROCK   WATER

PAN-MINUS BLUE (0.5 to 0.7 microns)

GRASS   HARDWOOD   CONIFER   SHRUB   ROCK   WATER

THERMAL INFRARED (8 to 14 microns)

GRASS   HARDWOOD   CONIFER   SHRUB   ROCK   WATER

SIDE-LOOKING AIRBORNE RADAR (1-3 cm)

(top) Conventional panchromatic minus blue aerial photography. (middle) Thermal infrared imagery. (bottom) Side-looking airborne radar (SLAR) imagery. The same 6 spots have been annotated on each of these 3 image types in order to highlight the fact that the 6 most important resource categories in this area exhibit different multiband ''tone signatures'' when photographed in these 3 bands. (American Society of Photogrammetry.)

Recent flights in the Explorer series launched in 1958 demonstrate the existence of a sharply defined boundary, known as the *magnetopause*, separating the earth's magnetosphere from the solar wind. It normally lies at 40,000 miles above the earth's sunlit side and is only about 60 miles thick. As we saw in Chapter 4, the solar wind of particles, traveling at between 200 and 400 miles per second, cannot penetrate the magnetopause but instead divides and flows around it, forming a "wake" extending from the earth's dark side at least a quarter of a million miles into space. Associated with the magnetopause, and lying slightly above it, is a shock wave produced when the supersonic solar wind particles are suddenly decelerated to subsonic velocities as they impinge on the earth's field.

Since the solar wind particles are unable to penetrate the magnetopause, it is difficult to understand how solar disturbances can be related to terrestrial phenomena. The best one can say at present is that the region of the impact of the solar wind on the magnetopause must be one of high turbulence. The magnetosphere's boundary is buffeted with particular violence when gusts of solar particles arrive, producing violent disturbances in the general field just inside the magnetopause. These result in all of the observed phenomena on the earth if the disturbances eventually penetrate into the atmosphere.

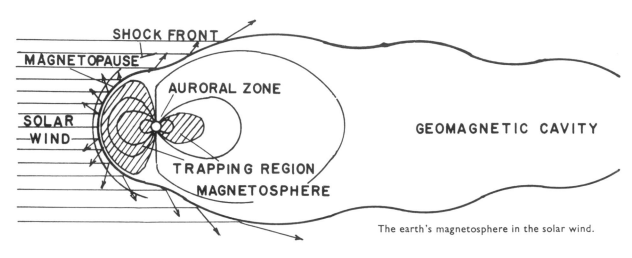

The earth's magnetosphere in the solar wind.

SECTION THREE

The Moon

CHAPTER **15** **LUNAR PORTRAIT**

Since time immemorial the moon has excited man's curiosity and awe. He has speculated about its markings and motions, and has made it the center of his superstitions and primitive religions. The ever-present light and dark markings on its surface, so easily discerned with the naked eye, are revealed through the telescope to be craters and mountains, plains and valleys in profusion, affording the amateur astronomer endless hours of fascinating viewing. The light areas are found to be mottled with thousands of craters of various sizes and shapes. They range in size from the great ringed Clavius, which might be called a ringed plain and which is 146 miles across, to the smallest that can be detected with great telescopes on the earth. Over 30,000 have been counted on the surface turned toward us. Smaller craters partially obliterate larger ones, showing that they were formed later, whether by meteor impact or volcanic action. They display an infinite variety of forms. Some are saucer-shaped, others have flat floors; some have central peaks; others, none. Some, like the great craters Tycho and Copernicus, are at the centers of complex systems of radiating rays, as though a meteor impact or volcano threw lunar dust and gas thousands of miles across the lunar surface. These rays are most conspicuous when the moon is full; they fade into comparative obscurity as the sun's rays fall on the moon's surface more and more obliquely as the moon moves into the quarter phase. It is then that shadows cast by crater walls reveal the full detail of their structures, and measures of shadow lengths can be used to calculate elevations.

The approximately circular, darker, lunar plains, or *maria*, appear at first glance to be quite flat and featureless. However, closer examination reveals markings not unlike flow patterns, as well as partially drowned hills and craters. One can easily imagine a giant meteor breaking through the moon's still tenuous outer crust in an early stage in the moon's evolution, and a great upwelling of lavalike material that flowed out over the surrounding terrain obliterating everything in its path.

The stark crisp sharpness of shadows cast by lunar relief has long suggested that the moon has little or no diffusing atmosphere. This lack has been confirmed by every recent attempt to detect an atmosphere, and indeed, we should not have expected one. The moon's surface gravity is only one-sixth that of the earth, and it can be shown that this is not enough to retain any but the heaviest gases. This means that most of the weathering processes that have so drastically modified the earth's surface features are not operating on the moon. Its remote past history is revealed in its topography, if we could only interpret it. This provides the single most important scientific justification of the series of Apollo explorations of

LUNAR DATA

Distance of moon from earth
Greatest .... 252,710 miles
Least .... 221,463 miles
Mean .... 238,857 miles
(60¼ times earth's radius)
Equatorial horizontal parallax at
mean distance ....57' 03"

Apparent angular diameter
Minimum ....29' 21"
Maximum ....33' 30"
Mean ....31' 05"

Eccentricity of orbit 1/18

Diameter of moon .... 2,160 miles

Volume ....1/49 that of earth

Mass ....1/81 that of earth

Mean density ....3/5 that of earth
(3⅓ times that of water)

Surface gravity ....1/6 that of earth

Velocity of escape................. 1½ miles per second

Approximate temperature
At noon .... 100° C.   212° F.
At midnight .... —150° C.   —238° F.

Shape
Moon is not quite spherical. Greatest diameter, directed toward the earth, is about 2/5 mile longer than the equatorial diameter at right angles to it and 1⅓ miles longer than the polar diameter.

Revolution and Rotation
Synodic month (from one
new moon to the next) .... 29^d 12^h 44^m 2^s.8
Sidereal month (true period of
revolution around earth) .... 27^d 7^h 43^m 11^s.5
Period of axial rotation .... 27^d 7^h 43^m 11^s.5

Librations
While in the long run the moon keeps the same face toward the earth, it is oscillating slightly, so that 59% of its surface has been seen from the earth.
Portion always visible 41%
Portion never visible 41%
Portion alternately visible and invisible 18%
The oscillations are called librations and arise from three causes:

(1) Libration in latitude comes from inclination of 6° 41' between moon's equator and plane of its orbit.

(2) Libration in longitude is due to the fact that the moon's angular motion in its elliptical orbit is variable, while the motion of rotation is uniform. Libration amounts to about 7¾° at each of eastern and western edges of the moon.

(3) Diurnal libration is caused by earth's rotation. At moonrise we see one degree farther over the western edge and at moonset one degree farther over the eastern edge of the moon than we see when the moon is overhead.

Inclination of orbit to ecliptic ....5° 08'

Period of revolution of nodes............... 18.6 years

Daily retardation in crossing meridian
Minimum .... 38 minutes
Maximum .... 66 minutes
Average.......................................50½ minutes
(Average retardation of moon's daily rising and setting is also 50½ minutes, but the range is much more variable, depending largely on the latitude. In latitude 40° the range is from 13 to 80 minutes. In higher latitudes it is still greater.)

Average velocity of moon around earth
Linear ....2,287 miles an hour
or 3,350 feet a second
Angular..13°.2 a day
or 33' an hour
(Moon moves in one hour a distance about equal to its own diameter.)

Brightness
Full moon gives about 1/465,000 as much light as the sun. If the sky were filled with full moons tangent to each other (about 77,000 of them), the total light received would be ⅙ as much as the sun gives. If the sky were illuminated uniformly with a brightness equal to that of the full moon, it would give as much light as about 98,000 full moons, which would be about ⅕ as much light as the sun gives.

Photographically, full moonlight is only about 1/650,000 as bright as sunlight. Since ordinary photographic plates are sensitive to blue and ultraviolet light, the ratios indicate that the moon's surface reflects less blue light than it would if it reflected all the wave lengths with equal efficiency. Thus the color of the lunar surface would be located toward the longer wave length end of the spectrum, i.e., yellow or red.

Light of full moon is about ¼ as bright as that of a standard candle at a distance of one meter.

Light of full moon varies nearly 30% with the changes in its distance.

Although the lighted area of the moon at first quarter or last quarter is apparently half that of the full moon, it is only about one ninth as bright. Most of this difference is due to the rough character of the moon's surface, which causes it to be somewhat darkened by shadows, except at full moon.

Stellar magnitude at greatest brilliancy .....—12.6

Albedo, or average fraction of light
reflected... 0.07

Craters
Total number visible from earth........Over 30,000
Three types:
(1) **Mountain-Walled Plains.** These are large, shallow, craterlike depressions in mountainous regions. Their diameters vary from about 40 to 150 miles. Typically they are polygonal (hexagons most common). Their origin is uncertain. Some lunar students believe

64

them to have been caused by impact. Others believe that in general they resemble the volcanic caldera of the earth. External walls are either low or missing. The Ptolemaeus wall is a typical example.

(2) **Ringed Plains.** These features exhibit definite explosive characteristics, due either to impact or to volcanic action. They have external walls. Usually their floors are less smooth than are those of the preceding class. Their diameters vary from about 10 to 60 miles. They are more nearly circular than are the mountain-walled plains. They may be found in any sort of terrain. Some of them brighten under a high sun much more than the neighboring regions. The ringed plains Copernicus, Tycho, Aristarchus, and Proclus are examples. They have radiating bright streaks called *rays*. Others are not so bright and have no ray systems. Eratosthenes is an example of the latter.

(3) **Craters.** This term in general discussions is applied to any craterlike formation. More technically it applies to features similar to the ringed plains but of diameters between 4 and 10 miles. Those with diameters less than 4 miles often are called *craterlets*. Tiny openings, such as observed on the tops of domes, usually are called *pits*.

There are many large features that exhibit the characteristics both of mountain-walled plains and of ringed plains. Arzachel is an excellent example of such hybrids.

Rays

These are bright streaks related to those ringed plains that brighten under a high sun. They vary from tiny, dagger-shaped, elementary ones to complex rays, which, in major systems, may be hundreds of miles long.

Maria (singular mare)

This is the Latin word for "seas." Maria are rather smooth, dark, usually depressed plains. On them may be found many scattered craters and ridges. Except for the conspicuous exceptions of Maria Crisium and Smythii, the major ones are interrelated. They are by far the largest of all lunar features and are easily visible to the naked eye.

Rills (clefts)

These are very narrow fissures in the lunar surface. Some are hundreds of miles long. Often they form rill systems. The finest systems are those related to the craters, Triesnecker, Hyginus, Ariadaeus, and Ramsden.

---

the moon's surface. The wealth of materials that have been brought back, and the numerous experiments that were carried out on the moon, have already augmented immeasurably our store of knowledge regarding its surface, its internal structure, and its probable past history. See also Chapter 20.

Evidence is mounting that the moon is far from being a static, unchanging object. In 1958 the Russian astronomer N. A. Kozyrev observed for two hours a brightening over the central peak in the Crater Alphonsus, and succeeded in photographing its spectrum. The results strongly suggested an ejection of gas in a volcanic eruption. Other observers have reported occasional changes in the distinctness of the view of the surface details in that same crater. "Moon quakes" are being recorded by seismographs left behind by the astronauts. It is certain that the surface of the moon is being continually bombarded by meteors since there is no atmosphere to burn up the incident meteors as is the case for the earth (fortunately for us!). This continual bombardment is gradually stirring the surface dust found by the astronauts. Their footprints will remain for thousands of years, but eventually they will disappear.

The unique double craters, Messier and W. H. Pickering, photographed during the Apollo VIII mission. Early astronomers were tricked by the shifting shadows of the double craters into thinking that the single crater they believed to be on the spot was undergoing changes. (National Aeronautics and Space Administration.)

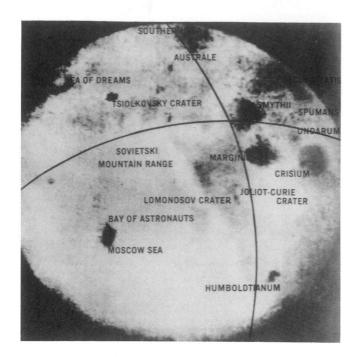

SOUTHERN SEA
AUSTRALE
SEA OF DREAMS
TSIOLKOVSKY CRATER
SOVIETSKI
MOUNTAIN RANGE
MARGIN
LOMONOSOV CRATER
BAY OF ASTRONAUTS
MOSCOW SEA
HUMBOLDTIANUM
FECUNDITATIS
SMYTHII
SPUMANS
UNDARUM
CRISIUM
JOLIOT-CURIE
CRATER

The farside of the moon, left, a composite of photographs taken by Soviet rocket. Even in this early, fuzzy glimpse it is apparent that the farside, to the left of the nearly vertical line, has many fewer maria than the side that faces earth. Below, chains of small craters along the rill that extend from the large crater of Hyginus, are indicative of lunar volcanism; so many evenly-spaced craters along the rill could hardly have been produced by impact. Dark material on mare to south and east of Hyginus may have erupted from that crater. (Sovfoto; National Aeronautics and Space Administration.)

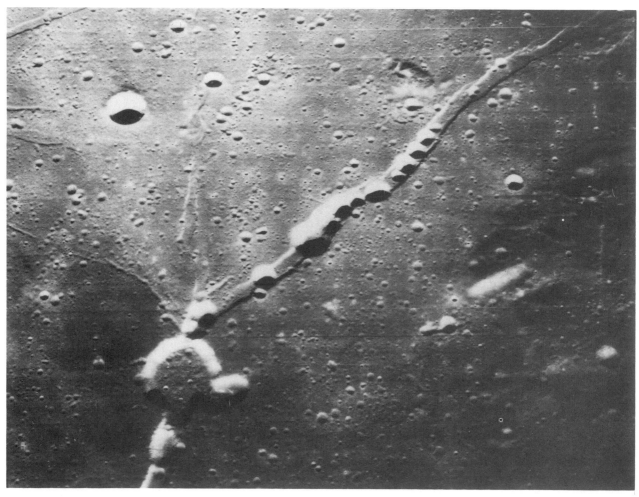

Close-up of lunar features (right). Mare Nubium, which occupies lower part of picture, received its name (meaning "Sea of Clouds") from numerous "ghosts" of former features seen on its surface. Just above Mare Nubium and at edge of black shadow is a large irregular depression called "Deslandres." Below it and at shoreline Mare Nubium is the Straight Wall (whitish line) at the center of the ruins of a similar and equally large mountain-walled plain. (Photo made at Cassegrain focus of 60-inch telescope at Mount Wilson and Palomar Observatories.)

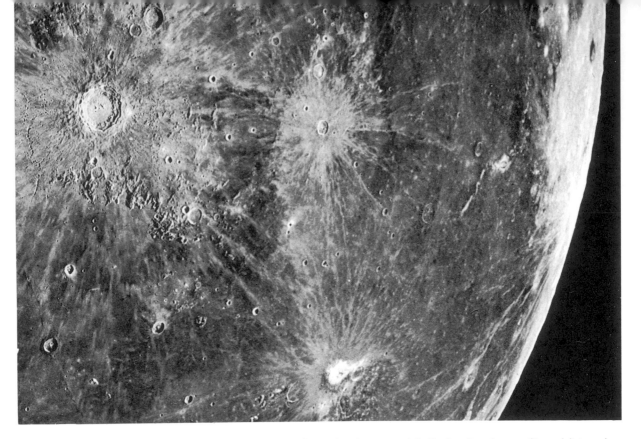

Three great ringed plains and their interlocking ray systems. Copernicus is at upper left, Kepler directly east of it, and Aristarchus (brightest area on moon) at bottom of picture. Large mountain-walled plain, Grimaldi, is at extreme upper right. Dark area at right is part of Oceanus Procellarum, that at lower left is part of Mare Imbrium. (Lick Observatory.)

Clavius, the largest mountain-walled plain (foreground). Note that it lies at the extreme eastern end of a much larger, very shallow, and older depression, which can be observed only when sun is very close to horizon. Note craters on rim of Clavius. Southern limb of moon is at top of picture. (Mount Wilson and Palomar Observatories, 60-inch telescope.)

Lunar photograph made at Lick Observatory. High magnification of lunar detail reveals that "crater walls" often are not walls, but are polygonal ranges of mountains. Diameter of plain inside Ptolemaeus, largest of the three "craters" in line (lower middle), is about 90 miles. Smallest details visible in plains are about two miles in diameter. When shadows are at left, these small round markings are craterlets. Some small markings show the bright side at left and therefore represent ordinary hills.

Ptolemaeus. Diameter of the floor of this typical mountain-walled plain is about 90 miles. At one time floor must have been molten rock. Fairly large crater on floor is Lyot. It was formed later. Approximately 150 small craters can be counted on floor, if one uses the original negative. This floor is an unusually smooth part of the lunar surface. The picture was made just before sunset. (Mount Wilson photo, 60-inch telescope.)

CHAPTER 16   THE CHANGING MOON

In a well-known passage from Shakespeare's "Romeo and Juliet," Juliet says to Romeo:

"O! swear not by the moon, the inconstant moon,
   That monthly changes in her circled orb,
   Lest that thy love prove likewise variable."

Indeed, the moon *is* inconstant, changing in appearance and position in the sky from night to night. Half of the moon is in sunlight and half is in darkness. Varying amounts of the moon's sunlit hemisphere are turned toward the earth as the moon revolves around the earth. These are the moon's phases.

The moon is new when it passes the sun. The dark hemisphere is turned completely toward the earth, and we cannot see the moon at this phase unless it happens to go directly in front of the sun. Usually the moon passes a little above or below the sun, because the moon's orbit is inclined at an angle of about five degrees to the earth's orbit around the sun. But at least twice a year the moon goes in front of the sun, causing a partial or a total eclipse.

The group of nine photographs show the moon with the south side at the top, the way it appears in the sky, not to the naked eye but through a telescope. It is, of course, inverted by the telescope. The first view was taken about two days after new moon. The second view shows the moon about two days later, when Mare Foecunditatis appears just above Mare Crisium.

It takes the moon about seven days to go one-quarter of the way around its orbit. The phase of the moon at this position is first quarter, shown in the fourth view. The lighted part of the moon appears as half a circle. The terminator, or boundary between day and night on the moon, appears as a straight line. The most striking views are obtainable at this phase and at last quarter, when the lunar mountains near the terminator cast long dark shadows, which give a fine effect of contrast with the bright parts.

The fifth view shows the moon, about half-

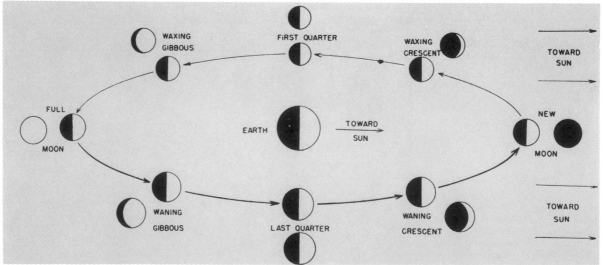

The phases of the moon.

  The moon is shown in eight different positions in its orbit, one-half of the moon being in sunlight and one-half in darkness. The outer figures show the phases as seen from the earth. These phases represent the varying amounts of the moon's sunlit hemisphere which are turned toward the earth as the moon revolves around it.

70

way between first quarter and full moon. The large crater a little right of center is Copernicus. To the north of it is a large dark area, Mare Imbrium, and on the nothern edge of that is a dark crater, Plato.

The sixth picture of the group was made about two days after full moon. Full moon occurs when the moon is opposite the sun in the sky and, therefore, has all its lighted half toward the earth. Its phase then is a little less than 15 days or about one-half of the period of $29\frac{1}{2}$ days from one new moon to the next. The prominent crater in the upper part of the picture is Tycho.

The last three pictures show the waning moon. The ages represented are about 20, 24, and 26 days. The crater near the eastern edge is Grimaldi.

Lady in the moon. If page is held at arm's length, drawing on bottom will help you find lady in photograph of moon at top. To find her in sky, look at moon when it is full or within a few days before full moon.

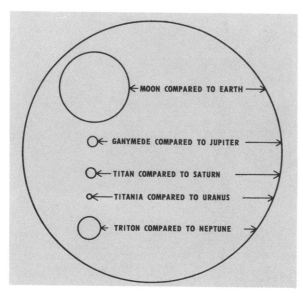

Sizes of satellites compared to their primaries.

The moon has a diameter of 2,160 miles, a little more than one-fourth that of the earth. Its distance from the earth averages less than a quarter of a million miles. If there were beings on other planets, they would probably refer to the earth and the moon as the "double planet." Most of the planets have satellites, but in no case is there one that, in proportion to the size of the primary, is nearly as large as our moon. Jupiter has two satellites about the same size as ours and two which are half again as large, but Jupiter's diameter is eleven times that of the earth.

When Venus and the earth are closest together, the earth would appear about six times brighter to an imaginary inhabitant of Venus than Venus at its best does to us. The moon would appear from Venus as bright as Jupiter at its best does to us. The moon would look yellow and the earth bluish. As seen from Venus, the maximum distance separating the two bodies would be half a degree, so that they could certainly be described as a double planet.

The moon's period of revolution around the earth, from one new moon to the next, has an average length of about $29\frac{1}{2}$ days, one day less than the average length of a calendar month. This means the date of any phase of the moon occurs an average of one day earlier from one month to the next. This period of 29½ days is called the *synodic* month and is the interval during which the moon goes through all its phases and makes a complete revolution with respect to the sun. Since the earth has moved forward nearly 30 degrees in its orbit during this time, the *sidereal* month, measured with respect to the stars, is a little more than two days shorter than the synodic month. Its average length is about 27⅓ days.

Because the moon's eastward motion among the stars is 13° a day, and the sun's is 1° a day, the moon moves eastward 12° each day with respect to the sun. The earth rotates eastward and makes the whole sky seem to move westward each day at the rate of 15° an hour, or 1° in 4 minutes. Thus the moon crosses the meridian about 50 minutes later each day. The length of the lunar day averages 24 hours 50 minutes. The excess over the 24 hours ranges all the way from 38 minutes to 66 minutes. This is due mainly to the changing speed of the moon in its elliptical orbit and the inclination of its path to the celestial equator.

the autumnal equinox, is called the harvest moon. For several evenings the moon rises only a little later each night, so that there is full moonlight in the early evening for an unusual number of evenings.

One of the most interesting facts about the moon's motion is the true shape of the moon's orbit in space with reference to the sun. If the earth did not go around the sun, the moon's orbit in space would be a small ellipse with the earth near the center of it. But the earth travels with a speed of 18½ miles a second around the sun. This speed is about 30 times as great as that of the moon around the earth. The distance of the moon from the earth is only about 1/400 of the distance of the sun. Therefore, the resulting path of the moon in space is one that deviates very slightly from the orbit of the earth. The accompanying diagram, which is made to scale, shows how the moon winds in and out along the earth's yearly path around the sun. One might think, in visualizing the moon's path around the earth as the earth travels around the sun, that part of the moon's orbit would be convex toward the sun. But the diagram illustrates the surprising fact that the moon's orbit in space is always concave toward the sun.

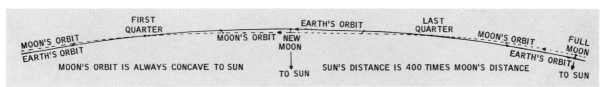

Moon's orbit compared to the earth's orbit.

The average delay in the moon's daily rising and setting is also the same 50 minutes, but the retardation is much more variable than that of the meridian transits. It depends on the latitude of the observer and on the position of the moon in its path. The minimum delay in rising occurs when the moon is near the vernal equinox. The moon's path there makes the smallest angle with the eastern horizon, and the moon is moving rapidly northward.

When the moon is near the vernal equinox and also is near the phase of full moon, the small delay in its rising becomes especially noticeable. The moon is very bright and rises in the early evening. This occurs in the autumn when the sun is near the autumnal equinox, for when the moon is full, the sun must be in the opposite part of the sky. The full moon occurring nearest September 23, the date of

We commonly say that one body revolves around another, such as the earth around the sun or the moon around the earth. However, that statement is not quite true. Both bodies revolve around their common center of gravity. If one is much more massive than the other, the center of gravity of the system is located inside the more massive one. Thus the center of gravity of the earth-moon system is about 3,000 miles from the earth's center, or about 1,000 miles below the earth's surface. Since the moon is 240,000 miles from the center of gravity and the earth's center is 3,000 miles from it, the moon's mass is about 1/80 the earth's mass.

The result of the monthly motion of the earth's center around a point 3,000 miles away from it is a slight alternate eastward and westward displacement in the sky of every object viewed from the earth as compared with the place the object would have if there were no such motion. This motion of the earth can be measured by observing during the month the apparent motion of the sun or of one of the nearer planets or asteroids.

The moon revolves around the earth in an ellipse, which has a greater eccentricity (is more flattened out) than the elliptical orbit of the earth around the sun. The mean distance of the sun from the earth is 93,000,000 miles, and the distance varies by $1\frac{1}{2}$ million miles on either side of this mean. If we divide $1\frac{1}{2}$ by 93, we get an eccentricity of 1/60 for the earth's orbit. In round numbers, the mean distance of the moon from the earth is 240,000 miles. Because of the gravitational pull of the sun, the moon's orbit is not of fixed shape. Its eccentricity varies from 1/15 to 1/23, averaging about 1/18. The greatest distance that the moon gets from the earth is about 252,700 miles. Its closest distance is about 221,500 miles. We may easily remember that it is only about 10 times farther to the moon than it is around the earth.

The sun and the moon appear to be about the same size, each one having an angular diameter, as measured from the earth, of about half a degree. That is a result of the curious coincidence that the sun is both 400 times larger than the moon and 400 times more distant. It is what makes the beautiful solar eclipses possible, with the disk of the sun covered and its corona revealed. But when the moon is too far away, its disk is too small to cover all of the sun and an annular or ring-shaped eclipse results.

Since the moon's distance from the earth varies more than the sun's, its angular diameter has a greater range. This is of importance to the marine navigator, who measures with his sextant the altitude of the edge, or limb, of the sun or moon. He must make a correction that will give him the altitude of the center of the disk. Usually he measures the lower limb and then the semidiameter must be added. This amounts to 16′ for the sun, with a variation of not more than 0′.3 on either side. The moon's semidiameter varies from about 15′ to 17′.

Since the earth's linear diameter is about four times the moon's, its angular diameter observed from the moon would be about four times the moon's angular diameter observed from the earth. Since the latter figure is $\frac{1}{2}°$, the other one will be 2°. The accompanying diagram shows the sizes of the

earth and the moon and the distance between them in true proportion. Lines drawn from the ends of a diameter of the earth make an angle of 2° at the moon.

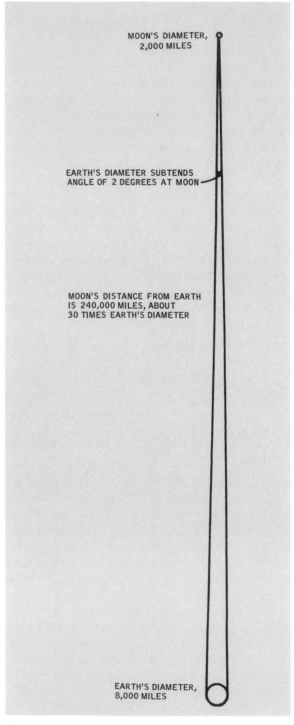

Sizes of the earth and the moon, and the distance between them in proportion.

74

If the moon were on the horizon of an observer at the North Pole of the earth, it would also be visible on the horizon of an observer at the South Pole. However, it would not appear in the same direction in space. There would be a displacement of 2°. This change in apparent direction of a body due to change of the point of observation is called *parallax*. The effect is easily seen by holding a pencil at arm's length and looking at it with one eye

the center of the earth to the moon passes through the observer.

Although the visible lighted portion of the moon is constantly changing shape, the same hemisphere of the moon is always turned toward the earth, and no changes have been found in the details of its surface. It is commonly believed that the moon does not rotate on its axis, but if that were so, we would be able to see all around it. It rotates on its axis at

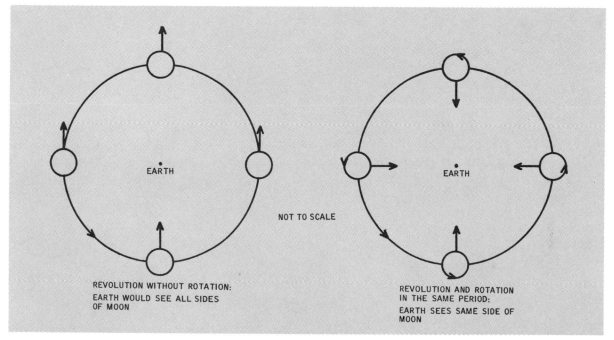

REVOLUTION WITHOUT ROTATION:
EARTH WOULD SEE ALL SIDES
OF MOON

NOT TO SCALE

REVOLUTION AND ROTATION
IN THE SAME PERIOD:
EARTH SEES SAME SIDE OF
MOON

The effect of revolution and rotation on the moon.

and then with the other. Instead of using the earth's diameter for measuring parallax, astronomers use its radius. Therefore, the moon's parallax is about 1°. It is the angle at the moon subtended by the equatorial radius of the earth. It varies from 54′ to 61′.

The navigator must allow for parallax when measuring the altitude of the moon. (Other celestial objects are too far away for parallax to have any appreciable effect.) Altitudes of the moon must be reduced to the center of the earth. Parallax differs with the altitude of the moon. It is the angle filled by the radius of the earth at any place on the earth as seen from the moon. When the moon is on the horizon, that radius is perpendicular to the horizon and the parallax is a maximum of about 1°, called the *horizontal parallax*. When the moon is in the zenith, the parallax is zero, because a line from

the same average rate that it revolves around the earth.

That can easily be shown in the accompanying two diagrams. Suppose that an arrow is erected perpendicularly upon the moon's surface. If there were no rotation, the arrow would always point in the same direction in space, as shown in the first diagram. All sides of the moon would be turned toward the earth.

The true situation is shown in the second diagram. The arrow is always pointed toward the earth, illustrating the observed fact that the same side of the moon is always turned earthward. It is seen that during one revolution of the moon around the earth the arrow points in all directions. It sweeps over a whole circle. Thus the moon rotates on its axis in exactly the same time it takes to revolve around the earth.

This is why the face of the "man in the moon" (or perhaps it is a lady—see photograph, page 72) is always toward us. Although the back of his head has never been seen, except in photographs from artificial satellites and by the astronauts, from the earth we do get glimpses behind his ears and a little over the top of his head and under his chin. The motion of the moon is not what it would be if it were attached to the earth by a rigid bar. During one revolution the moon does not keep

Map of the moon as seen in a telescope. The image is inverted.

101 LEIBNITZ MOUNTAINS	107 CARPATHIAN MOUNTAINS	113 JURA MOUNTAINS
102 ROOK MOUNTAINS	108 TAURUS MOUNTAINS	114 PICO, MOUNT
103 CORDILLERA MOUNTAINS	109 HAEMUS MOUNTAINS	115 PITON, MOUNT
104 D'ALEMBERT MOUNTAINS	110 ALPS	116 ALPINE VALLEY
105 PYRENEES MOUNTAINS	111 TENERIFFE RANGE	117 APENNINE MOUNTAINS
106 APENNINE MOUNTAINS	112 STRAIGHT RANGE	(117, 106) APENNINE MOUNTAINS

The moon as it appears in a telescope. Its normal image is inverted by the instrument. (Lick Observatory.)

exactly the same face toward the earth, but rotates independently of its orbital motion. With reference to the center of the earth the moon's face is continually oscillating slightly.

Since the moon turns on its axis in 29½ days, an observer on the moon would see the sun rise in the east and take about two weeks of our time to move across the sky and set in the west. In the middle of the long day on the moon the temperature reaches a maximum of about 212° Fahrenheit. After the sun sets there are about two weeks of night, during which the thermometer goes down to about 238° below zero. This range of 450° is made possible by the absence of any appreciable atmosphere. The air surrounding the earth acts as a protective blanket to keep out some of the sun's heat during the day and to hold in some of the earth's heat during the night. Also the long day and the long night on the moon make greater extremes of temperature possible. Thus we see another respect in which the moon changes.

If there is no atmosphere on the moon, there can be no water on its surface, for water would evaporate and form an atmosphere of water vapor. Also, observations with the naked eye show that there can be no bodies of water of any size in the equatorial regions of the moon. If there were any, they would occasionally be in a position to reflect sunlight to the earth. Such reflections would be very conspicuous, but have never been seen. Early observers thought that the dark regions on the moon were seas and called them by the Latin word for sea, which is "mare." However, observations with the telescope show that the moon has no water.

The lack of air and water makes the moon quite different from the earth. It never rains or snows. There are no winds. Rainbows and auroras do not exist. Sunrise and sunset are without color. There is no twilight, no sound to be heard.

The moon has been called an astronomer's paradise, in spite of the fact that it would be a very unpleasant place to live. The stars and planets shine by day as well as by night. They shine in a black sky with far greater splendor than we can imagine. Eclipse expeditions would not be necessary, for the solar corona is always observable. The spectra of the sun and stars could be examined to their fullest extent, with a resulting increase in knowledge about the composition of these objects.

There are a number of superstitions about the moon. The word lunacy comes from an early belief that the changes of the moon caused a kind of intermittent insanity. Even today one hears the statement that the weather will not change until the moon changes. People who say that ignore the fact that the moon changes the same way at the same time for all the world, and their statement implies the simultaneous occurrence of the same type of weather all over the earth. Such a uniform state never occurs. Also, records show that storms and weather changes have no relation to the position of the moon in its orbit.

Another error is the idea that the attitude of the crescent moon determines the weather. Some people believe that when the horns point up, the bowl of the moon holds water and there is no rain. When one horn is much lower than the other, water spills out of the tipped bowl and there is plenty of rain. Other people insist that when the horns are up the full bowl of water indicates rain, and when the bowl is tipped it holds no water and dry weather is in store. The truth is that the tilt of a line between the horns has no relation to the weather. It varies with the latitude and tends toward being horizontal at the equator and vertical at the poles. If the pointing of the horns controlled the weather, wet and dry regions would extend in belts around the earth. Such is not the fact. Also, at any intermediate latitude the horns may point nearly straight up at one time of the year and be nearly horizontal six months later. This is due to the changing angle that a line from the sun to the moon makes with the horizon.

Other superstitions about the moon have to do with farming. Potatoes should be planted during the dark of the moon, because they grow under the ground where there is no light. Shear sheep under a waxing moon, because the new wool will also grow fast. However, investigations by scientists show there is no truth in any of these beliefs. The moon has no relation to anything that affects the growth of crops.

In this connection it is interesting to point out the great difference in opinion as to exactly what periods are meant by "dark of the moon" and "light of the moon." Some define the first as the few days when the moon is too near the sun to be seen at all and the second as the two or three days on either side of full moon. Others accept this meaning of dark of the moon, but say that light of the moon covers all the rest of the lunar month.

Still others define light of the moon as the period of waxing moon, from new moon to full moon, and dark of the moon as the period of waning moon, from full moon to new moon. From these conflicting definitions, it is evident that a success-ful crop will result from planting at any time of the month, provided the conditions of soil and air are suitable and reasonable care is taken. Farming by the moon has no support in logic or in the statistics of accurate records.

CHAPTER 17   TIDES AND THE MOON

One of the standard ways of testing an individual's powers of reason is with a multiple-choice test. Imagine the following question and apply your usual methods of logic: "Check the true answer.... The daily high tides that the moon produces on the oceans of the earth occur near the time when (a) the moon is highest above the horizon; (b) the moon is near the horizon; (c) the moon is farthest below the horizon." The same choice of conditions might be offered for the low tides.

To create the greatest interest, the contestant must feel that there is a danger of choosing incorrectly and that there is enough difficulty to flatter him if his choices are correct. The statements used in the illustration are excellent ones for this purpose. With respect to high tides almost everyone, even if he has not studied astronomy, will choose (a). However, for the low tides there will be quite a split in choices between (b) and (c).

The truth in each case will surprise many. For the high tides, both (a) and (c) should be chosen. For the low tides, the correct choice is (b). The incorrect answers afford an excellent illustration of a peculiarity of our psychology of learning. It is probable that a superintelligent child would not answer incorrectly. We may miss the true answer because we are relying on our customary methods of thought and observation.

Consider any forces that you apply to an object. Pull on it or push it with your hand. You apply force on it solely through contact with the surface. All mechanical forces are applied in such a way. All forces that you can see being applied are applied only to surfaces. Now in the light of this consider the tides. Let the earth be a solid ball. The water may be represented by a soft rubber skin surrounding it in close contact but not cemented down.

Imagine yourself as the moon, and with your hands reach out to pull the "earth" toward you. You touch the loose rubber and through it exert a pull on the "earth." The rubber gives and piles up on the side toward you as a high tide. It stretches thin on the opposite side as a low tide. Your reasoning based on this observation causes you to check (a) for high tides and (c) for low tides.

The flaw in the reasoning is that you have used an incorrect analogy. All the forces you ever have watched have acted to make your answer correct. However, electric, magnetic, and gravitational forces act in a different manner. The moon uses the last of these to cause the tides. The best mechanical analogy would be to think of the moon as attaching a thread to each tiny particle of our earth, fastening it to the particles near the center just as easily as to the surface ones. The moon must then be supposed to pull on these threads and to pull hardest on the shortest ones. Let us consider what happens to three particles A, B, and C, as shown in the first diagram. A is pulled the hardest, B the least. A is pulled more toward the moon than is C; therefore, the distance AC increases. C is pulled more than B; therefore, the distance CB also *increases*.

In the second diagram, let A be a particle of water in the ocean and B one in the solid ground almost beside it. C is at the center of the earth, almost 4,000 miles farther from the moon. D is a solid particle at the bottom of the ocean on the far side, and E is a particle of ocean water beside it. D and E are 4,000 miles farther from the moon than point C, and 8,000 farther than points A and B.

The moon pulls A and B most strongly, C the next, and D and E the least. If the solid part of the earth actually is rigid, the three points B, C, and D

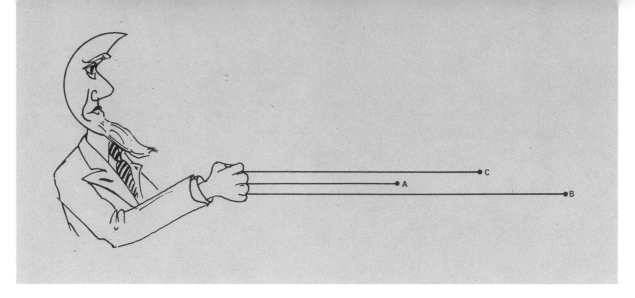

A mechanical analogy of the pull of the moon on the earth.

must all be pulled the same distance by the moon. This will be the average pull that is exerted on the three of them. That average is less than the pull on B and greater than the pull on D. As a result, the water at A is pulled away from the solid earth to give a high tide when the moon is overhead. But on the opposite side, the solid earth is pulled away from the water at E to give a high tide there as well. The moon is not overhead there, however, but directly underneath. We now have two high tides, so low tides must occur halfway between them, at places where the moon is on the horizon.

This explanation is the beginning of the study of the tides. Carrying through the study with all the complications occurring on the earth was almost a lifetime work for a very famous astronomer, George Darwin, son of the still greater Charles.

The depth of the ocean, the presence of large masses of land (especially continents), the shape of the coast, and other local factors greatly affect the actual times of high and low tides. Such local conditions also give us the extremely high tides of the Bay of Fundy, which average about 50 feet. In contrast, Mediterranean islands record tides of about two feet.

In addition to these local effects are the tides produced by the sun. The same diagrams may be used, merely substituting the sun for the moon. The sun pulls the whole earth far more strongly than does the moon, but because of the sun's great distance in relation to the distance between points on the earth, the difference of its pull at the points A and C or C and E is only about a third as great as is the difference of lunar pulls between

The moon's pull.

The moon's pull creates high tides at point A and point E.

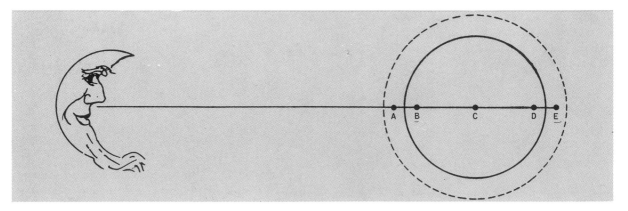

80

these points. The total pull has no effect on tides; it is the *difference* of pulls. So the lunar tides are about three times as great as the solar ones.

At times of new moon *and* full moon, both the sun and moon produce high tides at the same two places on the earth. As a result, the high tides are unusually high and the low ones unusually low at these phases of the moon. Such tides are called *spring* tides. At the times of half moon, the tidal range is least.

CHAPTER **18** THE LIFE STORY OF THE MOON

Most of the objects studied by man lose a great deal of their romantic interest when they are considered from the scientific standpoint. This is because man has a habit of building his romance around false premises. In our most serious studies we dissect, step by step, our ideas of the human body, of the objects around us, and even of the atoms, until we can see in every case a logical system replace the fanciful ideas previously held. Perhaps a new romance creeps in with this further knowledge, but it is so different from the old that only a few can partake of it.

In the case of the moon, however, the interest has clung around a reality—the pale, beautiful light that it reflects from the sun to dissipate the darkness. The moon's changing form and rapid motion among the stars remain just as interesting to us after we understand the laws that guide them. The moon is many times closer even than the asteroid that excited the scientific world a few years ago. Our sun appears at a vast distance when compared to the moon—400 times as far away. The moon moves with the earth as we swing in our gigantic orbit around the sun. Science even adds to the romantic aspect, for it tells us that last year the moon was closer to the earth than it is today, that 100 years ago it was even closer, and that if we go back a few hundred million years, we find the moon very, very close to the earth. When there was a clear patch in that early sky, the huge moon, covering many degrees in angular diameter, must have been a sight that brought awe to almost any creature. It seems a pity that man could not have been here to see it at that time.

As we regress in time and find the moon closer and closer to the earth, we find also an increase in the tides that it raised on the earth. When its distance was only half what it is today, the tides were eight times higher, and when it was half that distance, they were 64 times the height of present tides. Such tides would have made navigation across the sea almost impossible, even had there been a civilized race to build the ships. And the effect of the tides upon the earth certainly would have caused stresses to be liberated and tremendous earthquakes to occur, wherever the crust was at all weak. The change in the distance of the moon from then to now was caused by the friction of the tides slowing down the rotation of the earth. The earth, therefore, lost energy. But this energy, except for an almost infinitesimal amount that escaped as heat, could not leave the earth-moon system. The only way it could be used was in lifting the moon away from the earth. No energy can be destroyed. That is the law of *conservation of energy*.

This slowing down of the earth's rotation must also be considered in a life history of the moon. Yesterday the earth turned faster on its axis than it does today. When we go back a mere 3,000 years, we find the change large enough to cause eclipses to occur several hours from the time that today's rotation would predict. If we go back far enough, we find an epoch when the day was just a few of our present hours in length and when the month was almost equal to the day.

George Darwin, son of the great biologist, was the greatest authority the world has known on tides and their effects on the history of the earth. He carried the earth-moon system back to an early stage when the two bodies were close together and the tidal force was tremendous. He also speculated as to how the moon was formed. He considered any rotating fluid that departs slightly from being a perfect gas. When such a body rotates fast enough, a neck, or constriction, develops. This causes the body to assume the shape of an hourglass. Any rotating body that is shrinking in size must speed its rotation. If it rotates still faster, the two parts separate, and the tidal effects will push them farther and farther apart. Darwin believed that this probably was the process by which the moon was formed.

At first most astronomers were inclined to accept his conclusions. But in time, objections arose. It was demonstrated that the earth never did rotate quite fast enough to produce the separation.

In 1951 Gerard P. Kuiper developed a hypothesis for the evolution of the solar system. He postulated condensations of gas in a primeval, turbulent nebula. These condensations were far larger than are the planets of today, but were a first stage in their development. He called them protoplanets. He explained the formation of satellites by the assumption that there could be two or more centers of condensation in one protoplanet. The smaller ones would become moons. Today, nearly all astronomers favor this hypothesis as a general explanation for the evolution of planets and moons.

Our own moon, however, is a freak. Most other moons in the solar system revolve about their planets almost in the planes of the planets' equators and have orbits much less eccentric than that of our moon, which moves about the earth nearly in the plane of the earth's orbit about the sun. Furthermore, in respect to the size of its primary, our moon is far bigger than any other moon of the solar system. Accordingly, it is suggested that the earth and moon were formed out of the primeval nebula together as a double planet.

The experiments conducted by the astronauts on the moon during their moon walks together with studies of the structures and ages of rocks they brought back with them, have resulted in a wealth of information regarding the present condition of the moon and its past history. Evidence is mounting that the interior of the moon is very hot, and has been very hot since early in its formation some 4.5 billion years ago. The solid crust, with the outlines of surface features we know today, was formed early in the moon's evolution.

As we look into the future, the moon will get farther and farther away, and the earth's rotation will become slower and slower. The months will get longer and longer, as will the days, until the time comes when the month is almost twice as long as it is now and when the day will have increased to equal the month. The day will catch up with the month, partly because of additional tidal effects from the sun. If in this future period there were no disturbing factor, the two bodies would swing around each other as long as they existed, in unchanging orbits, each rotating on its axis in the month. However, tides caused by the sun will slow the day still more so that it will become a little longer than the month. With this transposition, the process that has pushed the two apart will be inverted, the day and the month will shorten and the two bodies will draw together, but always the day will remain the longer. (This has occurred already in the case of the planet Mars and its inner satellite.) Our moon will come back, ever closer, toward the earth, but the two will never collide. When they get close enough, the tidal forces will be so great that the moon will not be able to stand the stress and will break up into a swarm of moonlets somewhat similar to those that make up the rings of Saturn. The tides from such a ring will be negligible, for they will be exerted from each separate particle scattered around the path. The system will have reached stability, except for the continuing effects of the solar tides.

No one knows how long it will take all these things to come to pass. Long before that time it seems possible that the oceans will have frozen—even the air will have frozen—and our sun perhaps will have become a cold, dark body. Of course, that last is a guess. Under such conditions the actions of the tides will be many times slower than they are today. All we can say is, if our system lasts long enough, these things will occur. The time required will be many times longer than the earth's present age.

CHAPTER 19

ANCIENT MEASURES OF THE DISTANCE TO THE MOON

Whenever astronomers wish to determine the distance of objects they use instruments that measure accurately very small quantities. Often they make photographs and measure displacements on them to an accuracy of 1/50,000 of an inch or even better. We are so used to modern instruments that rather quickly determine small angles for us that we forget such instruments have been in use only for a very few generations and that long before their invention astronomers made rough measurements of the distances and sizes of the sun and moon. During the third century B.C. Aristarchus proved that the sun is much larger than the whole earth. Before that it had been guessed that it was a glowing rock perhaps as big as Greece. Practically no instruments were used in the first determination of the relative distances of the sun and moon. Aristarchus merely determined, as accurately as possible, the time of half moon. After that it required only a few minutes of arithmetical work to get an answer.

When the moon is at either of the positions M_2 or M_4 of the diagram we have half of the bright side turned toward us. In other words, we have half moon. At M_1 we have new moon, and at M_3 full moon. Both waxing and waning half moon occur closer to new moon than to full moon. The farther away the sun might be in comparison to the distance of the moon, the more nearly would the half moons occur halfway between new and full moon. Actually the sun is many times farther away

than the moon, so that the angle SEM_2 is very nearly a right angle. If it were possible to tell within a few seconds when half moon occurs, this simple triangle relationship would tell with great accuracy how many times farther away the sun is than the moon.

Aristarchus timed the half moon and computed the sun's distance from his measures. It seemed to him to be unbelievably great. Probably he tried to be conservative and took the sun's distance as being the least value permitted by his observations. He accepted, as a final value, that the sun is twenty times as far away as the moon. Of course, we know that it is 400 times as far away.

As soon as it was known that the sun is much farther away than the moon, it became possible to use eclipses of the moon to get the moon's distance and size in comparison with that of the earth. Lunar eclipses are caused by the full moon getting into the shadow of the earth. At the distance of the moon the earth's shadow is much bigger than the moon, since the earth is far the larger of the two bodies. When the moon is moving into the earth's shadow, the edge of the shadow can be seen on the moon as an arc of a circle. Comparing the sharpness of curvature of the shadow's edge and of the edge of the moon, one can find how many times greater is the diameter of the shadow than the diameter of the moon. Aristarchus found the ratio to be about 8/3. He then used the value of the sun's distance, i.e., 20 times that of the moon,

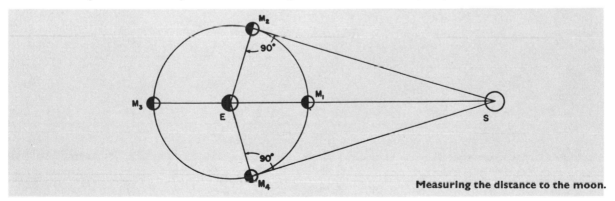

Measuring the distance to the moon.

83

Shadow of the earth.

If Aristarchus had not been misled by an effect of the brightness of the moon, he would next have obtained a fairly accurate estimate of the moon's distance. But bright objects appear to us larger than they actually are. It is hard to realize that the moon appears no bigger than a quarter held nearly ten feet away. Aristarchus guessed the apparent diameter of the moon to be two degrees instead of a quarter of this amount and, therefore, found it to be four times closer than it actually is.

With all the inaccuracies of this pioneering piece of research, he found the diameter of the sun to be nearly seven times as great as that of the earth. He could not believe that such a huge body, whose distance was at least 150 times the diameter of the earth, could move around the earth every day. Such a supposition seemed impossible. Therefore, the earth must spin daily about an axis. If the earth had one motion it might as well have others. The next simple conclusion was that the earth, as the smaller body, moves around the sun in the course of the year. To the objection that such a motion should displace the stars, he replied that the stars are immeasurably farther away than the sun. In all this he anticipated the work of Copernicus many centuries later. Unfortunately Aristotle had stated that the earth is fixed, and Aristotle's teachings were for nearly two thousand years to carry an authority almost equal to that of Scripture. The truer conclusions of Aristarchus were almost forgotten.

and by a fairly simple calculation from plane geometry determined that the earth's diameter is about three times the diameter of the moon. Fortunately the error in his first experiment did not affect this one very much. Actually the moon's diameter is more nearly a quarter than a third of ours, but the result is surprisingly good, considering the conditions under which he worked.

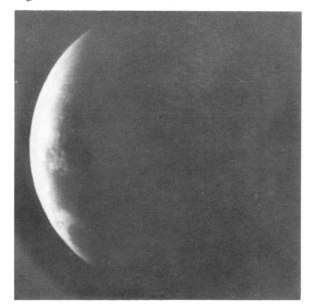

Two of a sequence of photos of the moon as it became totally eclipsed March 13, 1960. (U.P.I. from U.S. Naval Observatory.)

84

Fog, clouds, and eclipse of the sun at midday combine to give this impressive picture of Stockholm, Sweden. That crescent "moon" hanging in the sky is really the sun being swallowed by the moon. (U.P.I.)

The progression of a partial lunar eclipse over New York City on July 15, 1954, is shown in this photo made by the American Museum of Natural History. (U.P.I.)

Unmanned Surveyor and Lunar Orbiter missions preceded the Apollo-manned landings on the moon. This photograph by Lunar Orbiter V, from an altitude of about 136 miles, shows the zone near the young crater Tycho, selected for the landing of Surveyor VII. Material in the hummocky target area apparently was ejected from the Crater. (National Aeronautics and Space Administration.)

CHAPTER **20** EXPLORING THE LUNAR SURFACE

The series of Apollo landings on the moon have provided dramatic demonstration that mankind is no longer restricted to the earth. This grand adventure of the human spirit would have been eminently worthwhile even if the astronauts had done nothing more than prove man's ability to travel in space and return safely. However, the experiments they conducted or set up together with the observations they made and the wealth of rock and soil samples they brought back with them have gone far toward answering many questions about the moon's surface and past history. Additional questions that remain regarding the details of the lunar origin must await explorations made when the space shuttle becomes available in the 1980s.

Before the Apollo landings and the unmanned Surveyor and Russian Luna landings that preceded them, it had been a matter of debate whether the dust layer believed to be covering the moon might be so deep and soft that astronauts and instruments would sink below the surface and disappear forever. The first landings dispelled this fear. The footpads of lunar modules sank one to three inches into the surface, and astronauts had no trouble walking and hopping around in the 1/6 gravity of the lunar environment, although they frequently had trouble keeping the dust off themselves and their equipment. They found that the brownish, slightly gray, lunar soil ranged in depth from zero around the rims of craters to many feet within craters and in the lunar highlands. They could readily insert staffs, poles, or core tubes down a few inches, but then the resistance to penetration increased significantly. In the vacuum of the moon, the soil possesses a small amount of cohesion; holes made by core tubes were left intact when the tube was removed. The soil is made of grains that range in size from silt to fine sand. Under the microscope these grains show many evidences of melting induced by strong and intense shock. They take the form of glass dumbbells, teardrops, or nickel-iron spherules. It suggests that they may have been formed when meteorites crashed into the surface, melting the surface material and throwing molten droplets out in all directions, to solidify into the grains. About two percent of the lunar soil has come from meteorites which are raining down continuously onto the surface in the absence of a blanket of air, which, fortunately for us on the earth, burns up all but the largest bodies before they reach the earth's surface. As a consequence of this bombardment the lunar soil is being continuously stirred to a depth of about five inches. It has been estimated that this layer is "turned over" twice every million years. Thus those footprints left by the astronauts will survive for hundreds of thousands of years before they are eventually obliterated.

This photograph from the Apollo VIII spacecraft, obtained with a long-focus lens, looks south toward the large crater Goclenius. An unusual feature of this crater is the prominent rill that crosses the crater rim. (NASA.)

The crater Copernicus, taken from Lunar Orbiter II. (NASA.)

Colonel Edwin E. Aldrin, walking toward the Apollo XI module after having deployed the seismic experiments package to his right. (National Aeronautics and Space Administration.)

Apollo XV astronauts used Lunar Rover to explore region around Hadley Rill in 1971. First telescopic view down the winding rill (above) revealed rocky sides and faint horizontal layering but no signs of flow along the bottom. (National Aeronautics and Space Administration.)

Astronomers speculated that the Alpine Valley was formed by a meteorite slashing at a low angle through the moon's Alps Mountains until close-up photographs, such as this Lunar Orbiter V view, showed it to be a wide, irregular valley, with a rill wiggling down its center. (National Aeronautics and Space Administration.)

The detailed study of the 850 pounds of lunar material brought back by the astronauts has shed much light on the past history of the moon. It turns out that there are two main kinds of rock. One type is a fine or coarse-grained crystalline igneous basalt, whose internal structure contains gas cavities indicating that the rocks crystallized from melts. They are like lavas from terrestrial volcanoes except that they contain no water. Their presence predominantly in the lunar maria suggests that lava welled up from the lunar interior following the impact of large asteroid-size projectiles that blasted out the round lunar basins. The other type of rock is composed of highly compacted soil, called *breccia* by geologists. The breccia were probably formed at the surface of the moon as the result of violent and sudden compressions in the shocks when large meteorites struck the surface. Breccias predominate in the lunar highlands, those regions scarred by thousands of craters.

Since there is no water or atmosphere on the moon, it might be concluded that weathering processes are nonexistent. However, weathering changes can take place. For instance, the exposed upward-facing surfaces of rocks showed small pits with an average diameter of about one millimeter that were lined with glass. The supposition is that these pits were blasted out of the rock's surface when struck by small meteorites. The exposed rock surfaces were also rounded just like terrestrial rocks when subjected to sand blasting. More dramatic events can also take place. Astronauts on two Apollo missions saw bright flashes on the moon, which have been attributed to the impact of meteorites. Some of the conspicuous craters are comparatively recent. It has been estimated, for instance, that the crater Copernicus is only 600 million years old, and that Tycho, the crater that has the conspicuous ray system radiating from it so noticeable at full moon, is only 200 million years old. Within historical times, the chronicles of Gervase of Canterbury include a report that on the evening of June, 18 1178, when the moon showed a very thin crescent, the horns were cut apart by an obscuration that must have occurred on the moon itself. From the medieval description, it has been suggested recently that this was dust and debris thrown up in a meteorite impact that blasted out the Crater Giordano Bruno. This must be a very young crater, because it shows a beautiful and extensive ray system that extends out from the crater over 800 miles across the lunar surface.

Isotope dating techniques have been used to calculate the length of time that has elapsed since the various lunar rocks were formed. The results have been quite surprising. All of the rocks were crystallized over a relatively short period of time early in the moon's history, between 3.3 and 4.0 billion years ago. It had been hoped that Apollo XIV at Crater Fra Mauro might find among the debris in its vicinity older rocks that had been blasted out of Mare Imbrium when a meteorite impact produced the upwelling of molten material forming the circular "lunar sea." However, these supposedly older rocks were found to be only slightly older than "younger" rocks. These ages have suggested a very dramatic early history for the moon. It is hypothesized that the moon is about 4.6 billion years old, and that during the first 600 million years of its existence it was bombarded by numerous projectiles as large as ten kilometers in diameter. The impact that formed the Imbrium basin occurred about 4 billion years ago, and was so violent that debris was scattered over much of the lunar surface. These severe impacts had the effect of resetting the isotopic clocks of rocks now found on the surface. Subsequent collisions became less frequent and less violent as the aging solar system lost most of its initially large population of meteoritic objects through such "clean-up" processes. The surface of the moon has thus remained virtually unchanged over the past 3.3 billion years, with the possible exception of more recent impacts such as the ones mentioned above.

Four of the Apollo landings established seismometers that form a seismic network that has been monitoring moonquakes. Their results indicate that moonquakes occur at surprisingly great depths, between 800 and 1,100 kilometers below the surface, much deeper than on the earth. They seem to have their centers along two distinct belts, one running north to south for about 1,000 kilometers, and the other of equal length running northeast to southwest. Their explanation is unknown. They do not correspond in location with the pattern of lunar maria. From their depth, it has been inferred that the moon has an unexpectedly thick crust, with a rigid mantle beneath the crust, and, possibly, a partially molten core. From the length of time it takes seismic waves to traverse the lunar interior, it has been calculated that the crust is about 60 kilometers deep, since there is a sharp

This closeup feature of the Apennine foothills shows Silver Spur in the Hadley Delta region. Its face shows a ledged structure. The massive ledges appear to be the ends of strata dipping gently toward the left, which may represent stratified rock of lava flows. The depth of the face of Silver Spur which is shown in this view is about 800 meters (½ mile). (NASA.)

change in the speed of these waves at that depth. The crust probably consists of two layers; the outer 20 kilometers are believed to be fractured rock. Seismic sheer waves are damped out in the deep interior, just as on the earth, suggesting a partially molten core, which would require a temperature of about 1,500°C near the moon's center.

The shape of the moon has been derived from laser altitude determinations at the landing sites of Apollo XV and XVI. They show that the center of mass of the moon is about 2 kilometers closer to the earth than the center of the moon's figure. This may be completely explained by the presence of low altitude maria on the moon's near side and their virtual absence on the far side. The crust appears to be considerably thicker on the far side, which may have inhibited the formation of maria when the surface of the moon was bombarded early in its history.

One of the widely reported discoveries by Apollo XVII was the presence near Shorty crater of orange-colored soil, which might have been colored with rust from a recent volcanic event. However, analysis of this orange soil showed that it was a red-tinted glassy material similar to the basalts, with ages around 3.7 billion years.

The first astronauts to return from the moon were subjected to an elaborate isolation procedure in the remote possibility that they might have returned with lunar organic materials that might be inimical to life on the earth. These fears were completely dispelled, and later astronauts were not subjected to this isolation. Indeed, none of the rock or soil samples that have been brought back show any evidence of microscopic living, previously living, or fossil material. In a way, this is a disappointment to those advocating widespread life in the universe, but it is not surprising that living organisms never had a chance to evolve in the extremely hostile lunar environment. On the other hand, terrestrial bacteria have survived on the moon. They were carried there aboard Surveyor III, and were recovered, still alive after several years, by the Apollo XII astronauts.

SECTION FOUR

Eclipses

CHAPTER **21** **ANCIENT AND MODERN ECLIPSE OBSERVATIONS**

Almost 5,000 years ago the Chinese are believed to have started making predictions of eclipses. The occasional darkenings of the sun and moon had frightened them. Their explanation, like that of many other early peoples, was extremely naïve. For them two dragons lived in the sky at opposite points, and if the sun happened to be near enough to either of these dragons at the instant of new moon, the dragon ate it. Their fears that the sun would be entirely lost to them caused them to make a very noisy ceremonial, which always sickened the dragon and caused it to vomit the sun, leaving it for terrestrial illumination. The Chinese were not alone in this belief. The symbol used today for these points in the sky is a dragon from which the head and tail have been cut. We received this symbol from the Greeks, the belief in the dragon being common to these two widely separated peoples.

The Chinese may have learned how to predict eclipses as early as 2800 B.C., more than 2,000 years before the same method was first learned in Greece. They found that after an interval of about 18 years an eclipse took place again and that they could depend with fair certainty on the recurrence of the eclipse. Indeed, through many centuries there would be no failure of it at all. Rather early the Chinese established an observatory, and one of the duties of the astronomers was to make forecasts of these obscurations in order that preparations

could be made to combat them. These astronomers were made princes in the land. They were given wealth and honors and palaces, and many wives. Neither before nor since has the world so honored astronomers. But there was a catch: if they failed, their heads were cut off. Records show that this was the fate of the astronomers Hsi and Ho, who became drunk and failed to announce the approaching phenomenon.

Other records from history tell of battles suspended as the moon crossed between the earth and sun, and even of a treaty of peace made during an eclipse. Anyone who has watched a total eclipse of the sun would be certain, even without historical references, that among primitive people the spectacle must have been frightening. Even an astronomer, who is trained to make the calculations and knows to the second when the eclipse will end, cannot help but feel awed as the light in the sky takes on an eerie aspect and queer shadows dance across the ground, racing past his feet with tremendous speed.

Not only is an eclipse one of the greatest spectacles known to man, but there are few phenomena of the sky that have an equal scientific value. When the main disk of the sun has been blotted out, the upper atmosphere of the sun appears in streamers, sometimes millions of miles long. Only during these few precious moments have astronomers any unimpeded opportunity to make a study of this

Total eclipse of sun, September 10, 1923, near a time of sunspot minimum. At such times the corona usually forms these long streamers, paralleling solar equator. (Dr. Joaquin Gallo, Director, National Astronomical Observatory of Mexico.)

upper atmosphere, although an invention by the French astronomer B. Lyot does provide imperfect observations at other times. If it were not for total eclipses of the sun, science might never have been aware of the sun's atmosphere.

With the darkening of the sky during an eclipse, the brighter stars appear, and astronomers photograph them quite near to the limb of the sun. They measure the star positions with extreme accuracy, and from certain small residuals they compute the effect of gravitation on the light. In 1912, in his theory of general relativity, Einstein showed that two effects operate to deflect a beam of light from a Euclidean straight line when the light passes close to a large mass. First, there is a deflection toward the large mass caused by its acceleration of gravity, as though the light beam possessed weight. The deflection of a horizontal beam near the surface of the earth is negligible, but one passing close to the much larger mass of the sun is changed in direction by 0.87 seconds of arc. Second, an

equally large deflection in the same direction is produced by space itself which is curved in the vicinity of large masses. The total resultant deflection for a beam passing close to the surface of the sun is 1.75 seconds of arc. This is the expected shift in the position of a star close to the limb of the eclipsed sun. Though the shift is small, careful measurements of stellar positions made at several eclipses during the past forty years have confirmed Einstein's prediction. These measurements have been used as further evidence to confirm the relativity hypothesis.

The study of the sun's upper atmosphere and the measurement of the sun's effect on the light from stars are merely the two principal reasons why astronomers make long journeys—sometimes halfway around the world—to spend perhaps two or three minutes observing solar eclipses. Sometimes, after they have made the journey, clouds interfere. The lure of eclipse expeditions is one of the most romantic aspects of astronomy.

CHAPTER 22 CONDITIONS THAT PRODUCE ECLIPSES

A combination of factors determines where, when, and how often eclipses occur; how long they last; and whether they are partial or full. These are explained in part by the six diagrams accompanying this chapter. The first shows the phases of the moon, the second and third explain solar eclipses, the fourth explains lunar eclipses, and the fifth and sixth are maps of the paths of the solar eclipses of June 8, 1937 and July 20, 1963. In using them, it is necessary to remember that the diameter of the sun is 109 times that of the earth, and the diameter of the moon is a quarter of the earth's. This does not mean, however, that the illustrations are distorted, for we are looking at the three bodies from points not very far from the earth.

The first figure shows the orbit of the moon around the earth, with the moon at four positions. The light from the sun is represented as parallel rays. When it is at the new moon position, the moon is almost between us and the sun. Whenever it gets exactly on the line that joins any point on the earth to the sun, the sun must be either entirely or partly hidden from an observer at that point. Figures 2 and 3 show this effect. If, at the time of full moon (see Fig. 4) the moon cuts this same line, it must pass into the shadow of the earth and fail to receive the sunlight. It is evident, therefore, that we can have eclipses of the sun only at times very near to that of the new moon, and eclipses of the moon only at the time of full moon.

The diameter of the sun is approximately 400 times that of the moon, but the sun also is approximately 400 times as far away as the moon. Therefore, the two bodies generally appear nearly the same size in the sky.

However, their sizes may vary, depending on the position of the moon. The moon moves around the earth in a fairly eccentric oval path called an ellipse. At one time of the month the moon is only a little more than 221,000 miles away, but two weeks later has moved 31,000 miles farther from

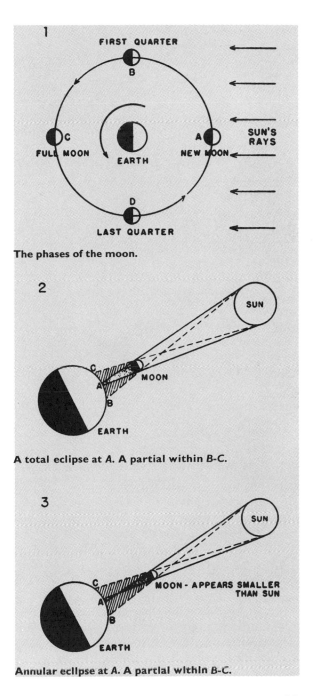

1

The phases of the moon.

2

A total eclipse at A. A partial within B-C.

3

Annular eclipse at A. A partial within B-C.

99

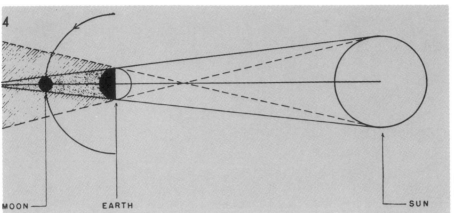

Full moon eclipsed by shadow of the earth.

us. When the moon is closest to the earth, it appears considerably larger than the sun; when it is farthest it appears smaller.

If the moon moved around the earth in exactly the same plane that the earth moves around the sun, we would have an eclipse of the sun at each new moon and an eclipse of the moon at each full one. If there were a large angle between the planes of the two paths, eclipses would be very rare, for the moon would nearly always pass north or south of the sun. But the angle of inclination is very small—a little more than five degrees. With this value, it is impossible for us to go a whole year without having the moon hide the sun, at least partially, on two occasions. There may be as many as five new moons at which this occurs. Eclipses of the moon are more rare, and it is not uncommon for none to occur during a whole year. We could never, under any conditions, have more than three eclipses of the moon in one year. Furthermore, when five eclipses of the sun occur, it is impossible within the same year to have three eclipses of the moon because the eighth eclipse, under the most favorable conditions, would occur early in the next year. A very interesting group of eight eclipses started with a partial eclipse of the sun on January 5, 1935, and ended with a total eclipse of the moon on January 8, 1936. In 368 days there were five eclipses of the sun and three of the moon, the eighth occurring at approximately the minimum possible interval.

Several factors influence the length of time that the sun may be hidden totally from any point of the earth. As seen in the second diagram, the earth's surface intersects the cone of the dark shadow of the moon. If the moon's distance from the sun is such that it appears exactly the size of

the sun, the vertex, or point, of this cone is on the surface of the earth. In such a case the moon hides the sun for but an instant. This occurred during the eclipse of 1930, the path of which crossed California and Nevada. The sun's distance from the earth, like the moon's, varies, but the change in its apparent diameter is small enough to have a much less important effect on eclipses than does the variation of the moon's diameter.

In the third diagram the moon approximates its farthest distance from the earth. The dark part of the shadow, or *umbra*, comes to a point before reaching the earth, and the observer at A will see the moon as smaller than the sun. Therefore, he will behold a thin ring of sunlight completely around the dark center. Such an eclipse is called an *annular* eclipse and is extremely beautiful, but it has no great scientific value, for if even the smallest fraction of the bright surface of the sun remains unhidden, the corona and the stars are not satisfactorily observable.

If the sun is at its greatest distance from us and the moon is closest, and if the shadow falls near the equator of the earth, an eclipse must last the maximum length of time. This is a little more than $7\frac{1}{2}$ minutes, and in the eclipse of the 8th of June, 1937, this condition was very closely approximated. The earth is farthest from the sun near the first of July and throughout the month of June is near this maximum distance; the conical lunar shadow also is close to its greatest length. The moon was closest to the earth that month on the day of the eclipse, and also the eclipse path crossed the equator. All three conditions were unusually favorable.

There are two principles involving the motion of the eclipse shadow across the face of the earth that make the equator a particularly favorable vantage

100

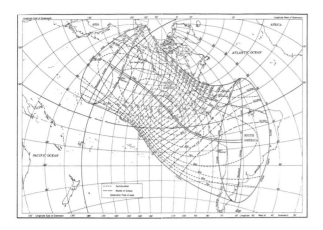

Figure 5.
Annular solar eclipse of September 11, 1969.
(*American Ephemeris and Nautical Almanac.*)

Figure 6.
Total solar eclipse of June 30, 1973.
(*American Ephemeris and Nautical Almanac.*)

point. The chief cause for the shadow of the moon to move across the earth is the motion of the moon in its orbit (see Fig. 1), which causes the shadow to move from west to east. As a result of this, the eclipse must begin at some point at sunrise and must end at some point much farther east at sunset. The second cause is evident when one considers what would happen if the moon could be stopped and held motionless in its orbit. Then the turning of the earth eastward would cause the point of darkness to travel in the opposite direction on our surface, namely from east to west. This latter effect is a much smaller one than the first and only counters a rather small fraction of the eastward motion of the shadow. It has, however, its greatest effect near the equator, where the daily spinning of the earth moves its surface eastward most rapidly.

About 18 years and 10 days after an eclipse takes place, there will be another very similar one, occurring, however, not exactly at the same point in the orbit of the moon. Continued repetition

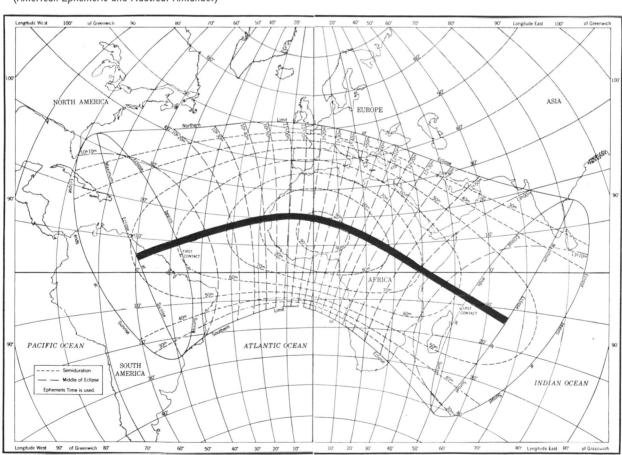

results in a family of eclipses, beginning with a small partial eclipse of the sun near one of the poles of the earth. This pole is the south pole if the moon is moving south in the sky at the times of the eclipses. The next eclipse, 18 years later, will hide a little more of the sun; the next, still more; and so on, until finally a total eclipse of the sun occurs near that pole. Then the path of totality marches northward through succeeding centuries until after more than 600 years it occurs near the equator. At this time the longest possible eclipses occur. From now on, as the path of totality continues to the other hemisphere, the eclipses get shorter and shorter until the path ends in partial eclipses at the pole opposite the one where the family entered 1,300 years before.

This interval of 18 years, 11⅓ days (10⅓ days when there are 5 leap years) is called the saros and is described and illustrated in Chapter 23. The eclipse of June 8, 1937 was near the center of a

TABLE I
Principal Total Solar Eclipses (1980–2000)

DATE	DURATION OF TOTALITY (MINUTES)	WHERE VISIBLE
1980 Feb. 16	4.1	Central Africa, India
1981 July 31	2.0	Central Asia
1983 June 11	5.2	Indonesia
1984 Nov. 22	2.0	New Guinea
1988 Mar. 18	3.8	Indonesia, Philippines
1990 July 22	2.5	Finland, Northeastern Asia
1991 July 11	6.9	Hawaii, Central America, Brazil
1992 June 30	5.3	South Atlantic Ocean
1994 Nov. 3	4.4	South America
1995 Oct. 24	2.2	Southern Asia
1997 Mar. 9	2.8	Northeastern Asia
1998 Feb. 26	4.1	Columbia
1999 Aug. 11	2.4	Central Europe, Southern Asia

series. It was followed by those of June 20, 1955 and June 30, 1973. The next one, on July 11, 1991, is listed in Table I, which includes examples of two other saros cycles. These are the pairs of eclipses of 1980 and 1998 and of 1981 and 1999.

Since a lunar eclipse occurs during full moon, which is opposite the sun, it can be observed at any one instant by all of the night half of the earth. Also, since a total lunar eclipse can last for more than an hour, it can be seen from a little more than half of the earth's surface.

Table II lists ten total lunar eclipses visible in the lower 48 United States from 1982 to 2000. It is based on Oppolzer's *Canon der Finsternisse* published in 1887. Most of the predictions are probably accurate to within five minutes. The times are Pacific standard time. For other standard times, add one hour for Mountain, two hours for Central and three hours for Eastern.

The times given cover only the interval of total eclipse. To find the approximate time when the moon enters the umbra, subtract one hour from the time when totality begins. Similarly, to find when the moon leaves the umbra, add one hour to the time when totality ends.

Four of the eclipses are not visible from coast to coast. The 1986 and 1993 ones are visible only within approximately the Pacific standard time zone. Those in 1992 and 1996 are limited approximately to the Eastern standard time zone.

TABLE II
Ten Total Lunar Eclipses (1982–2000) Visible in United States (Pacific Standard Time)

DATE	BEGINS		ENDS	
1982 July 5/6	22ʰ	39ᵐ	00ʰ	21ᵐ
1982 Dec. 30	02	53	03	59
1986 Apr. 24	04	10	05	18
1989 Aug. 16	18	15	19	53
1992 Dec. 9	15	06	16	20
1993 June 4	04	11	05	49
1993 Nov. 28	22	01	22	51
1996 Apr. 3	15	27	16	51
1996 Sep. 26	18	17	19	29
2000 Jan. 20	20	02	21	26

At first thought the prediction of eclipses of the sun and of the moon appears to be a very complicated procedure. There are, however, several things that aid the investigator. The solutions of other problems have given us, with extreme accuracy, the positions of the sun and moon for any desired instant. An eclipse of the moon can occur only when the sun and moon occupy opposite positions among the stars. In such a case the full moon passes through the shadow of the earth. An eclipse of the sun can occur only when the sun and moon appear to occupy the same position in the sky. In this case the moon passes between the earth and the sun. Diagrams explaining these facts can be found in the preceding chapter.

Modern accurate predictions of total and of annular solar eclipses tell us many things. They give us the instant when the eclipse is first seen at sunrise and the exact position on the earth for observation at that instant. They trace accurately the narrow path of totality eastward from this point to where it leaves the earth at sunset. They give the times when the shadow of the moon passes each point along the path. They mark out the northern and southern boundaries of the path with accuracy. They state the number of seconds that the sun is hidden. They indicate also the area on the earth within which the sun is partly hidden. Such a calculation is too difficult for the amateur sky-watcher.

It is possible, however, for a person with no mathematical knowledge other than addition and subtraction, plus an understanding of our calendar, to make quite usable predictions, even for eclipses that will occur several centuries in advance. He not only can tell the day on which such eclipses will take place, but their approximate positions on the surface of the earth. Such simple methods were used by the Chinese perhaps 5,000 years ago for predicting eclipses.

On the accompanying map of the sky has been traced the ecliptic, the path the sun appears to follow around the sky in one year. Also on this map has been traced the path of the moon among the stars for April, 1939. The two paths follow each other rather closely, and the moon is never more than a few degrees away from the ecliptic. It will be observed that one half of the moon's path is south of the ecliptic and that the other half is north of the ecliptic. The point where the moon crosses the ecliptic going northward is called the *ascending node*, and the point where it crosses to the south of the ecliptic, the *descending node*. Obviously eclipses can occur only when the sun and moon are at or near these nodes at the same instant. Usually at the time of new or of full moon, the moon is too far north or south of the sun for an eclipse to be produced. At the time of new moon the sun may be as much as $18\frac{1}{2}°$ away from the node and still be partly hidden. Thus there is a critical zone on the ecliptic of a length twice the $18\frac{1}{2}°$, straddling each node. If a new moon occurs with the sun within this critical zone, one or another of the three kinds of solar eclipses must occur. Due to the fact that the sun travels less than twice the $18\frac{1}{2}°$ in a month, it is impossible for it to traverse the critical zone without at least one solar eclipse taking place, and two may occur, spaced one month apart. Thus, as we mentioned earlier, in any given year, it is impossible for us to have less than two eclipses, and four may occur. Indeed, it is sometimes possible to have as many as five, since the eclipse year is about half a month shorter than a calendar year, so that the sun could make a return to a node before the end of the year if its former visit took place early in January.

A short arc from the April, 1940, path of the moon is shown on the map, almost paralleling the 1939 path and containing the ascending node. A mere glance will show that the ascending node in April, 1940, is west of the April, 1939, position by about 1/18 of the distance around the sky. If on a

103

given day the sun is at the ascending node of the moon's orbit, it will meet that node again before it has traveled completely around the ecliptic. This is due to the fact that during the interval the node will have moved backward by approximately the amount shown on the map between April, 1939, and April, 1940. The sun will come again to the node after 346.62 days. This interval is defined as an *eclipse year*. The interval from new moon to new moon is called the *synodic month* and is on the average 29.530588 days long.

The discussion to follow, concerning cycles in which eclipses occur, can be applied equally well to both lunar and solar eclipses. In order to avoid complication, however, only the solar will be considered. Since each lunar eclipse is observable over more than half of the earth, the study of such eclipses is even simpler than the case discussed.

Let us assume that an eclipse occurs today and that the sun and moon are exactly at one of the nodes, say the ascending node. After 346 days and a fraction the sun will be at this node again, the node having moved backward slightly. However, this interval is not an exact number of synodic months. In other words, we will not have a new moon when the sun again passes this node. The nearest new moon will occur after about 354.3 days, or $7\frac{3}{4}$ days too late. During this $7\frac{3}{4}$ days the sun will travel eastward $7°.6$ and, therefore, will be only that distance from the node at the new moon 354 days after the first one. We must have either a *total* or an *annular eclipse* at any new moon for which the sun is within $9°55'$ of the node and may have such an eclipse with it as much as $11°50'$ away. An annular

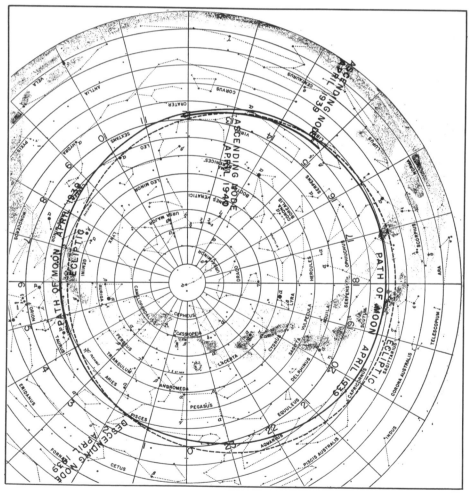

Path of the moon.

Path of the moon during April, 1939, with short arc of the April, 1940, path near the ascending node. The backward shift of the node during the year is apparent.

eclipse occurs when the positions of the sun and moon are those that produce a total eclipse, except that the moon is far enough from the earth to appear smaller than the sun.

Note the two requirements of a solar eclipse: there must be a new moon, and the sun must be near a node. Therefore, an eclipse must follow a previous one after any interval of an exact number of eclipse years that is also an exact number of synodic months. If this equality be exact, the two eclipses will trace out paths on the earth almost exactly similar in shape and in the same latitudes. If also the interval is an exact number of days, the paths will be practically identical on the earth.

Of course, such exact equalities never occur. However, whenever the equalities are fairly close, eclipses must repeat. Some equalities are nearly enough exact that for long runs of more than a thousand years they may be used without any check-up for prediction purposes.

An example will make this clear. Suppose that after some given number of eclipse years, say ten, following an eclipse exactly at a node, there is such a near equality that the sun is only one degree away from the node at the instant of new moon. Let us begin with some partial eclipse that occurred with the sun $18\frac{1}{2}°$ ahead of the node. Due to this near equality, the sun must fall back by one degree each "ten" years. It will require 370 eclipse years for it to fall back 37° so that no more eclipses of the series can be produced after a ten-year interval. One could predict, if such an equality were true, that there would be a solar eclipse each 3,466.2 days through an interval of 370 eclipse years.

The simple ratio used does not produce an equality but a great many other such near ones do exist, some of them far more exact than in the example cited.

The shortest fairly exact equality consists of four eclipse years, or 47 synodic months.

47 synodic months = 1,387.93764 days
4 eclipse years = 1,386.48012 days

The sun comes to the node less than one and one-half days too early with respect to the nearest new moon. The cycle will be repeated twelve times after the first eclipse occurs before the sun has fallen back too far to give a solar eclipse. During this interval it may be used with confidence to predict solar eclipses.

The cycle can be used either at the ascending or the descending node. Suppose it is the former. The first eclipse will occur with the sun beyond the node. That means that the new moon is north of the ecliptic. The eclipse must, therefore, be seen from near the earth's North Pole. Since the sun falls back each time, the second eclipse will not be quite so far north. The next one will be still farther south until the thirteenth one will be seen near the South Pole. For eclipses at the descending node this cycle reverses the progression from the South Pole to the North Pole.

The 47 synodic months of this cycle are nearly equal to 1,388 days. If they were exactly equal to an integral number of days, each eclipse would be due south or north of the preceding one, depending on

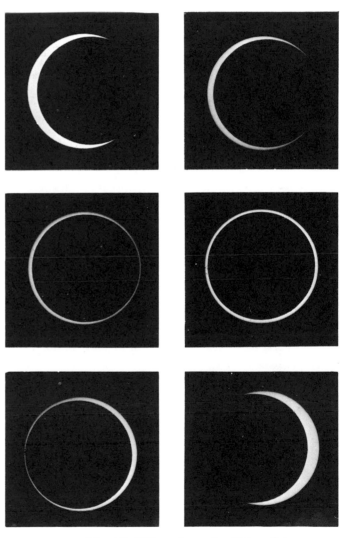

Annular eclipse of April 7, 1940, as observed at Mulege, Baja California. (Griffith Observatory and Pomona College.)

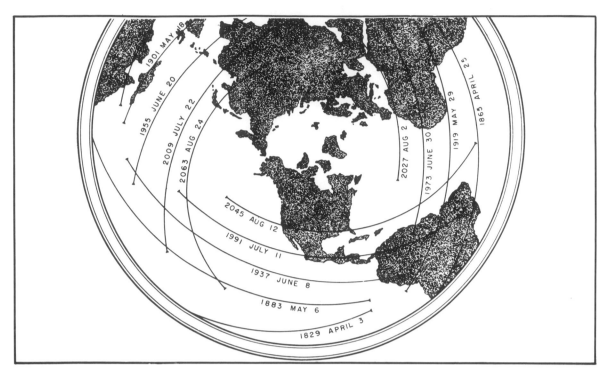

Saros cycle.
Two hundred years of the saros cycle, which contains the great eclipse of June 8, 1937. The saros cycle causes eclipses to recur after 6,585.32 days.

which node produced the eclipse. The fraction of a day causes each eclipse to fall east of the preceding one by about one-fourteenth of the circumference of the earth.

The next equality is the one that gives the very famous cycle called the *saros*. This was discovered in early times. The Chaldeans appear to have used it during their later years. Also, the Babylonians may well have had a knowledge of it. Thales, the Greek scientist, used it to predict the most famous eclipse of history and by it to stop a war. The still earlier Chinese used either it or some similar cycle. This equality is:

19 eclipse years   = 6,585.78 days
223 synodic months = 6,585.32 days

For the saros, the equality is far more exact than for the shorter cycle. In the saros cycle the sun gains on the node, but little more than a quarter what it lost in the previously mentioned four-year cycle. As a result, instead of a dozen cycles carrying the progression from pole to pole, a number approaching 81 is required. More than 1,300 years elapse from the time the first eclipse is observable at one pole till the last one appears at the opposite pole.

The first eclipse is always a very slight partial one. For each following one more of the sun is hidden until a total or an annular occurs. These two types often are grouped as *central* eclipses. After a time the eclipses again become partial and are less and less important till finally the cycle ends with a slight partial eclipse seen near the opposite pole from where the first occurred.

In the saros cycle the difference in equality of a fraction of a day causes each eclipse to fall west of its predecessor by approximately a third of the way around the earth. After three saros intervals the eclipse falls again near the same longitude but is now quite noticeably farther north or south. This relationship is so exact that the two centuries of these eclipses shown on the accompanying map fall beautifully into three groups. One of the eclipses shown is the very long one of June 8, 1937.

Other fairly good cycles are 23, 42, and 61 eclipse years. The last one is even more exact than the saros.

61 eclipse years   = 21,143.82 days
716 synodic months = 21,143.90 days

This interval equals 58 years minus 26.1 days and also minus one day for each leap year in the

106

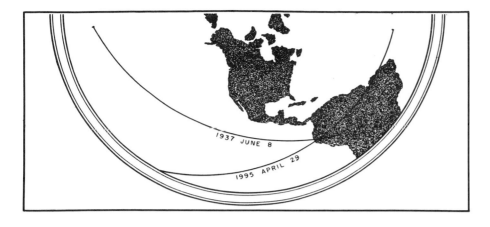

Eclipse of 1937.
The June 8, 1937 eclipse and the one which follows it in the 21,143.90-day cycle. The 1995 eclipse is annular.

interval. The last map shows again the great eclipse of June 8, 1937. It must be repeated with great exactness in 1995. There are 14 leap years; therefore, it must fall 40.08 days earlier, or on April 29, 1995. The eight one-hundredths of a day earlier causes it to be displaced eastward by one-tenth of the way around the earth.

From every standpoint that has been considered this cycle appears to be far better than the saros. There is, however, another factor to be included. As the moon moves around the earth in the course of the month, its distance from the earth varies. At one point, called *perigee*, it is closest to the earth, being only 221,463 miles away. About two weeks later it is at *apogee* and is 252,710 miles from the center of the earth. If the moon is near perigee it appears to be bigger than the sun and may produce a total eclipse. If it is near apogee it appears smaller than the sun and at best can produce only an annular eclipse. The interval from perigee to perigee is called the *anomalistic* month and consists of 27.554550 days.

If an eclipse cycle is also equal to an almost exact number of anomalistic months, it will produce many total eclipses in succession or else similar successions of annular eclipses. This condition does hold for the saros, which equals 238.99 anomalistic months. As a result all of the eclipses shown in the map for the saros *family* are total eclipses. The 58-year cycle equals 767.34 anomalistic months. For this cycle the first eclipse is the best total eclipse observable in many centuries, but the 1995 eclipse is an annular one. It is fairly easy to determine which will appear, though it cannot be done by the simple additions that have been used here. Probably the saros is the more useful cycle.

If cycles several centuries long are considered, even closer relationships can be found than those exhibited. It is, however, necessary to consider cycle lengths of more than 300 eclipse years to obtain more exact relationships than that given by a 61-year cycle. In the above-300 neighborhood are:

$$3{,}803 \text{ synodic months} = 112{,}304.826 \text{ days}$$
$$324 \text{ eclipse years} = 112{,}304.890 \text{ days}$$
$$\text{and } 4{,}519 \text{ synodic months} = 133{,}448.727 \text{ days}$$
$$385 \text{ eclipse years} = 133{,}448.712 \text{ days}$$

The longer of these equals 4,843.074 anomalistic months. It, therefore, is useful through so many thousands of years that the variations of the mean orbits of the earth and moon become serious. Longer cycles would be vitiated by these orbital uncertainties. The simple method has been driven to the absolute' limit of its application. This last cycle tells us that a total eclipse of the sun following the June, 1937, eclipse will occur near the equator on October 21, 2302. Not even the great work of the Austrian astronomer Oppolzer extends far enough to show it. However, carrying backward, it tells that one did occur on January 15, 1572, by the Julian calendar. Oppolzer shows it on the proper map.

SECTION FIVE

The Planets

CHAPTER **24** **PLANETARY PORTRAIT**

Most people seem to feel that astronomy is strictly a science of the sky totally removed from everyday life. Indeed, when reminded of the facts, many seem surprised that the day and the month and the year, as well as the light from the sun and from the moon, are all astronomical subjects. When the telescope is pointed at the other planets and details begin to show upon them, the average watcher breaks out in rather eager questioning: Do you think there is life there? Does it have people? Would they be like us? Could we live if we went there? Where did the planets come from?

Questions like these have been asked for generations. Three hundred and seventy years ago the telescope had just been invented and had begun its work of scanning the skies. Compared with the telescopes of later centuries, Galileo's best telescope was a very poor instrument. It merely brought objects to an apparent distance $\frac{1}{33}$ of that for the unaided eye. The later telescopes of that century were somewhat more powerful. However, only one object was close enough to the earth to be seen in much detail through those telescopes. That object was the moon. As a result of the restless desire of the human mind to find other abodes of life, the astronomers of that time looked at the plains of the moon and called them "seas." They imagined they saw clouds and atmospheric effects, and they pictured the moon as a world that might be delightful. That same century, however, this optimistic picture faded. The moon was shown to have neither air nor water and to be an absolutely dead world.

The same telescopes that showed the true nature of the moon began to reveal details on other planets and raised anew the question of extraterrestrial life. Since then, astronomers have gathered much data bearing on such speculations. In this chapter we describe the general characteristics of the planetary system, their distances, motions, sizes, and brightnesses, and then in subsequent chapters look at each of the planets in turn in an attempt to answer these questions.

The distances of planets from the sun can be most easily remembered by expressing them in terms of the earth's distance from the sun as a unit. This is called the *astronomical unit* and is equal to 93,000,000 miles. Mercury and Venus are closer to the sun than the earth, their mean distances being about 0.4 and 0.7 of an astronomical unit, respectively. Mars is about 1.5 units away. Rounding off the numbers of astronomical units for the outer five planets, we get: Jupiter, 5; Saturn, 10; Uranus, 20; Neptune, 30; and Pluto, 40.

The sidereal period of a planet is its true period of revolution around the sun, from a star to the same star again, as seen from the sun. The periods range from 88 days for Mercury to 248 years for Pluto. An interesting relation exists between the distances and periods of the planets. Known as Kepler's third law, it states that the cubes of the mean distances of the planets from the sun are

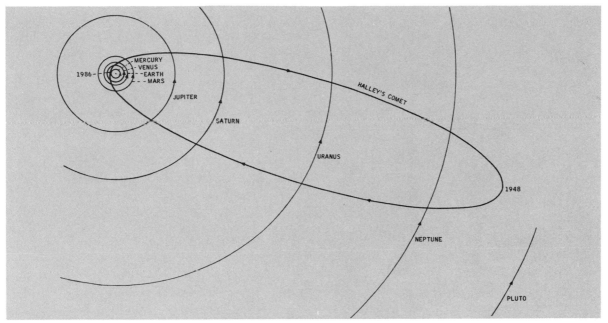

Orbits of the planets and of Halley's Comet.

proportional to the squares of their sidereal periods. This simply means that in Table I the cube of any number in the column of distances expressed in astronomical units is equal to the square of the corresponding number in the column of sidereal periods. For example, if we use approximate values for Neptune, the cube of 30 astronomical units (30x30x30) is 27,000, and the square of 165 years (165x165) is 27,225. These relationships are a result of Newton's law of gravitation.

The mean distance of Pluto from the sun is almost exactly 100 times the mean distance of Mercury. Hence the period of Pluto is about 1,000 times the period of Mercury. It is also interesting to notice that the mean orbital velocity of Mercury is about 10 times that of Pluto, since this quantity varies inversely as the square root of the mean distance.

The synodic period is the interval between two successive times when the planet occupies the same position in relation to the sun, as seen from the earth. In the cases of Mercury and Venus, the synodic period is the interval in which one of them gains a lap on the slower-moving earth. In the cases of the planets outside the earth's orbit, the synodic period is the interval in which the earth gains a lap on one of them.

Mercury travels so fast that it gains a lap on the earth in less than four months. The earth and Mars are so evenly matched that it takes the earth over two years to overtake Mars, after starting out even with it. Since the sidereal period of Jupiter is about 12 years, it goes about 1/12 of the way around the sun in one year. Therefore, the earth passes it again after going around 1-1/12 times, which takes about 13 months. The synodic period approaches a year as we go to the slowest moving planets. Since Pluto moves only $1\frac{1}{2}°$ in a year, the earth overtakes it in only $1\frac{1}{2}$ days more than a year.

The planets all move in ellipses. The eccentricity denotes the degree of flattening of the ellipse. It may have any value between 0, when the figure is a

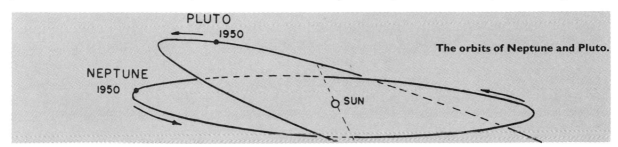

The orbits of Neptune and Pluto.

109

PLANETARY PORTRAIT

TABLE I. THE ORBITS OF THE PLANETS

| Planet | Mean Distance from Sun | | | Period of Revolution | | Mean Orbital Velocity | | Eccentricity | Inclination to Ecliptic |
	Astronomical Units	Millions of miles	Millions of kilometers	Sidereal (Years)	Synodic (Days)	Miles per second	Kilometers per second		(Degrees)
Mercury	0.39	36	58	0.24	115.88	29.8	47.9	0.206	7.0
Venus	0.72	67	108	0.62	583.92	21.8	35.1	0.007	3.4
Earth	1.00	93	150	1.00	....	18.5	29.8	0.017	0.0
Mars	1.52	142	228	1.88	779.94	15.0	24.1	0.093	1.9
Jupiter	5.20	484	779	11.86	398.88	8.1	13.1	0.048	1.3
Saturn	9.54	887	1,427	29.46	378.09	6.0	9.6	0.056	2.5
Uranus	19.18	1,785	2,870	84.0	369.66	4.2	6.8	0.047	0.8
Neptune......	30.06	2,797	4,496	164.8	367.49	3.4	5.4	0.012	1.8
Pluto	39.46	3,670	5,906	247.7	366.74	2.9	4.7	0.249	17.1

Note: Three of the elements seldom needed except by astronomers are omitted.

TABLE II. SIZES, MASSES, AND RELATED DATA OF THE PLANETS

| Planet | Angular Diameter (Seconds of Arc) | | Equatorial Diameter | | Volume | Mass | Density | Surface Gravity | Velocity of Escape at Surface | |
	Maximum	Minimum	Miles	Kilometers	(Earth=1)	(Earth=1)	(Water=1)	(Earth=1)	Miles/Sec.	Km./Sec.
Mercury	12.7	4.7	3,025	4,868	0.06	0.05	5.4	0.4	2.7	4.2
Venus	64.5	9.9	7,600	12,200	0.92	0.8	5.3	0.9	6.3	10.2
Earth	......	......	7,927	12,757	1.00	1.0	5.5	1.0	7.0	11.2
Mars	25.1	3.5	4,200	6,800	0.15	0.1	4.0	0.4	3.1	5.0
Jupiter	50.0	30.8	89,000	143,600	1347	318	1.3	2.6	37	57.5
Saturn...........	20.6	14.9	75,000	121,000	771	95	0.7	1.2	22	33.1
Uranus	4.2	3.4	29,200	47,000	51	15	1.6	1	13	21.6
Neptune........	2.4	2.2	27,700	44,600	43	17	2.2	1.41	15.5	24.6
Pluto	*	*	4,200*	6,800*	*	0.002	*	*	*	*

Note: *Figures for Pluto are unknown.

TABLE III. OTHER PLANETARY DATA

Planet	Period of Rotation	Inclination of Equator to Orbit	Oblateness	Albedo	Magnitude When Brightest	Number of Satellites
Mercury	59 days	<28	0	0.07	—1.2	0
Venus	243 days retrograde	88°	0	0.76	—4.3	0
Earth	23ʰ 56ᵐ	23½°	1/297	0.40		1
Mars	24 37	25	1/192	0.15	—2.8	2
Jupiter	9 50-56	3	1/15	0.44	—2.5	15
Saturn.................	10 14-38	27	1/9.5	0.42	—0.4	17
Uranus	10 49	98	1/14	0.45	+5.7	5
Neptune..............	15 48	29	1/50	0.52	+7.6	2
Pluto	6ᵈ 10ʰ	*	*	*	+14.5	1

Note: The albedo of a planet is the ratio of the total amount of sunlight reflected from the planet, in all directions, to the amount that falls on it. This cannot be determined as accurately as most of the other planetary data.
* Figures for Pluto are either unknown or very uncertain.

110

circle, and 1, when it becomes a parabola. The orbits of Venus and Neptune are most nearly circular. Pluto has the most eccentric orbit of all the planets. Its aphelion distance (farthest from the sun) is about 4,580,000,000 miles. Its perihelion distance (nearest the sun) is about 2,760,000,000 miles. The sum of these two distances is 7,340,000,000 miles, which is the length of the major axis of the planet's orbit. Half of the major axis is the mean distance, which is 3,670,000,000 miles. The aphelion distance is about 910,000,000

Pluto's minimum distance from the sun is about 60,000,000 miles less than Neptune's maximum distance. If the orbits of Neptune and Pluto were in the same plane, they would intersect. But because of their inclination, they do not come closer than about 240,000,000 miles to each other and they do not interlock.

Mercury's orbit has the second largest inclination (7°) to the plane of the earth's orbit, Pluto's being 17°. Thus Mercury, the nearest planet to the sun, and Pluto, the farthest, have orbits that are

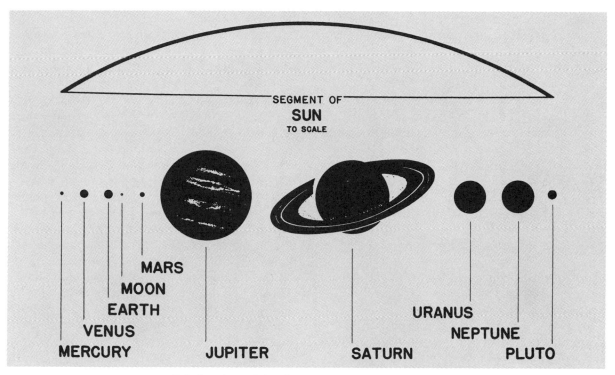

SEGMENT OF
SUN
TO SCALE

MARS
MOON
EARTH
VENUS
MERCURY   **JUPITER**   **SATURN**

URANUS
NEPTUNE
PLUTO

Relative sizes of the planets, the moon, and the sun.

miles greater than this mean distance, and the perihelion distance is the same amount smaller. The eccentricity is the fraction obtained by dividing the 910,000,000 miles by the mean distance of 3,670,000,000 miles. The result is about 0.25.

Although Pluto's mean distance from the sun is the greatest of all the planets, part of its orbit lies closer to the sun than the orbit of Neptune. The perihelion distance of Pluto (2,760,000,000 miles) is about 37,000,000 miles less than the mean distance of Neptune (2,797,000,000 miles). Also it happens that the perihelion of Pluto lies in nearly the same direction from the sun as the aphelion of Neptune.

the most eccentric and that have the largest inclinations.

Table II lists the sizes, masses, and related data of all the planets except Pluto, for which the figures are too uncertain to be included. The apparent size of a planet depends on its real size and its distance from the earth. Venus comes closer to the earth than any other planet. When it is at inferior conjunction (on the same side of the earth as is the sun, and between earth and sun) it is about 26 (93 — 67) million miles away and has an angular diameter of 64.5 seconds, or a little over one minute, of arc. When it is at superior conjunction (on the same side of the earth as is the sun, but beyond the sun) it is about 160 (93 + 67) million miles away. This is about 6 times the minimum distance, and hence its angular diameter is about 1/6 of a minute of arc, or almost 10 seconds, at this position. The planet whose angular diameter varies the most is Mars. At the most favorable opposition (when it is on the opposite side of the earth from the sun) its distance from the earth is 35 million miles. At conjunction (Mars can only be at superior conjunction) its distance is almost 250 million miles, or seven times greater. Thus its angular diameter varies by seven times, from 3.5 to 25 seconds of arc. Jupiter, the largest planet, has a maximum angular diameter of 50″, a little less than that of Venus, but its minimum angular diameter is 31″, or three times the minimum of Venus.

The planets that are listed in Table II can be grouped in four pairs, as far as the real sizes are concerned. Jupiter and Saturn are the largest, having equatorial diameters of 89,000 and 75,000 miles, respectively. Uranus and Neptune are each a little over 27,500 miles in diameter. Venus and the earth are often called twin planets, differing in size by only about 300 miles. Mercury's diameter is about 1,000 miles less than that of Mars, which is about half as great as the earth's.

Since the volume of a sphere is proportional to the cube of the diameter, we find a much greater difference in the volumes of the planets than in their diameters. Jupiter is bigger than all the rest of the planets put together, having a volume 1¾ times that of Saturn and 1,347 times that of the earth. Jupiter is also the most massive, but it has only 318 times the earth's mass. Thus its density is only ¼ that of the earth. The earth has the highest density of all the planets, 5.5 times that of water. Nearly as dense are Venus, Mars, and Mercury. The four giant planets have low average densities because of their extensive atmospheres. Saturn's density is the only one less than that of water.

The figures in the column headed "Surface Gravity" give the relative weights a person would have on the surface of each planet. It may be a surprise to learn that he would weigh the same on Uranus that he does on the earth and that he would weigh only 2.6 times as much on Jupiter. This is because the surface gravity depends not only on the mass of the planet but also on the distance from the center to the surface. This force varies directly as the mass and inversely as the square of the distance.

Thus Uranus has a mass 15 times that of the earth but a diameter ratio of about 79 to 310. The relative weights are, therefore, 1 over 79 squared and 15 over 310 squared, which are almost the same. Jupiter's diameter is 11 times that of the earth. If we divide 121 (the square of 11) into 318 (the relative mass of Jupiter, we get 2.6. Thus a 100-pound boy on the earth would weigh 260 pounds on Jupiter.

The velocity of escape is the initial speed an object must have at the surface of a planet in order to overcome the pull of gravity and get away. At the earth's surface it is seven miles a second. This is the speed a projectile would need to escape from the earth if it were shot from a gun and then had only the force of gravitation acting on it. A rocket does not have to develop a speed of seven miles a second to escape from the earth. Its propulsive force does not stop at the beginning of the motion, as in the case of a gun, but continues as long as its fuel lasts. The problem is to provide enough fuel.

In order to find the velocity of escape for another planet, we can use the relation that the square of this velocity is proportional to the mass divided by the radius. For example, Neptune's mass is about 16 times the earth's and its radius is about 4 times greater. 16 divided by 4 equals 4. The square root of 4 is 2. Hence the velocity of escape for the planet Neptune is twice that of the planet earth. Its velocity equals fourteen miles per second.

Table III gives some other miscellaneous information about the planets. Mercury's period of rotation is close to 59 days, shorter than its 88-day period of revolution around the sun. Recent radar results show that Venus has a retrograde rotation with a 243-day period, in resonance with the relative orbital motions of Venus and the earth. Mars has very nearly the same period as the earth, while Jupiter, Saturn, and Uranus rotate in about ten hours. Neptune's period is nearly 16 hours. In the cases of Jupiter and Saturn the period of rotation is not the same at all latitudes, being shortest at the equator.

It is well known that the earth's equator is tilted at an angle of $23\frac{1}{2}°$ to the plane of its orbit around the sun. Three other planets have similar but slightly larger tilts: 25° for Mars, 27° for Saturn, and 29° for Neptune. The equator of Uranus has the very high inclination of 98°. Another way of stating this is that the tilt is 82° and the planet is rotating backward. Its five satellites are also revolving in the same direction and in the same plane as its equator.

112

The oblateness of a planet describes how much its shape differs from a sphere. It is found by dividing the difference between the equatorial and polar diameters by the equatorial diameter.

Saturn has the largest value. Its equatorial diameter is 75,000 miles and its polar diameter is 67,800. The difference is 7,200 miles. This goes into 75,000 miles about 9.6 times, so that the oblateness is 1/9.6.

The column of magnitudes shows that three planets can appear brighter than the brightest star, Sirius, whose magnitude is −1.6. Venus is always brighter than Sirius, Jupiter is most of the time, and Mars is occasionally. Venus, at a magnitude of −4.3, is 18.8 magnitudes brighter than Pluto, whose magnitude is −14.5. This corresponds to a ratio of brightness of about 32,000,000.

In 1972 a computer analysis of small irregularities in the period of Halley's comet suggested the existence of a remote planet called Planet X, about twice as far from the sun as Neptune. Planet X has not been found, despite its large predicted diameter, and it is unlikely that it exists.

ORIGIN OF THE PLANETS

Ever since man recognized that the earth was accompanied in its orbital motion around the sun by sister planets, he has been speculating on their origin. As explained in a later chapter on life in the universe, the circumstances surrounding the original formation of our planetary system have an all-important bearing on the question of the uniqueness of intelligent life on the earth. If the planets were born as a consequence of an accidental cataclysm, then planetary systems associated with stars other than the sun would be extremely rare, and life elsewhere would be unlikely. On the other hand, if contemporary theories are correct, the solar system evolved as a natural step in the evolution of the sun. So there is no reason why many, if not all, stars like the sun should not possess an accompanying family of planets, with the result that planets with favorable surface conditions, moderate temperatures, an atmosphere, and water should be extremely numerous.

The current intensive research on the origin of the solar system will produce sweeping modifications of current theories, but it is believed that these modern theories are beginning to reveal some of the events that must have occurred during the formation of the solar system. It is interesting that these theories are revivals of some early hypotheses by the philosopher Kant in 1755 and the mathematician Laplace in 1796, who suggested that the solar system was formed out of the same contracting nebula, or cloud, that resulted in the formation of the sun. Difficulties with these early, oversimplified models led to the abandonment of the "nebular"

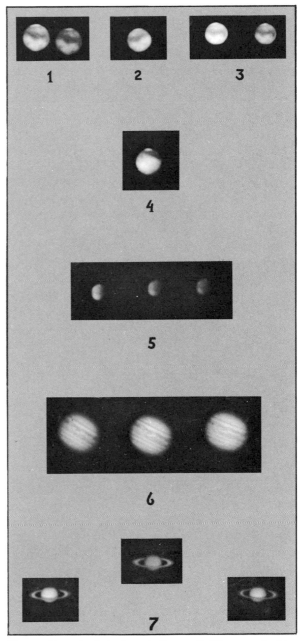

Planets: Mars, 1—5; Jupiter, 6; Saturn, 7. (Photographed by George W. Bunton and George Herbig with 12-inch refractor of Griffith Observatory.)

hypothesis until its recent revival by several experts in the field. As an example of the current thinking, we might briefly outline the protoplanet hypothesis of G. P. Kuiper. According to this theory, the Laplacian presolar nebula is replaced by a protosun surrounded by a highly flattened rotating nebula that might have developed into a double star if the nebula had been somewhat more massive. If the nebula had contained about a hundred times as much material as is presently found in the planets, then when condensations formed in the nebula they would have been able to resist the tidal disruptions by the gravitational field of the sun. These condensations, or protoplanets, formed solid cores, by a sedimentation process, surrounded by gaseous envelopes. Thus one can explain why all the planets revolve around the sun in the same direction, with orbits all nearly in the same plane. Some protoplanets, as they condensed, may have become gravitationally unstable in their turn, leading to the formation of protosatellites. Initially, all of these condensing objects had compositions similar to the sun, which is composed primarily of the light gas hydrogen. However, as the protosun contracted and heated, its increasing radiation eventually succeeded in dispersing all of the remaining gaseous parts of the nebula, and also the gaseous envelopes of the nearer planets, *i.e.*, Mercury, Venus, the earth, and Mars. Even the more remote planets lost much of their lighter gases, but enough was retained by them to explain the marked difference in structure between the terrestrial and the Jovian planets.

PHOTOGRAPHING THE PLANETS

It is a common belief that satisfactory photographs of the planets can be made only with the largest of instruments. The photographs on page 107 disprove this idea. They were made with the 12-inch visual refractor of the Griffith Observatory, using commercial photographic filters. This telescope, through which the public may observe the sky free, six evenings a week, was made by the Zeiss Optical Company and is of excellent quality, both optically and mechanically. However, as compared with modern research instruments, it is of only moderate aperture.

The first photographs using the filters were made on July 21, 1939 when Mars was close to us. Each of them exhibits some of the broader "canals," in addition to the polar cap and the large dark region. The second photograph was made on July 17, and the third several weeks after opposition, on August 17. The fourth photograph, made on August 5, shows a different part of the surface and also the shrinkage of the polar cap. All of these are good, but those of July 21 are surprisingly so and will compare favorably with photographs made though much larger telescopes. The three pictures shown as No. 5 were made on November 28, when Mars was at quadrature (at right angles to the sun) and much farther from the earth than during the summer.

The three photographs numbered 6 were made on September 15 and will be recognized at once as of the planet Jupiter. Each is rich in detail, showing plainly the six belts across the planet.

The topmost photograph of Saturn was made on October 22. The two lower ones were made on September 15. Cassini's division, between the outer rings, shows plainly on the originals. The bulge at the equator is quite noticeable, as it also is in the Jupiter photographs.

CHAPTER **25** **THE MOTIONS OF THE PLANETS**

The time required for a planet to make one revolution around the sun is called its sidereal period. The earth's period is one year and that of Venus is 225 days. We may compare the two planets to runners who start off on a race when they are lined up with the sun. This line-up is called inferior conjunction. Venus is on the inside track, which is much shorter than the earth's. Also, the speed of Venus is about 22 miles a second, as compared with the earth's speed of $18\frac{1}{2}$ miles a second.

114

Therefore, Venus overtakes the earth at regular intervals. The length of this interval is called the *synodic period*.

The sidereal period of Venus, consisting of 225 days, is one day more than 32 weeks, whereas one year is one day more than 52 weeks. Thus we can see that the period of Venus is very nearly equal to $\frac{32}{52}$, or $\frac{8}{13}$ of a year. If it makes a trip around the sun in $\frac{8}{13}$ of a year, the number of trips it will make in 1 year is 1 divided by $\frac{8}{13}$. This equals $\frac{13}{8}$, or $1\frac{5}{8}$. Therefore, it gains $\frac{5}{8}$ of a lap on the earth in 1 year. The number of years it will take to gain a whole lap is 1 divided by $\frac{5}{8}$. This result is $\frac{8}{5}$ or 1.6 years, the synodic period of Venus.

This is not the ordinary method for determining the synodic period, and usually cannot be used. However, the sidereal periods of these two planets are such that this simpler way can be more easily explained. The standard method will now be given.

Let S be the synodic period, E the sidereal period of the earth, and P the sidereal period of another planet. If all periods are expressed in days, then $1/E$ is the fraction of a circumference moved over by the earth in one day. Similarly, $1/P$ is the fraction of a circumference moved over by the planet in one day. The difference, $1/P - 1/E$, is the amount that either of the inferior planets, Mercury and Venus, gains on the earth each day. The difference, $1/E - 1/P$, is the amount that the earth gains each day on any of the superior planets, which are outside the earth's orbit.

An inferior planet will gain a whole revolution on the earth in a synodic period of S days. It gains $1/S$ of the circumference in one day. Therefore, we have for an inferior planet, $1/S = 1/P - 1/E$. Similarly, for a superior planet, $1/S = 1/E - 1/P$. If we use the periods to the nearest tenth of a day for Venus and the earth, we get

$$1/S = 1/224.7 - 1/365.3$$
$$\text{Then } 1/S = 140.6/82{,}082.9$$
$$\text{Finally,} \quad S = 583.9 \text{ days}$$

This synodic period of 583.9 days is equal to 1.60 years. This checks the result obtained by our first method.

It is interesting to note that Venus completes 5 synodic periods in 8 years, since 5 times 1.6 equals 8. The dates do not agree perfectly, because the synodic period is not exactly 1.6 years. However, it is only about $7\frac{1}{2}$ hours less than 1.6 years. Therefore, in 8 years the corresponding configurations come about 60 hours, or $2\frac{1}{2}$ days, earlier.

In the accompanying diagram full lines have been drawn connecting simultaneous positions of the earth and Venus at 10-day intervals. Dotted lines have been drawn from the sun to the two planets at their positions on September 5, 1954, and January 25, 1955. These are the times when Venus is at its greatest elongation, that is, when there is a maximum value of the angle at the earth formed by lines from the sun and Venus. This angle is about 45°. When Venus is east of the sun, it should be looked for soon after sunset above the western horizon. When it is west of the sun, it will appear just before sunrise above the eastern horizon.

After passing greatest eastern elongation, Venus appears to move westward relative to the sun and passes between the earth and the sun at inferior conjunction. It thus changes from an evening star to a morning star. It would be too confusing to show the positions of the planets during a complete synodic period, and so they have been shown for one sidereal period of Venus.

Since Venus does not revolve around the sun in exactly the same plane that the earth does, it usually does not pass directly in front of the sun. Such a passage across the sun's disk is called a transit. It is not an eclipse, because Venus appears as only a small black dot on the solar disk. Its angular diameter at that time is only about 1/30 that of the sun. It covers up only about 1/1,000 of the area of the sun's disk. The last two transits of Venus occurred in 1874 and 1882, and the next two will occur in 2004 and 2012. Transits of Mercury occur more often. One took place in 1940, and the next two were in 1953 and 1957.

The relative positions of Venus and the earth during one synodic period are indicated in the other diagram, which has been simplified by keeping the earth's position fixed. As Venus revolves around the sun, its sunlit hemisphere is presented to the earth in varying amounts. It thus shows phases resembling those of the moon. These phases are indicated at the left and at the right of the diagram. They are seen only in a telescope, and were first observed by Galileo in 1610. Venus shows the full phase at superior conjunction, the quarter phases at greatest elongations, and the new phase at inferior conjunction. Usually a very thin crescent remains at the new phase, because the planet crosses a little above or below the sun.

As Venus moves from superior to inferior conjunction, its distance from the earth decreases from 160 million miles (93 million + 67 million, the sum of their mean distances from the sun) to 26 million miles (93 million — 67 million, the difference in mean distances). This decrease in distance causes its apparent diameter to increase about six times. As it approaches the earth, its apparent brightness increases until the crescent phase is reached. Greatest brilliancy occurs about five weeks preceding and five weeks following the time of inferior conjunction.

At greatest brilliancy, the sunlit crescent of Venus sends to the earth nearly $2\frac{1}{2}$ times more light than the smaller, fully illuminated disk at superior conjunction. Venus at its best appears about 15 times brighter than Sirius, the brightest star. Viewed from Venus, the earth at its best would appear about six times brighter than Venus ever appears to the earth. This is because the earth presents all of its illuminated hemisphere toward Venus when the two planets are closest.

The five views of Venus taken at the Lowell Observatory show the phases and the great range in apparent diameter. The angular diameter varies from about 10″ to 65″. Mercury also shows the same phases in the telescope, but the angular diameter and its range are not as great. The angular diameter varies from about 5″ to 13″.

Mercury is visible to the unaided eye only during the intervals when it is far enough away from the sun in the sky. These intervals vary from a week to two weeks or more, and there are usually six of them each year. Each one is centered around the time of greatest elongation. Mercury takes only three months to revolve around the sun, but it requires nearly four months to gain a lap on the earth. The average interval from a greatest eastern elongation to the next one or from a greatest

Phases of Venus, showing great range in apparent diameter as planet passes from superior to inferior conjunction. (E. C. Slipher at Lowell Observatory.)

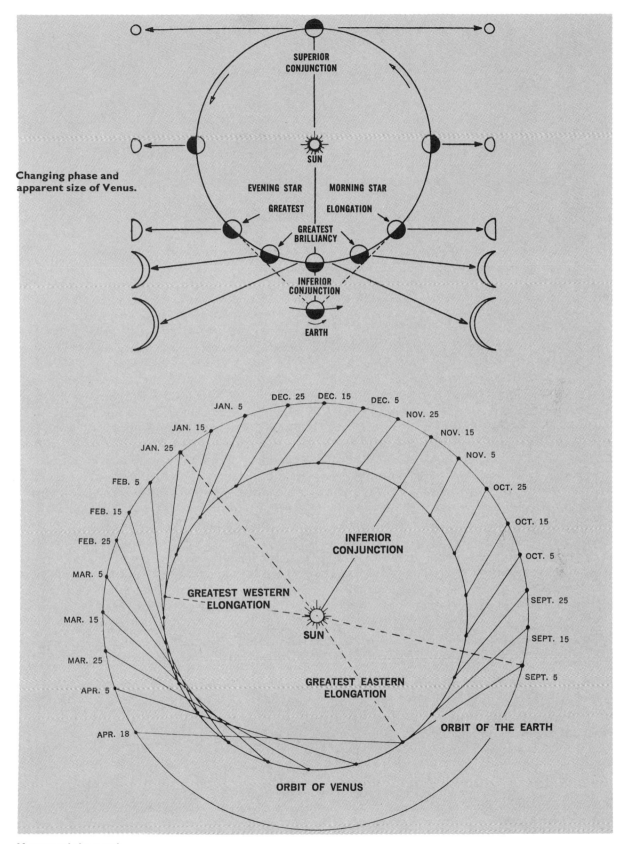

Changing phase and apparent size of Venus.

SUPERIOR CONJUNCTION

SUN

EVENING STAR

MORNING STAR

GREATEST

ELONGATION

GREATEST BRILLIANCY

INFERIOR CONJUNCTION

EARTH

DEC. 25 DEC. 15 DEC. 5
JAN. 5 NOV. 25
JAN. 15 NOV. 15
JAN. 25 NOV. 5
FEB. 5 OCT. 25
FEB. 15 OCT. 15
FEB. 25 OCT. 5
MAR. 5 SEPT. 25
MAR. 15 SEPT. 15
MAR. 25 SEPT. 5
APR. 5
APR. 18

INFERIOR CONJUNCTION

GREATEST WESTERN ELONGATION

SUN

GREATEST EASTERN ELONGATION

ORBIT OF THE EARTH

ORBIT OF VENUS

Venus and the earth.

Positions of Venus and the earth at ten-day intervals during one revolution of Venus around the sun from September 5, 1954, to April 18, 1955, an interval of 225 days.

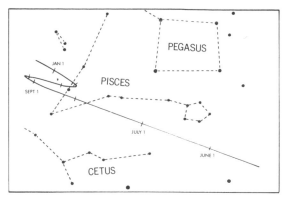

The path of Mars among the stars during the latter part of 1973.

TABLE I

Oppositions of Mars

YEAR	DATE	DISTANCE FROM EARTH (MILLIONS OF MILES)	CONSTELLATION
1982	Mar. 31	60	Virgo
1984	May 11	50	Libra
1986	July 10	39	Sagittarius
1988	Sep. 28	37	Pisces
1990	Nov. 27	49	Taurus
1993	Jan. 7	58	Gemini
1995	Feb. 12	62	Leo
1997	Mar. 17	61	Virgo
1999	Apr. 24	54	Virgo

western elongation to the next one, is equal to the synodic period.

The most suitable times to observe Mercury are when it is an evening star in the spring and when it is a morning star in the autumn. At these times the line joining the planet to the sun is most nearly vertical to the horizon. This gives the maximum distance above the horizon.

Those who have attended a planetarium show know how the planets swing back and forth in their courses among the stars. Most of the time the planets move eastward through the constellations in what is called *direct motion*. At intervals they appear to stop and then move backward among the stars in what is called *retrograde motion*. Again they appear to stop and resume their eastward motion, thus progressing around the sky in a series of loops.

The looped motions of the planets are easily explained by the Copernican theory. For example, the earth moves faster than Mars and overtakes it about every two years. When the earth is passing it, Mars appears to shift backward against the more distant background of stars. It is the same effect one gets as he drives along in a car and overtakes another car. The latter seems to move backward against the landscape. The same reasoning can be used to explain the retrograde motions of the other outer planets. This apparent backward motion occurs around the time when the planet is in opposition—that is, on the opposite side of the earth from the sun.

Because of the considerable eccentricity of the orbit of Mars, the interval from one of its oppositions to the next varies between about 2 years 1 month and 2 years 2½ months. Also, the distance from the earth to Mars at opposition ranges from 35 million miles to 63 million miles. Table I gives the approximate dates and distances of Mars at opposition from 1982 to 1999 and the constellation in which it is located on each of these dates.

The interval from one opposition of Jupiter to the next is a little over 13 months. This is the time required for the earth to gain a lap on Jupiter. On the average, the earth overtakes Jupiter about 34 days later from one year to the next. Since the planets move in ellipses and their orbital speeds are not quite constant, this value ranges from 30 to 38 days. It is shown in the following table, which gives the approximate dates of opposition of Jupiter from 1982 to 2000 and the constellation in which it is located on each of these dates.

The synodic period of Saturn is 378 days, or about 13 days more than a year. On the average, oppositions of Saturn occur 13 days later from one year to the next. This is illustrated in the next table, which gives the approximate dates of opposition of Saturn from 1982 to 2000 and the constellation in which it is located on each of these dates.

Conjunctions of Jupiter and Saturn occurred in 1941, 1961, and 1981. That such conjunctions must occur about every 20 years can be easily determined. Dividing Jupiter's period of 12 years into 360°, we find that it moves eastward among the stars 30° every year. Dividing Saturn's period of 30 years into 360°, we find that it moves eastward among the stars 12° every year. Thus Jupiter gains 18° on Saturn in one year. Therefore, it will gain 360°, or a whole lap, on Saturn in 20 years.

PLANETS AND PREDICTIONS

The planets and their motions across the sky are key elements in the body of investigation known as astrology. The early astrologers identified the planets with their pagan gods. In fact, the planets received their names due to a fancied resemblance to the gods of the Greeks and Romans. Saturn, the most distant of the planets, represented the

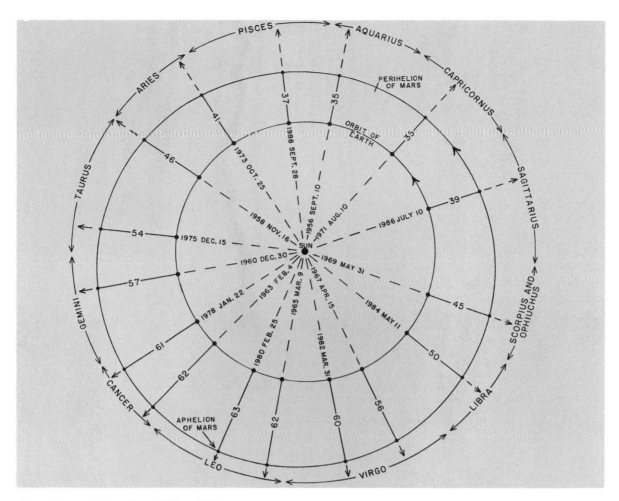

Oppositions of Mars from 1956 to 1988.

The constellation in which Mars appears and the distance in millions of miles from the earth to Mars are given for each opposition. Because of the eccentric orbit of Mars, these distances range from 35 million miles in 1956 and 1971, when Mars is near perihelion, to 63 million miles in 1980, when it is near aphelion.

oldest of the family. Jupiter, which is bright enough to rival Venus but which can remain in our sky throughout the whole night, naturally played the part of the ruler of the gods. Mars' red color could leave only one name possible for it. Venus, the most beautiful object in our sky, received her name quite naturally. Mercury, moving among the stars far more rapidly than any other object except our moon, could be named only for the god with the winged feet who was the messenger of the gods.

The early belief that planets had powers similar in character to those attributed to the gods whose names they bore is still present among some people today. How absurd it is to believe that large chunks of rock millions of miles away exercise influence on us because they are identified with the names of pagan gods who existed only in mythology.

The amounts of radiation received on the earth from stars and planets are exceedingly small. Their gravitational effects are so slight as to be negligible in comparison with those from nearby objects. It is true that the moon and the sun cause tides on the earth, but that does not prove anything in astrology.

Astrologers do not attempt to explain the mechanism for the transfer of planetary influence. They have no sound hypothesis that might serve as a basis for their speculations. They try to offset this lack by the authority of such terms as "cosmic vibrations,"—quite unknown to physicists and astronomers. If the mysterious vibrations travel with the speed of light, they reach the earth in a few minutes or hours from the planets, but they

119

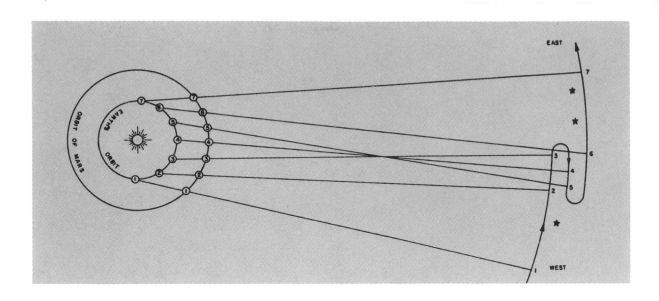

The Copernican system.

By placing the sun at the center and having the earth and the other planets revolve in circles around it, Copernicus was able to explain the backward motions of the planets much more simply than in the Ptolemaic system. This is illustrated here in the case of the earth and Mars.

The closer a planet is to the sun the greater is its velocity. The earth's speed is 18½ miles a second, while that of Mars is only 15 miles a second. As the earth overtakes Mars, the latter seems to move backward. This is shown by the lines which indicate the direction of Mars as seen from the earth. The stars are hundreds of thousands of times farther away than the planets and form a fixed background, against which Mars appears to move backward when the earth passes it.

The direct motion of Mars to the east is shown at positions *1*, *2*, and *3*, backward motion to the west at *4* and *5*, and direct motion to the east again at *6* and *7*. The wide loops in this path are not intended to indicate any change in distance of the planet from the sun. They are meant to show only the forward and backward motion in the sky.

| | TABLE II | | | TABLE III | |
| | Oppositions of Jupiter | | | Oppositions of Saturn | |
YEAR	DATE	CONSTELLATION	YEAR	DATE	CONSTELLATION
1982	Apr. 25	Virgo	1982	Apr. 8	Virgo
1983	May 27	Scorpius	1983	Apr. 21	Virgo
1984	June 29	Sagittarius	1984	May 3	Libra
1985	Aug. 4	Capricornus	1985	May 15	Libra
1986	Sep. 10	Aquarius	1986	May 27	Scorpius
1987	Oct. 18	Pisces	1987	June 9	Ophiuchus
1988	Nov. 23	Taurus	1988	June 20	Sagittarius
1989	Dec. 27	Gemini	1989	July 2	Sagittarius
1991	Jan. 28	Cancer	1990	July 14	Sagittarius
1992	Feb. 28	Leo	1991	July 26	Capricornus
1993	Mar. 30	Virgo	1992	Aug. 7	Capricornus
1994	Apr. 30	Libra	1993	Aug. 19	Capricornus
1995	June 1	Ophiuchus	1994	Sep. 1	Aquarius
1996	July 4	Sagittarius	1995	Sep. 14	Aquarius
1997	Aug. 9	Capricornus	1996	Sep. 26	Pisces
1998	Sep. 16	Pisces	1997	Oct. 10	Pisces
1999	Oct. 23	Aries	1998	Oct. 23	Aries
2000	Nov. 28	Taurus	1999	Nov. 6	Aries
			2000	Nov. 19	Taurus

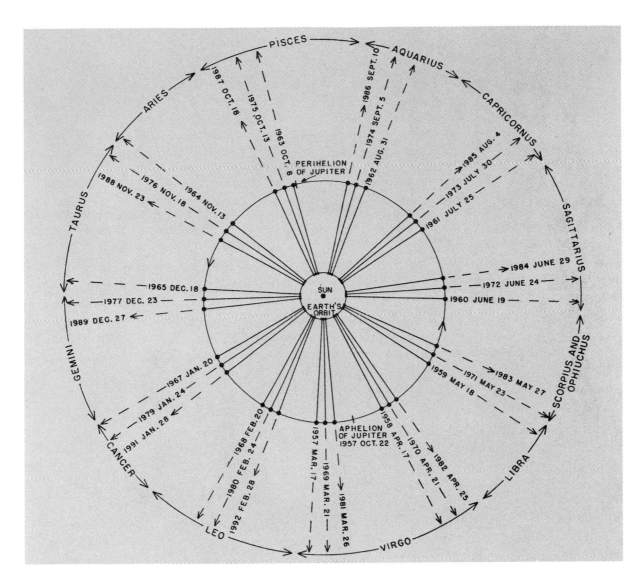

Oppositions of Jupiter from 1957 to 1992.

take hundreds and even thousands of years from the stars. Also, astrology requires that planets, which are much alike in their constitution, affect human affairs in different ways.

According to astrology, a physician who hastens birth for the benefit of the mother changes the whole life of the child. But even if we were affected by planetary influence, why is the moment of birth favored over the time of conception?

The rules by which astrologers interpret their horoscopes have not been derived from any known experiments or observations. Three planets—Uranus, Neptune, and Pluto—were discovered with the telescope. In each case the astrologers ascribed influences to the planet before preliminary observational tests of the influences could have been made, and even before accurate orbits could be assigned to the planets. Other influences were

added later. In one astrology book, we find that Uranus rules the railroads and Neptune governs aviation. Imagine the difficulty of trying to demonstrate the influence of Neptune on aviation.

The zodiac is an imaginary belt in the heavens 16° wide through which the sun's annual path runs centrally. It includes the paths of the moon and the principal planets. It is divided into twelve divisions, each 30° long, marked off eastward from the vernal equinox. These are called the signs of the zodiac. Their names were originally the names of the constellations occupying the divisions of the zodiac.

About 2,000 years ago the signs of the zodiac and the constellations coincided. Now, as a result of the precession of the equinoxes, the signs have separated about 30 degrees from their constellations. When the sun reaches the vernal equinox on March 21, it enters the sign of Aries, which is always the first 30° of the zodiac measured eastward from the vernal equinox. However, the sign of Aries is now in the constellation of Pisces. So when the sun is in the sign of Aries, it is in front of the stars of Pisces.

The astrologer completely disregards this precessional motion. In casting a horoscope, he imagines that the sky is arranged as it was 2,000 years ago. He pays no attention to the stars, but bases his predictions on the positions of the sun, moon, and planets with respect to twelve arbitrary divisions of an imaginary circle.

In low-priced horoscopes based only on the time of birth between two dates a month apart, only the position of the sun is considered. All persons born between March 21 and April 20, for example, are supposed to be influenced in the same way. Then on April 21 there is a profound change in the effects of the sun on people born after that date. This continues unchanged until May 21, when the sun enters the next sign. Even if the sun did have any influence, there is no more reason why it should change when the sun crosses an imaginary line in the sky than there is that the destinies of passengers on a ship should be changed when it crosses the 180th meridian in the Pacific Ocean.

According to several books on astrology, persons

Oppositions of Saturn from 1957 to 1986.

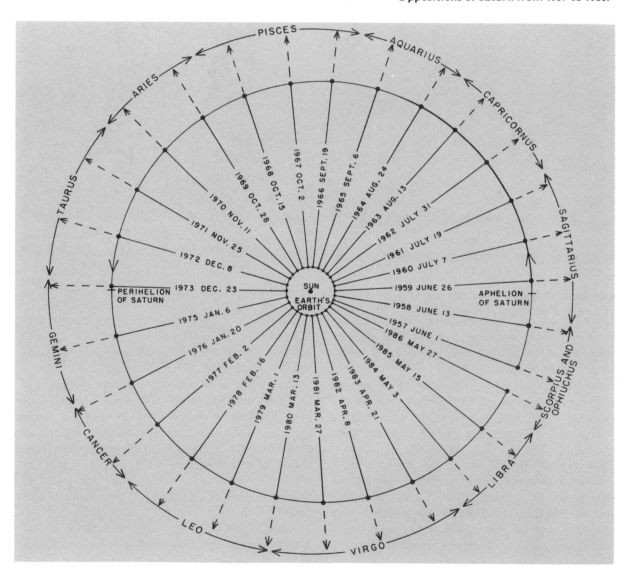

born from September 24 to October 23 should have musical ability. During that interval, the sun is in the sign of Libra ("The Scales"). Psychologist Farnsworth looked up the birth dates of 1,498 musicians and found that fewer were born under Libra than under any other sign except Scorpius. In fact, Libra and Scorpius were tied for last place as musician-makers. Thus in picking a musical sign, the astrologers could have made ten better choices than Libra, and could not have made a worse one.

Suppose that two people consult an astrologer for financial advice. He tells one client to buy a certain stock in the market and advises the other to sell the same stock. If the stock goes up, the man who bought it thinks astrology is wonderful and tells everybody about it. The man who sold the stock is ashamed to tell anyone of his experience. If the stock goes down, the purchaser keeps quiet and the seller recommends the astrologer, who thus wins either way.

Enough claims are made by an astrologer to cancel one another. A person will remember one thing mentioned by an astrologer that turns out to be true and will forget the ten things that are wrong. Here are a few astrological predictions that went sour: there would be no war in Europe in 1939; Hitler would be assassinated in 1940; the Maginot Line would remain impregnable; France would not be defeated.

In 1552 an astrologer predicted that King Edward VI of England would live beyond middle life, though after the age of 55 years, 3 months, 17 days, he would be the victim of various diseases. Within one year the king died at the age of 16. This same astrologer predicted that he himself would die on a particular day. He was obliged to commit suicide in order to maintain his reputation and that of astrology.

When newspapers devoting space to astrology are courteously asked why they represent themselves as dedicated to truth and enlightenment and then deliberately spread superstition, their answer usually is that the daily horoscopes amuse the ignorant, make them buy papers, and do no harm. But they do a great deal of harm. They keep alive a superstition through which fakers collect money by fraud.

CHAPTER **26** **SUNBURNED MERCURY**

Mercury is as bright as the brightest of the stars. Yet it has been seen by few people. The tiny planet stays so close to the sun in our sky that we are able to see it for intervals of only one to two weeks three times a year in the evening and at three other such intervals before sunrise. Only when it is at its extreme distance from the sun can it be above our horizon after the night becomes entirely dark. Always it is necessary to look for it in the twilight or the dawn. Although few modern people know this little world, it was known to the ancients before the beginning of history. The earliest observers, however, did not recognize it as one body that sometimes was east and at other times west of the sun, but thought there were two satellites of the sun, one a morning "star," the other an evening "star." The Egyptians called these two bodies "Set" and "Horus," the Hindus spoke of them as "Buddha" and "Raulineya," and the Greeks as "Apollo" and "Mercury." Tradition tells us that Pythagoras in the sixth century B.C. was the first to recognize the identity of the two bodies. It seems probable, however, that the fact had been recognized long before his time.

Mercury's mean distance from the sun is less than 4/10 of our own. It swings around the sun in a year that has only 88 of our days. At the end of one of our years Mercury is not far from the position that it occupied among the stars the year before. Moving around the sun four times to our one, it gains approximately three laps on the earth each year.

The diameter of Mercury is almost exactly 3,000 miles, not much more than a third of the earth's. We do not know accurately the amount of material in this little world. If the planet had a moon moving around it, the motion of that moon would tell us with great accuracy the gravitational force exerted

123

and therefore the mass of the planet. Mercury and Venus have no moons, and Mercury's mass must be obtained by observing the slight amount that it pulls its distant neighbor, Venus, from her path. These *perturbations* are very slight, and the uncertainty of their exact amount leaves a corresponding uncertainty as to the mass of Mercury. It is, however, somewhere in the neighborhood of 1/20 that of the earth. As a result of this small mass and diameter, a man on Mercury would weigh approximately one-third of what he does here—240 pounds on the earth would pull down a spring balance on Mercury only to the 80-pound mark.

Relative sizes of Mercury and earth. Heavy black line shows the great circle route from New York to Tokyo, which is 6,800 miles long.

The average density of the interior material of Mercury is almost exactly the same as that of the earth, which is rather surprising, since it is much lighter and the internal pressures are much lower. It must mean, therefore, that the central core of dense iron must be proportionately larger inside Mercury, possibly as much as 2000 miles in diameter.

As we saw in Chapter 24, Mercury rotates very slowly, taking about 59 days to turn once on its axis. This period of rotation has been found by bouncing radar pulses off its slowly rotating surface and then measuring the small Doppler shifts

that are consequences of the rotation. It is a very interesting period of rotation, because it is exactly ⅔ of the planet's period of revolution around the sun. It means that as the planet revolves around the sun once, it turns on its axis 1½ times. This must be a consequence of the tidal influence of the sun on the slightly bulged shape of the planet, which has slowed an originally rapidly rotating planet down until it became locked into this simple ratio of 1 to 1½.

The orbit of Mercury around the sun is quite far from a perfect circle; indeed, of all the planets, only Pluto has a more eccentric orbit. When Mercury is closest to the sun (at perihelion) it is 17 million miles closer than when it is farthest (at aphelion). As a consequence, the sunlit surface varies considerably in temperature, being a searing 400°C when closest, hot enough to melt tin and lead, dropping to 285°C when it is farthest.

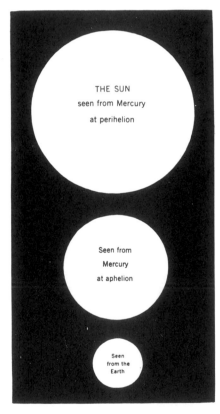

The effect of the sun on Mercury.

On the earth the changing distance from the sun throughout the year is too small to matter much. The sun, as observed from Mercury when nearest to that planet, gives about twice as much radiation as when farthest. This factor, almost negligible in the case of the earth, produces very marked seasons on Mercury.

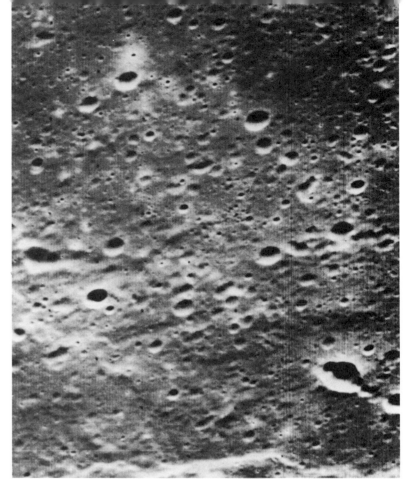

Eighteen pictures, taken at 42-second intervals by Mariner X's two TV cameras, were computer-enhanced at the Jet Propulsion Laboratory and fashioned into this photomosaic of Mercury. The pictures were taken during a 13-minute period when Mariner was 200,000 kilometers (124,000 miles) and six hours away from Mercury on its approach to the planet. About two-thirds of the portion of Mercury seen in this mosaic is in the southern hemisphere. The cratered surface is somewhat similar to the cratered highlands on the Moon. Largest of the craters are about 200 kilometers (124 miles) in diameter. (NASA.)

This photograph, taken only minutes after Mariner X's closest approach to Mercury, is one of the highest resolution pictures taken during the mission. Taken from a distance of 5,900 kilometers (3700 miles) from the planet, it shows a surface area of about 50 by 28 kilometers (31 × 17 miles). Evident are both primary and secondary impact craters; the secondary craters can be recognized as being less circular than the primary ones, and frequently form chain or cluster structures. (NASA.)

Drawings of Mercury made by Denning in 1882.

The photograph shows what is evidently a small, new crater (center of photo) near the middle of an older, larger crater. The young crater is about 12 kilometers (7.5 miles) in diameter. This picture, covering an area of 91 by 170 kilometers (65 by 105 miles) was taken from a distance of 20,700 kilometers (12,860 miles). (NASA.)

Measurements made with the great 210-foot radio telescope at Parkes, New South Wales, Australia, showed that the radio emission from the dark side comes from a surface whose temperature is near the freezing point of water. This suggests that there may be a tenuous atmosphere, carrying heat from the hot to the cold portion of the planet. All attempts to detect an atmosphere by making observations from the earth have failed, however. It remained for Mariner 10 to show that there was indeed a very slight atmosphere.

Some astronomers of the last century believed that they had observed a great many kinds of markings on the planet, and some very complicated drawings were made. One set of these, as reproduced on page 125, was made by Denning in 1882. We now know from Mariner X photographs that the dark line drawn parallel to the rounded

markably similar to that of the moon, pitted with thousands of craters that are undoubtedly the result of meteorite impacts. Many of the larger craters show central peaks, just as on the moon. The only significant difference between these craters is that on Mercury the craters are not as deep, probably as consequence of the planet's greater surface gravity. Some of the more recent craters are surrounded by brighter radial rays, although they do not extend as far from the crater as they do on the moon. One big difference is that Mercury does not have the numerous flat plains ("maria") that cover such large areas of the moon. Only one feature on Mercury comes even close to looking like a lunar plain. This is the Caloris Basin, about as large as the lunar Mare Imbrium, and probably also the result of an impact with quite a large body. What a violent event that must have been! It

Mercury's apparent size.

Changes of phase and of apparent size in various parts of Mercury's orbit. Note that when Mercury is nearest us most of the surface toward us is dark, and that when it is in the full-moon phase it is farthest from us. If this relationship between distance and phase were reversed, it would be much easier to observe Mercury's markings through our telescopes.

limb is merely an optical illusion. When one is straining his eye to observe an object that is hard to see, such spurious markings become quite common. Some astronomers thought they observed mountains on the planet, even mountains rising to unbelievably great heights. Schroeter (1745–1816) concluded that there was a range of peaks, which he called the Cordilleras, rising to heights of nearly 63,000 feet. He also concluded that the planet turned once on its axis in almost exactly the same period as our own earth. As we now know, his observations have since been entirely discredited as incorrect.

When Mariner X sailed past Mercury in 1974 it sent back 1,800 photographs of its surface. At its closest approach, it was within 500 miles of its surface, so that these pictures revealed surface detail in exquisite detail. They showed a surface re-

shook the entire planet to its very core, sending shock waves throughout its bulk to throw up ripples that we can see today on the surface opposite the Caloris Basin.

In addition to the craters, the surface of Mercury shows a number of steplike semicircular cliffs that are over a half mile high and, in some cases, are hundreds of miles long. There does not seem to have been any slippage along these cliffs, so that they are probably not the result of the movements of continental plates. The best guess is that they are the result of a slight shrinkage of the planet's surface area as the large inner core of iron shrank as it gradually cooled.

A spectrometer mounted on Mariner X succeeded in detecting a very tenuous atmosphere on Mercury. It reported the presence of nitrogen, oxygen, argon, xenon, carbon, and helium. It is rath-

The Mariner X Television-Science Team has proposed the name "Kuiper" for this very conspicuous bright Mercury crater (top center) on the rim of a larger older crater. (Jet Propulsion Laboratory/NASA.)

er surprising that there is any atmosphere at all, since the planet's high temperature and low surface gravity would allow these gases to escape, particularly helium. It has been suggested that they are actually transitory gases trapped out of the solar wind that is continually blowing past the planet. In the case of helium, it has been proposed that outgassing from the planet's interior is responsible for its presence in the atmosphere.

A further surprise was the detection by Mariner 10 of a very weak magnetic field around Mercury. Usually magnetic fields are found around rotating planets or stars, becoming the stronger the more rapid the rotation, but here is a field around a planet that is rotating extremely slowly. The best guess is that it is the very weak remaining ghost of a much stronger field with which the planet was endowed when it was rotating very much faster in its infancy.

If, at some time when the human race has developed far beyond our present stage, we should find it possible to consider colonizing Mercury, it would be necessary to protect ourselves from the intense radiation of the sun during the three-month-long "day." The "night" would not be nearly so inhospitable. Science fiction writers often have used the planet as a base of exploration and of mineral supplies for future generations. It is very difficult from a practical point of view to believe that such use will ever be made. The conditions against which explorers would find it necessary to protect themselves are much more rigorous than those on our moon. However, it would be very foolish to attempt to underestimate the extent of our future inventions. It must be considered as within the bounds of possibility that some day our race will make practical use of this tiny sunburned world.

CHAPTER **27** QUEEN OF THE EVENING SKY

Venus puts on an exciting evening show at intervals of a little more than a year and a half. At such times it becomes so bright that nearsighted people have mistaken it for the crescent moon. It even casts a noticeable shadow. When its evening brightness comes in December, hundreds of people phone observatories to ask whether the Star of Bethlehem has reappeared. All of these questioners are emphatic in stating that never before have they seen such a bright object in the sky. This planet even becomes visible in full daylight, and observers call to ask what sort of a ship is above them. During the war Venus excited the officers and men of a large army transport on the Indian Ocean who could not identify the object seen high in the sky shortly before noon.

At average intervals of 584 days this twin of the earth repeats its positions with respect to the earth and sun. There is a slight variation in the number of days, since the orbits of Venus and the earth around the sun are not perfect circles. This cyclic period is a little more than 19 months; therefore, if Venus should be bright in the winter evening sky this year, it would next show up the same way in the summer sky. There is a longer and

very exact relationship that is sufficiently accurate to hold quite well during a person's whole lifetime. Venus requires 224.701 days to go around the sun, and the earth requires 365.256 days. Venus goes around the sun 13 times in 2,921 days, and the earth 8 times in 2,922 days. If a man were to live to be a century old, predictions made by him in his youth for the phases of Venus during his old age would be in error by less than two weeks.

One of the diagrams of Chapter 25 shows Venus (a) at its new moon phase passing between us and the sun, i.e., at *inferior conjunction*, (b) appearing as a half moon through the telescope when in the western evening sky at *greatest eastern elongation*, and (c) as a half moon when in the eastern morning sky at *greatest western elongation*. The following diagram gives dates of the positions. Predictions may be made merely by adding 2,920 days to any of them. It can be fun to watch the fulfillments through the years.

In the evening sky near the end of winter Venus gives a far better show for us of the northern hemisphere than it can give when similarly seen in the summer. The reverse holds for its dawn display. Like all other planets, it follows very closely the

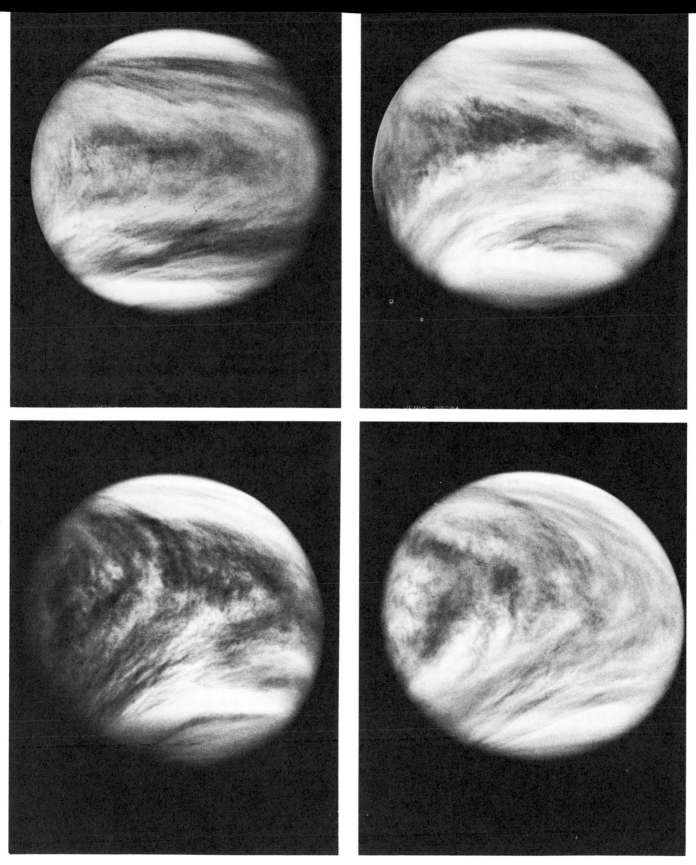

These four images of Venus were taken from February 2, 1979, to March 3, 1979. The first image (top left) shows a relatively dark, uniform band in the equatorial region. Superimposed on this are numerous, small-scale cellular features. The second image (top right) shows a dark, almost triangle shape in the middle and western portion (left) of the disk which is the tail of the dark Y. The third image (bottom left) shows a small wedge-shaped dark feature near the left hand side which might have been a Y feature that was not well developed. Cellular features appear throughout this same region. (NASA.)

same apparent path as the sun among the stars. It is about 47° from the sun at its greatest elongations. A greatest eastern elongation on the first of March would place it where the sun is after the middle of April. This is much farther north than the sun is on March 1st, and causes Venus to remain in the sky for more than four hours after sunset. When it is farthest east of the sun near September 1st, it is also far south of it and remains above the horizon for, roughly, only half that long after sunset. In the southern hemisphere the advantages are reversed so far as months are concerned, but still hold if we use the words late winter and summer instead of March and September.

Astronomers of a couple of centuries ago saw rather vague or perhaps even illusory markings on Venus, and from their observations concluded that Venus had a period of rotation approximately equal to that of the earth. Later observations disproved this short day and indicated that possibly Venus keeps one face to the sun continually, as the moon does toward the earth. This is almost correct. A slow rotation was dramatically confirmed in 1962 when radar pulses beamed toward Venus gave echoes from various parts of the rotating disk shifted in wavelength through the operation of the Doppler effect. An analysis of these echoes showed a rotational period of about 250 days, but most unexpectedly, in a *retrograde* direction. The most likely reason for this remarkable result is that the earth has affected Venus' rotation, acting most strongly when Venus is closest to us. The resonance period of these perturbations is 243.16 days, so close to the observed period that 243.16 days is now taken as the true period of rotation.

Venus has frequently been called the earth's "twin sister," since it is nearly the same diameter as the earth. It is a world for which surface weight is approximately 8/10 that to which we are accustomed and, therefore, one to which we could adjust rapidly. This planet receives twice the radiation from the sun that we do. If its surface were a perfect absorber and radiator, its surface temperature would be about 80°C, which is considerably below the boiling point of water. Here, therefore, is a world to which much consideration was given in the past when one speculated about life elsewhere in the solar system. Unfortunately, such speculations came to an abrupt end when recent space probes and radio astronomy observations indicated that the surface temperature is a searing 500°C.

Observations from the earth have shown that Venus is covered by clouds, since hazy markings appear when the planet is photographed in ultraviolet light. These markings change their pattern in a haphazard, nonperiodic manner, so they cannot be markings on a solid surface. The clouds reflect back to space about ¾ of the visible and infrared sunlight, and about ½ of the ultraviolet. The 16 space satellites sent to Venus by the Russians and the U.S. between 1961 and 1978 have

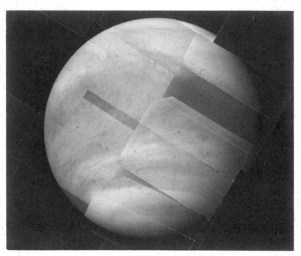

Cloud patterns, visible only in ultraviolet light, reveal the general circulation of Venus' upper atmosphere in this mosaic of TV pictures, taken by Mariner X in February 1974 from 440,000 miles away, a day after the probe's closest approach to the planet. (National Aeronautics and Space Administration.)

given us the best information about its atmosphere and surface conditions. They amply confirmed the ground-based observation that 90 percent of the atmosphere was made up of carbon dioxide, with traces of sulphur dioxide and water vapor. The clouds turned out to be largely sulphuric acid (H_2SO_4) particles. The Pioneer 2 probes that descended through the atmosphere to the solid surface found that the clouds started 29 miles above the surface and extended upward another 18 miles. The surface atmospheric pressure was a hefty 90 times that at the surface of the earth.

In 1974, Mariner X produced thousands of pictures of the cloud surface as it orbited the planet. The clouds were found to resemble quite closely cirrus clouds here on the earth, extending in long streaks from the equator toward the poles. A con-

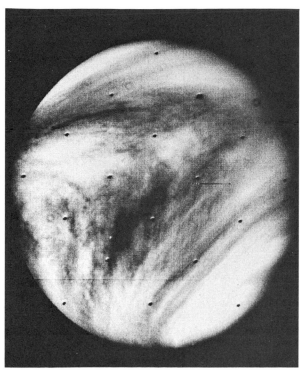

A photograph of Venus taken in the ultraviolet by the Mariner X spacecraft. The markings are clouds that can be detected only in the ultraviolet and are the subject of great curiosity among those who study this always mysterious planet. (NASA.)

spicuous mottling of the clouds along the equator suggested that this was a zone of strong vertical convection. A further elongated area of convection extended downwind from the subsolar point, showing that, when the sun was in the zenith, the strong heating produced vertical convection of the heated gas, and marked turbulence downwind. The clouds showed that there were strong winds in the upper levels of the atmosphere, reaching almost 200 miles per hour, analogous to very strong jet streams on the earth. On the other hand, the winds down at the surface were quite light; the Russian Venera 9 and 10 landers reported winds up to only 2 to 3 miles per hour. These winds, coupled with the fact that the cloud tops were almost 6°C hotter at the poles than at the equator, with a somewhat lower elevation, suggested that warm air rises at the equator, cools as it rises, then moves toward the poles, sinks, and heats up. The general circulation is much simpler than on the earth, where rapid rotation and the effect of the continents produce a very complex situation.

It may seem strange that we can say anything about the topography of the solid surface of Venus beneath those heavy clouds, but it has actually been mapped in surprising detail. This is because radar pulses beamed at the planet penetrate through to the surface and are reflected back; the timing and strength of the echoes as the planet slowly rotates eventually result in quite accurate maps of elevations and reflectivities of the surface. In addition to ground-based radar scans, the orbiting Pioneer 1 satellite carried a radar signal generator and receiver. The scans revealed a surface that is quite a bit less heavily endowed with highlands than is the earth, although there is one highland area equivalent in elevation and extent to Africa, and another the size of Australia. There are a number of large craters, some almost 200 miles in diameter. Most of the craters have central peaks, indicating that they are the result of meteorite impacts, although one huge peak, 450 miles across at its base, with a central depression 50 miles wide at its summit, is probably a giant volcano. Finally, the scans revealed a giant canyon stretching along the equator for 900 miles. In places it is 3 miles deep and 240 miles wide.

The views of the surface of Venus as seen by Venera 9 and 10. The view from Venera 9 (top) is nearly filled by a heap of rocks. Notice the sharp edges, which surprised the scientists' expectation that exposed edges would be well worn. In contract, the view from Venera 10 (bottom) shows a smooth surface. An important question now is the nature of the erosion that produced it. (USSR Academy of Sciences.)

131

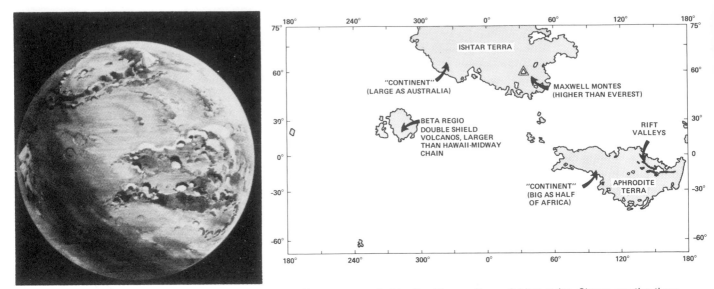

(left) An artist's conception of the surface of Venus as revealed by the Pioneer Venus Orbiter radar. Shown are the three continent-sized highland regions—Aphrodite Terra (foreground), Ishtar Terra (top) and Beta Regis (left). The elevation of the features is exaggerated on this drawing for artistic effect. (right) A preliminary map of the surface of Venus prepared from the Pioneer Venus Orbiter radar data. Note that the Mercator projection makes features near the upper latitudes appear much larger than they actually are. Maxwell Montes is about 11 kilometers above the mean altitude of the Venus plains. (NASA/Ames.)

Further information regarding the surface conditions was provided by four Russian landers that sent back pictures of the surrounding terrain before they burned up. The pictures showed some areas strewn with sharp-edged rocks and other areas that looked like lava beds. The landers found that only two percent of the sunlight falling on the planet filters down to the surface. In sum, a terrain that is inhospitable in the extreme, with searing heat, a suffocating atmosphere, dimly visible surface features, and acid clouds that forever block the view of the rest of the universe.

A surprise was the report of frequent lightning flashes, sometimes as many as 25 per second between three and seven miles above the surface. They are probably the explanation of a constant glow that was seen by two Pioneer probes that descended through the atmosphere of the planet on the night side.

Why are conditions on Venus so different from those on the earth? It is almost entirely due to its higher temperature. The atmosphere of the earth only contains one percent carbon dioxide, but there are great amounts trapped in terrestrial rocks and dissolved in sea water. If all this carbon

dioxide were released, there would be as much as in the atmosphere of Venus. The higher temperature of Venus has prevented the operation of those reactions that have trapped terrestrial CO_2. The great abundance on Venus of carbon dioxide, together with some sulphur dioxide, also helps to raise the surface temperature through what is known as the "greenhouse effect"; these gases allow much of the sunlight to pass through to lower layers, but prevent the longer wavelength radiation from these layers from getting out. The high abundance of free oxygen in the earth's atmosphere is due to the action of plant life in breaking up the chemical compounds containing that gas. Oxygen is a very active element, and under the probable high temperatures that exist in the early days of a planet, would form many compounds. On the earth, most of the carbon, aluminum, and iron exist in oxidized form. It seems probable that on almost any planet the oxygen would all be used in such a manner. The fact that oxygen is found on the earth seems, therefore, to be a result, not a cause, of life. The lack of it on the planet Venus gives additional strong evidence, if any is needed, that even plant life does not exist.

132

CHAPTER **28** THE MARTIAN DREAM

Dreams are the stuff of childhood, and when astronomy was young it had its share of dreams. It was only a little more than 300 years ago that modern astronomy was born. Early astronomers, seeing what appeared to be oceans and mountain ranges on the moon, dreamed of the time when telescopes would be powerful enough to watch the doings of the lunar population. This dream soon exploded, and others took its place, some being fulfilled beyond the wildest expectations and others falling silently by the wayside.

One of the most persistent dreams concerned the possibility of life on Mars. This dream stirred up a sharp controversy two generations ago, a controversy that has come down to the present without losing much of its keenness.

The dream began about one century ago, when telescopes were becoming just powerful enough to show some detail on Mars. Let us turn back to the pages of older astronomy books and try to experience the excitement of those earlier watchers of the sky.

We begin with Sir John Herschel, who shows a drawing he made in 1830. He says:

"The case is very different with Mars. In this planet we frequently discern, with perfect distinctness, the outlines of what may be continents and seas. Of these, the former are distinguished by that ruddy colour which characterizes the light of this planet (which always appears red and fiery), and indicates, no doubt, an ochrey tinge in the general soil, like what the red sandstone districts on the Earth may possibly offer to the inhabitants of Mars, only more decided. Contrasted with this (by general law in optics), the seas, as we may call them, appear greenish. These spots, however, are not always to be seen equally distinct, but, when seen, they offer the appearance of forms considerably definite and highly characteristic, brought successively into view by the rotation of the planet, from the assiduous observation of which it has even been found practicable to construct a rude chart of the

Drawing of Mars.
Mars as drawn by Sir John Herschel, August 16, 1830. Many years later he stated that he had never observed the color of the "seas" more distinctly than he did that night.

surface of the planet. The variety in the spots may arise from the planet not being destitute of atmosphere and clouds; and what adds greatly to the probability of this is the appearance of brilliant white spots at its poles—one of which appears in our figure—which have been conjectured, with some probability, to be snow; as they disappear when they have been long exposed to the sun, and are greatest when just emerging from the long night of their polar winter, the snow line then extending to above six degrees (reckoned on a meridian of the planet) from the pole."

The scene changes to 50 years later. It is now 1880, and larger telescopes are in use. The 26-inch refractor at Washington is typical. Asaph Hall, using it in 1877, had found the two tiny Martian moons that Gulliver had "seen" 150 years earlier in Dean Swift's great satire. Another great English astronomer had pondered the evidence and in a text had shown drawings with far more detail than in Herschel's picture. Sir Norman Lockyer summarizes the situation:

"In Fig. 59 are presented two sketches of Mars. Here we have something strangely like the Earth. The shaded portions represent water, the lighter ones land, and the bright spot at the top of the drawings is probably snow lying around the south pole.

"The two drawings represent the planet as seen in a telescope, which inverts objects, so that the south pole of the planet is shown at the top. In the upper drawing, which was made on the 25th of September, a sea is seen on the left, stretching down northward; while, joined to it, as the Mediterranean is to the Atlantic, is a long, narrow sea, which widens at its termination.

"In the lower drawing, made September 23rd, this narrow sea is represented on the left. The coast-line on the right reminds one of the Scandinavian peninsula, and the included Baltic Sea.

"Mars has not only land, water, and snow, like the Earth, but also clouds and mists, and

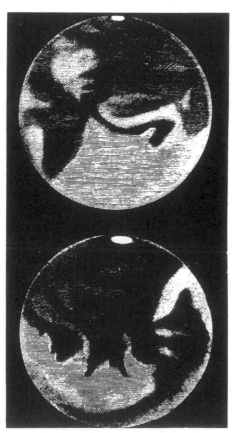

Drawings of Mars as published by Sir Norman Lockyer in 1880.

these have been watched at different times. The land is generally reddish when the planet's atmosphere is clear; this is due to the absorption of the atmosphere, as is the color of the setting Sun with us. Hence the fiery red light by which Mars is distinguished in the heavens. The water appears of a greenish tinge.

"Now, if we are right in supposing that the bright spot surrounding the pole is ice and snow, we ought to see it rapidly decrease in the planet's summer. This is actually found to be the case, and the rate at which the thaw takes place is one of the most interesting facts to be gathered from a close study of the planet. In 1862, this decrease was very visible. The summer solstice of Mars occurred on the 30th of August, and the snow-zone was observed to be smallest on the 11th of October, or forty-two of our days after the highest position of the Sun. This very rapid melting may be ascribed to the inclination of the planet's axis, the great eccentricity of its orbit, and the fact that the summer of the southern hemisphere occurs when the planet is near perihelion."

A few more years pass. The Italian astronomer Giovanni Schiaparelli has announced his discovery of the canals on Mars. The 36-inch Lick Observatory refractor has been put in operation, and in 1892, Samuel P. Langley, a great American astronomer, writes:

"By the industry of numerous astronomers, seizing every favorable opportunity when Mars comes near, so many of these features have been gathered that we have been enabled to make fairly complete maps of the planet, one of which by Mr. Green is here given.

"Here we see the surface more diversified than that of our earth, while the oceans are long, narrow, canal-like seas, which everywhere invade the land, so that on Mars one could travel almost everywhere by water. These canals have appeared to some observers to exist in pairs, or to resemble two close parallel lines; but this cannot be said to be at present wholly certain. The spectroscope indicates water-vapor in the Martian atmosphere, and some of the continents, like 'Lockyer Land,' are sometimes seen white, as though covered with ice; while one island (marked on our map as Hall Island) has

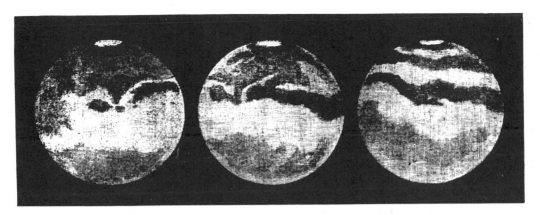

Mars as presented by Langley in *The New Astronomy*, 1892.

been seen so frequently thus, that it is very probable that here some mountain or table-land rises into the region of perpetual snow.

"The cause of the red color of Mars has never been satisfactorily ascertained. Its atmosphere does not appear to be dark enough to produce such an effect, and perhaps as probable an explanation as any is one the suggestion of which is a little startling at first. It is that vegetation on Mars may be *red* instead of green! There is no intrinsic improbability in the idea, for we are even today unprepared to say with any certainty why vegetation is green here, and it is quite easy to conceive of atmospheric conditions which would make red the best absorber of the solar heat. Here, then, we find a planet on which we obtain many of the conditions of life which we know ourselves, and here, if anywhere in the system, we may allowably inquire for evidence of the presence of something like our own race; but though we may indulge in supposition, there is unfortunately no prospect that with any conceivable improvement in our telescopes we shall ever obtain anything like certainty. We cannot assert that there are any bounds to man's invention, or that science may not, by some means as unknown to us as the spectroscope was to our grandfathers, achieve what now seems impossible; but to our present knowledge no such means exist, though we are not forbidden to look at the ruddy planet with the feeling that it may hold possibilities more interesting to our humanity than all the wonders of the sun, and all the uninhabitable immensities of his other worlds."

Langley was an astronomer, neither inclined to exaggerate nor to jump at conclusions. When he

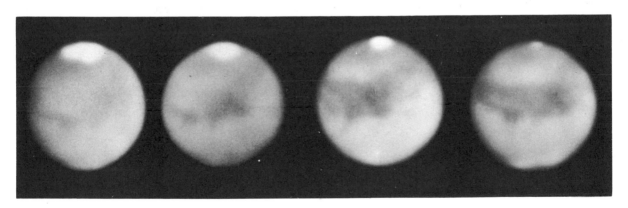

Changes in southern hemisphere during Martian summer. Seasonal dates on Mars in the four views correspond to our May 11, May 29, June 23, and July 31, with June 21 considered the beginning of summer, despite the fact that this is the southern hemisphere. (E. C. Slipher, Lowell Observatory.)

spoke with certainty of the seas he reflected the sober opinion of the majority of astronomers of his time.

So far as the existence of oceans is concerned, the dream had almost ended when Langley wrote his book, *The New Astronomy*. The dream had inspired a young man from one of the most prominent Boston families to turn from exploration and other avocations to astronomy as a serious lifetime work. He became convinced of the probability of life on Mars and, with his own fortune, erected a large observatory in Arizona. From that time the name Percival Lowell became almost synonymous with Mars. Lowell was one of the most colorful of all astronomers. He was a brother of Abbott Lawrence Lowell, president of Harvard, and of Amy Lowell, the poetess. After graduation from Harvard in 1876 he spent nearly twenty years in business, in travel, and in writing books on travel. The later years of his life were devoted almost exclusively to the study of the tiny planet. It is ironical that he should have observed the "canals" crossing the oceans and by such data have been one of those to prove the aridity of our neighbor planet. He studied the markings of the planet at every opportunity and, with his assistants, gave the world more knowledge of the surface detail than had all previous astronomers. Some of his conclusions aroused much opposition. According to him and his followers, the desert planet must conserve water as has never been done on earth. Irrigation canals lead water from the melting polar snows. Vegetation along their borders renders them visible. Their regularity proves the existence of intelligent life. Seasonal changes in the dark markings are considered as proof of vegetation.

Even before the recent satellite visits to the vicinity of Mars, earth-based observations over the past twenty years have dispelled the possibility of intelligent life on that planet and even cast serious doubts on the existence of vegetation as we know it on the earth. Its tenuous atmosphere produces a surface pressure only one percent as great as on the earth. Yellowish dust storms, fogs, and mists occasionally obscure surface features, and wind-shifted dust overlying dark areas change their shapes as seen from the earth. The spectroscope has shown that the atmosphere is almost totally carbon dioxide, with traces of water vapor and possibly nitrogen.

The nature of the polar caps is still a matter of some controversy. The weight of the evidence seems to support the suggestion that they are deposits of frozen carbon dioxide ("dry ice") with probably some frozen water added. However, the deposits must be extremely thin, probably only a small fraction of an inch thick, otherwise they would not change in size so dramatically with the seasons. The average temperatures on Mars are extremely low by terrestrial standards, though at midday they can reach 80°F. After sunset the temperature plummets dramatically on account of the tenuous atmosphere, reaching −95°F by the end of the Martian night. Under these rigorous conditions life, if it exists, must be extremely rudimentary. Plant life would be limited to primitive lichens or mosses. Even under our favorable terrestrial conditions, with hospitable oceans and an atmosphere with life-supporting oxygen, it has taken millions of years for intelligent life to evolve. On Mars this process would be slowed to such an extent that we should expect only the most primitive forms, if even that.

136

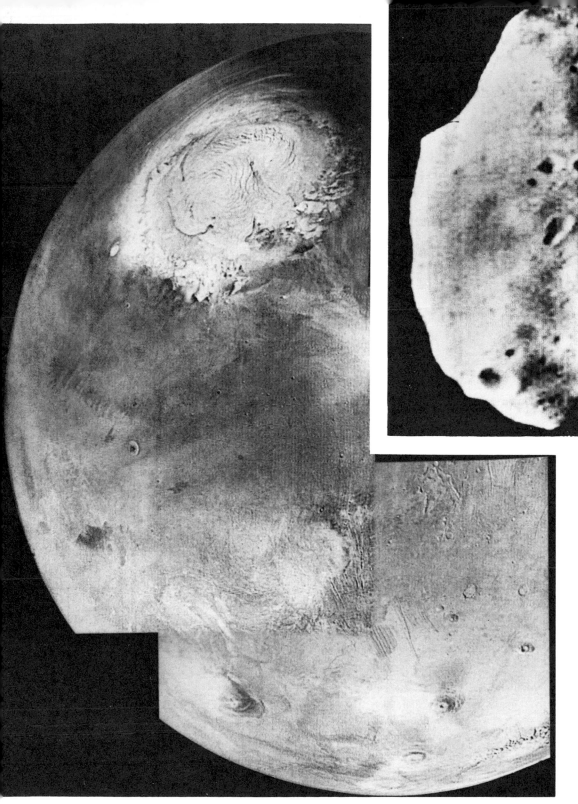

Mars and Phobos, the inner of its two moons, as photographed by Mariner IX. The mosaic of three photos of Mars, taken from 8500 miles away, shows the planet's northern hemisphere, from the polar cap to a few degrees south of the equator. Clearly seen in the bottom photograph are the huge Martian volcanoes, including Nix Olympica at the lower left, and the western end of the great equatorial canyon, lower right. The photograph of Phobos was taken from a distance of 3444 miles. Phobos appears to be about thirteen miles high and three miles long. The many craters suggest that it is very old. (Jet Propulsion Laboratory/National Aeronautics and Space Administration.)

137

All earth-bound attempts to map the Martian terrain faded into insignificance on November 13, 1971, when Mariner IX was successfully thrust into orbit around Mars after a 167-day flight, and when Viking 1 was more recently landed in the Plain of Chryse on July 20, 1976, after an eleven-month journey. Thousands of spectacular pictures from Mariner IX revealed a complex world far beyond our wildest expectations. The diffuse light and dark markings so painstakingly mapped by Herschel, Langley and Lowell have become gigantic inactive volcanoes, drifting sand dunes, flat plains, meandering river beds, and a great rift valley paralleling the Martian equator for 2,300 miles. Completely dwarfing the Grand Canyon, this valley is 150 miles wide in places, and reaches a depth of 20,000 feet. Branching serrations along its rim suggest tributory streams that may have flowed at an earlier stage in Mars's history. Nearly half the surface of Mars is covered with ancient lava flows from nineteen large volcanoes. The largest, Nix Olympica, covers an area 350 miles in diameter, over twice the diameter of Mauna Loa in the Hawaiian Islands.

A whole new era in astronomy opened up when Viking 1 made a perfect landing on the Martian surface. At this writing we can look forward to a continuing stream of fundamental information from its numerous instruments. However, already important discoveries have been made. While the Viking lander was making its nine-minute descent towards the surface, it measured the chemical composition of the tenuous Martian atmosphere to be about 3% nitrogen and 1.5% argon. The presence of nitrogen, an essential requirement for life, together with the probability that water exists locked into the soil as a permafrost has raised again the hopes of those attempting to find some sort of life on Mars.

Almost immediately upon landing, Viking 1

Nix Olympica, the gigantic volcanic mountain on Mars, is some six miles high and more than 300 miles across the base, making it twice as broad as the most massive volcanic pile on earth. The photographs in this mosaic were taken by Mariner IX in January 1972. (Jet Propulsion Laboratory/National Aeronautics and Space Administration.)

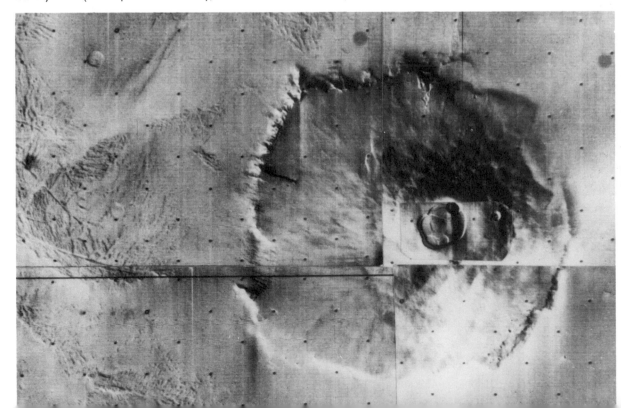

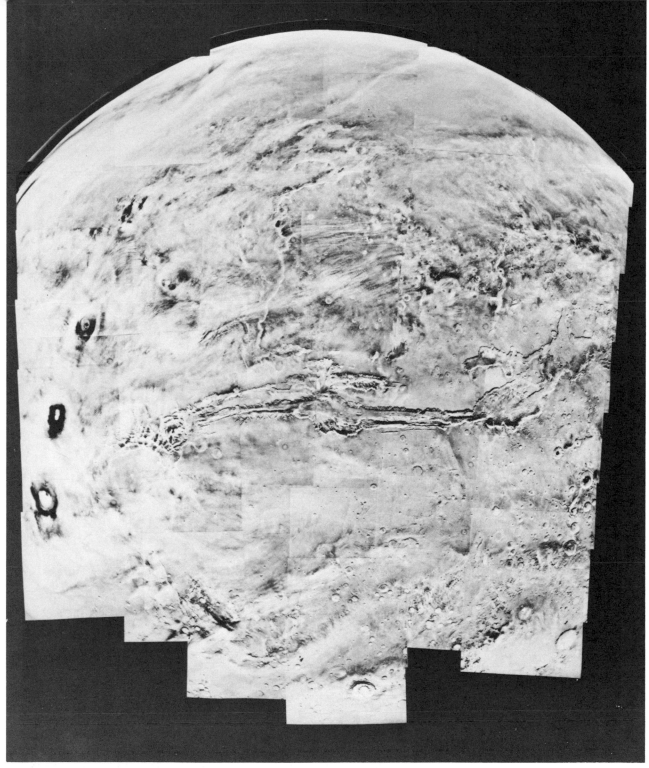

This mosaic of 102 photos of Mars was taken by Viking Orbiter I. Several prominent Martian features and at least two rare weather phenomena are visible. Valles Marineris, as long as the North American continent from coast to coast, stretches across the center. Three huge volcanoes of the Tharsis Ridge are at left: Arsia Mons, Pavonis Mons and Ascraeus Mons (from south to north). A sharp line, either a weather front or an atmospheric shock wave, curves north and east from Arsia Mons. This is the first time a feature like this has been seen. Four tiny clouds can be seen in the southern-most frame, just north of a large crater named Lowell. While the clouds are too close together to be resolved, even under high magnification, their shadows can be separated easily. The largest cloud is nearly 32 kilometers (20 miles) long. Measurements show the clouds' elevation as nearly 28 kilometers (91,000 feet). Such distinct cloud-shadow patterns are extremely rare on Mars. (NASA.)

139

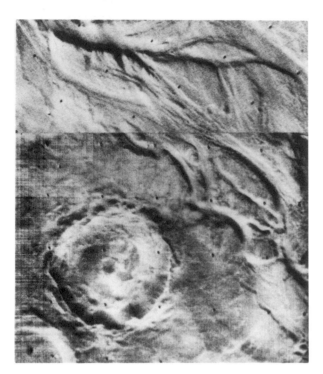 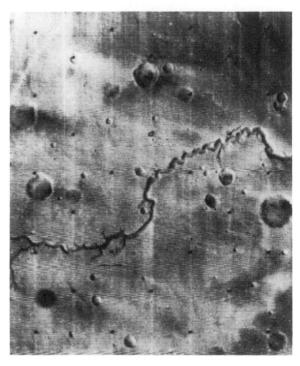

After mapping Mars, Mariner IX focused on details of special interest. The presence of water on Mars at some time in the past is suggested by the braided channel, above left, sweeping past a crater, 12 miles in diameter, and by the sinuous valley, above right. The entire valley, only part of which is shown here, is some 250 miles long. Below, when photographed by Mariner IX's narrow-angle camera, the black spot in the wide-angle photograph of the crater at left, turned out to be a field of dunes, spaced about one mile apart—evidence of strong winds blowing from a consistent direction. (Jet Propulsion Laboratory/National Aeronautics and Space Administration.)

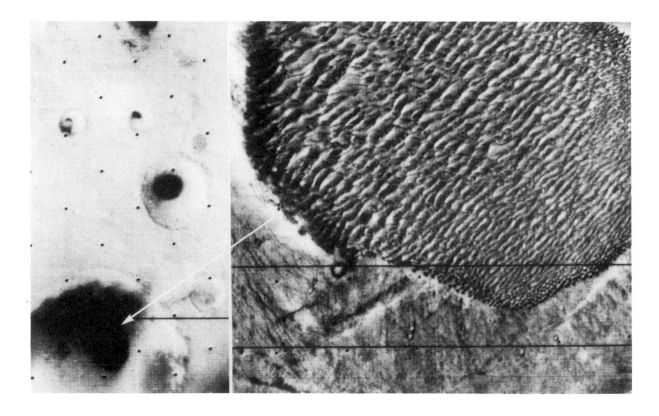

Above, first photograph by Viking 1 ever taken on the surface of the planet Mars, minutes after the spacecraft landed. The center of the image is about five feet from Viking camera #2. Below, this close-up but partial panoramic view of the Martian surface taken on the same mission shows that the soil consists mainly of reddish fine-grained material. A group of blue-black rocks near the horizon may be volcanic rocks recently excavated from the surface. (National Aeronautics and Space Administration.)

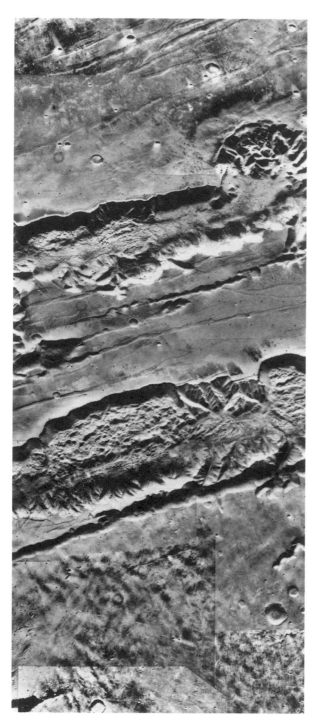

This is the best photograph of Deimos, Mars' tiny moon, obtained by Viking Orbiter I from a range of 3300 kilometers (2050 miles). At least a dozen craters are prominent—the two largest measuring 1.3 kilometers (0.8 miles) and one kilometer (0.62 mile) in diameter. Craters as small as 100 meters (330 feet) are seen to pock-mark the surface. Deimos is the smaller and outermost of Mars's two small satellites. (NASA.)

Probably the most photogenic feature on Mars is the enormous canyon—Valles Marineris—which cuts deeply into the surface and stretches nearly a third of the way around the planet. This photomosaic was made by Viking Orbiter I from an average range of 4200 kilometers (2600 miles). The principal canyon crosses the bottom half of the picture. The far wall of the main canyon shows several large landslides which probably formed in episodes and perhaps were triggered by Mars quakes. (NASA.)

began transmitting across 200 million miles of space almost unbelievably clear color pictures of its surroundings. It revealed a beautiful boulder-strewn terrain with a predominantly red color, similar to Arizona's Painted Desert. In various places dark gray patches may be exposed soil swept clear of overlying red dust. Boulders on a distant hillside bore horizontal bands that may be either evidence of sedimentation (requiring an ancient ocean) or, more likely, the effect of wind and sand erosion. The color of the atmosphere was initially thought to be bright blue, but subsequent calibration of the signals showed it to be pinkish, or even gray, probably as a consequence of wind-borne dust.

A miniature weather station on Viking 1 is transmitting data on atmospheric temperature, pressure, and wind velocities. Initial results showed light winds ranging from 2 to 15 miles per hour. Temperatures ranged downwards from 22 degrees below zero Fahrenheit in the late afternoon to 122 degrees below zero just after dawn. The atmospheric pressure was less than 1/200 that on the surface of the earth.

About a week after landing, Viking 1's balky scoop succeeded in depositing samples of Martian soil in an ingenious miniature laboratory designed to carry out crucial biology tests. The initial results proved to be tantalizingly baffling. The presence of oxygen in the samples, and the results of radioactive carbon tests were quite suggestive of the presence of microorganisms, but the results were also enough different from those achieved by earthly microbes that it was suggested that life-mimicking chemical reactions were going on. As of the end of August 1976, the situation was still unclear. It can be expected that further experiments will help to settle the matter.

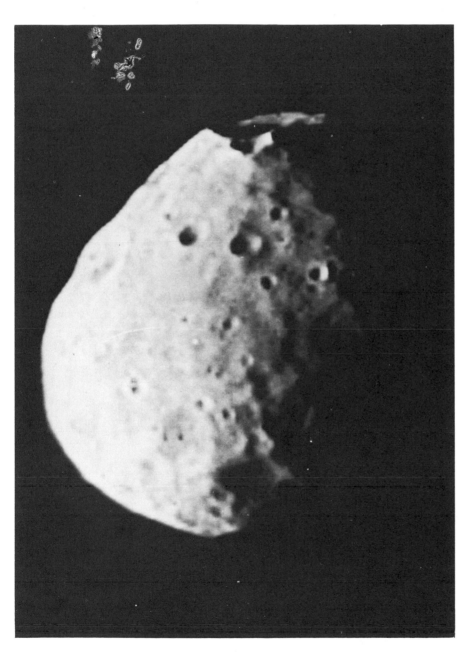

The first Viking Orbiter I picture of Phobos shows a heavily cratered side of the Mars satellite. The large crater near the North pole (top) is approximately 5 kilometers (3 miles) across while craters as small as a few hundred meters can be seen. The diameter of Phobos when viewed from this angle is about 22 kilometers (14 miles). (NASA.)

143

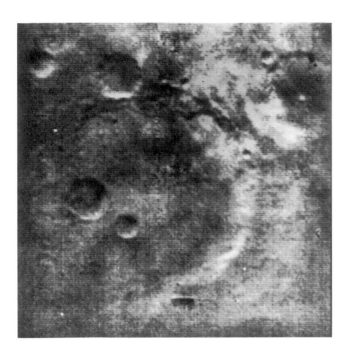

Progressive improvements in man's view of Mars: Above left, one of the first close-up photographs from Mariner IV in 1965. Above right, a 1969 photograph by Mariner VI. Below, mosaic of Mariner IX photos, made in 1972, shows almost a third of planet's equatorial region. The rift canyon parallel to the equator is 2300 miles long, 150 miles wide in places, and 20,000 deep—3-4 times deeper than Grand Canyon! In the upper left-hand corner of the mosaic is the huge volcanic mountain, Nix Olympica. (Jet Propulsion Laboratory/National Aeronautics and Space Administration.)

CHAPTER 30 JUPITER, THE GIANT PLANET

Jupiter, the largest of the planets, is a favorite object of amateur astronomers, for their telescopes reveal changing markings on its disk. Even small telescopes show certain belts parallel to its equator. Larger instruments bring to view a great deal of detail in these belts and show that they change markedly from time to time. From these changes, it has been known for a long time that what we are seeing is a cloud surface, but the details had to await the visits of the Pioneer and Voyager spacecrafts.

Even before these visits, however, earthbound studies had revealed a great deal of information about Jupiter. It was known, for instance, to be over 300 times as massive as the earth, and 11 times its diameter. It is in extremely rapid rotation, completing one revolution in $9^h 50^m$ at its equator. At higher latitudes, however, it rotates somewhat more slowly, taking $9^h 55^m$ to get around once. Its chemical composition is quite similar to what we would expect if it originated out of the same nebula that formed the sun. Spectroscopic observation of gaseous absorptions in the Jupiter atmosphere has shown it to be a poisonous, smelly atmosphere made up of methane, ammonia, great amounts of hydrogen, and traces of phosphine. From analogy with the sun, it is expected that there must be great amounts of chemically inert helium and neon also present. Since the planet is over five times as far from the sun as is the earth, its temperature at the cloud level is very low, about $-150°$. It is interesting, however, that this is higher than the temperature would have been if the sun were the sole source of heat. Evidently, there must be some heat from the interior. It is believed that this heat comes from a gradual compression and heating of the interior. Computer calculations of the interior conditions show that the pressure and temperature both increase inward. The increasingly compressed gas becomes liquid, and eventually, about halfway to the center, the hydrogen be-

comes a solid with all the properties of a metal. If there is any rocky core at all, it is small, probably no bigger than the diameters of any of the inner planets. The temperature of the gaseous atmosphere also rises rapidly, reaching over 700° before it turns into a liquid.

The Voyager flights to Jupiter, together with the earlier flybys of Pioneer 10 and 11 in 1972 and 1973, added tremendous amounts of detailed information that will take years to analyze. Voyager 1 was launched on September 5, 1977, sixteen days after Voyager 2, though its orbit was such that it arrived at Jupiter first, making its closest approach to the surface of the planet on March 5, 1979, when it was only 174,000 miles away. Voyager 2 arrived four months later, on July 9, when it was 400,000 miles from the surface of Jupiter. Together, the two spacecraft sent back 30,000 pictures that revealed the complex structure of the surface in brilliant detail.

The Voyager photographs confirmed the general structure of the belts that had been gleaned from earthbound observations. Two bright zones (called the north and south tropical zones) lie some 25° north and south of the equator. Their lower temperature relative to darker intervening belts suggests that they are composed of high clouds that probably are moving upward relative to the belts. By tracking small features as little as 120 miles in diameter, the photographs confirmed the expectation that the belts were produced by winds blowing parallel to the equator. This eastward or westward series of winds proves to be surprisingly complex, changing from one to the other in just three to 4° in latitude. At middle latitudes, speeds as high as 330 miles per hour have been measured, three times as fast as severe hurricane winds on the earth. Immediately over the equator there is an eastward wind blowing at 200 miles per hour, increasing rapidly to over 260 miles per hour at 8° latitude north or south. These rapid changes

A brilliant halo around Jupiter, the thin ring of particles, is portrayed here by Voyager 2 from the planet's night's side. This four-picture mosaic was obtained while the spacecraft was some 900,000 miles beyond the planet. The ring is unusually bright because of forward scattering from small particles within it. Similarly, the planet is outlined by sunlight scattered toward the spacecraft from a haze layer high in Jupiter's atmosphere. (NASA.)

in wind speed produce marked turbulence in the interfaces between zones, resulting in swirls that rotate clockwise in the northern hemisphere, counterclockwise in the southern.

Superimposed on the general pattern of belts, there are a number of small and very puzzling features. For instance, right over the equator there is a series of white features called plumes, which look for all the world like a series of comet tails chasing each other around the planet. Each plume trails out from a small, bright spot that is probably located at a point where there is rising gas. Twelve to fifteen of these plumes cover the entire equator. It is possible that they are produced by a rising and falling wave running right along the equator, the rising gas at each cycle producing the spot at the head of a plume. Other mysterious objects are a number of oval spots located mainly at higher latitudes. Some of these spots are white, others are brown. The former may be high altitude clouds, while in the brown spots we are looking deeper down into the atmosphere. They seem to be semipermanent features of the atmosphere, lasting for

fifty years or more in the case of the white spots, and one or two years for the brown.

The most famous of the spots on Jupiter is the Great Red Spot in the southern hemisphere, which has been seen through telescopes for at least 150 years. It is enormous, extending at least 25,000 miles from east to west, and for 6,000 miles from north to south. The Voyager photographs of its internal structure show that it is in slow counterclockwise rotation, taking six days to rotate once. A conspicuous white feature within the Great Red Spot may lend some support to the traditional theory that it is produced by gases thrown upward as lower-lying winds impinge on a protuberance on a solid, possibly icy, surface.

After the Voyager spacecraft had passed the planet, and looked back at its dark, night side, they found a number of conspicuous, diffuse aurorae. Bright by terrestrial standards, these seemed to occur in at least three layers, about 400, 900, and 1,500 miles above the cloud tops. The largest that was photographed ran in an arc that was 18,000 miles long. They each varied in brightness on time

Jupiter, its Great Red Spot and three of its four largest satellites are visible in this photo taken by Voyager 1, at a distance of 28.4 million kilometers (17.5 million miles). The innermost large satellite, Io, can be seen against Jupiter's disk; to the right is Europa; barely visible at the bottom left is Callisto. (NASA.)

(*left*) The major components of the Voyager spacecraft are identified in this drawing. (*right*) Voyager 1–Jupiter encounter. This view from the planet's north pole shows the trajectory's closest approach to the planet and selected satellites, satellite periods.

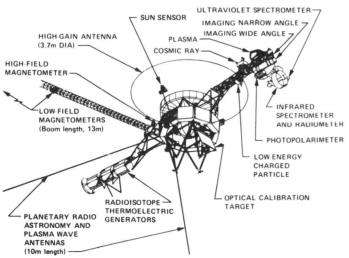

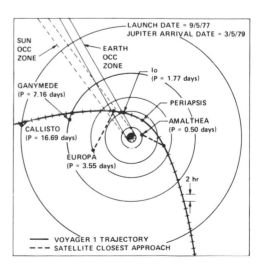

147

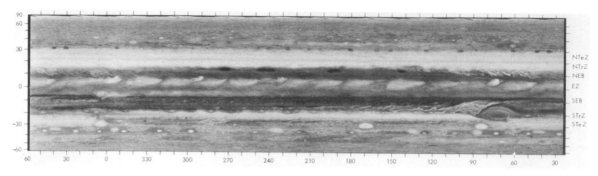

Jupiter's zones and belts are evident in this image of Jupiter produced by combining 10 images. The abbreviations on the right edge of the photograph identify the major belts (dark features) and zones (light features): NTeZ, North Temperate Zone; NTrZ, North Tropical Zone; NEB, North Equatorial Belt; EZ, Equatoria Zone; SEB, South Equatorial Belt; STrZ, South Tropical Zone; and STeZ, South Temperate Zone. North is at the top. (Jet Propulsion Laboratory/NASA.)

scales of less than one minute. Just as is the case with terrestrial aurorae, they were located predominantly in the polar regions, in zones extending from the poles equatorward to latitudes of 60°. A surprise was the photography of a series of lightning bolts, which seemed to occur in clusters that were fairly uniformly distributed over the entire planet. Their strengths were comparable to those of very strong terrestrial lightning bolts observed near the tops of massive thunderheads.

Exciting as these discoveries were, they were completely overshadowed by the discovery that Jupiter has a ring system of its very own. Admittedly, it is not very bright (otherwise earthbound observers would have seen it years ago), and does not contain very much material; estimates put its

Two closeup views of the Great Red Spot. (*above*) This Voyager 1 photo shows the Great Red Spot to be a tremendous atmospheric storm, twice the size of earth, that has been observed for centuries. It rotates counterclockwise with one revolution every six days. Wind currents on the top flow east to west, and currents on the bottom flow west to east. The large white oval is a similar, but smaller, storm that has existed for about 40 years. (NASA.) (*below*) This Voyager 2 image shows several distinct changes in the Jovian atmosphere around the Great Red Spot. The white oval beneath it in the first picture has moved farther around Jupiter, and a different white oval is shown here. The cloud regions around the Great Red Spot have changed, and the white zone west of the Great Red Spot has narrowed. (NASA.)

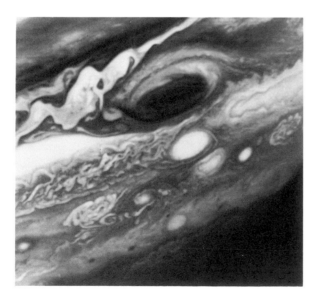
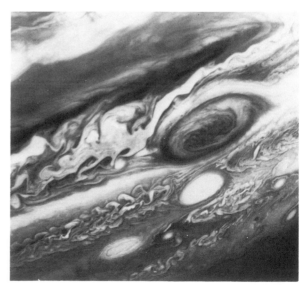

148

total mass as less than twice the mass of the tiny satellite Amalthea, Jupiter's innermost satellite. It is brightest at its outer edge, which is 35,000 miles above the cloud tops of the planet. It fades off at lower elevations, but may extend all the way down to the planet. Just as is the case with the rings of Saturn, it lies immediately above the planet's equator and is undoubtedly composed of particles in orbit around the planet. One of the reasons why terrestrial observers failed to see the ring around Jupiter is that it is much brighter when viewed from outside the planet's orbit, that is, when looking in the direction of the sun. We know this because Voyager 2 found the ring to be brighter when it was leaving Jupiter and on its way toward the more distant Saturn. This is not what one would expect if the ring were made up of small rocks or chunks of ice, since in that case their sunlit sides would be brightest. The conclusion is that the ring must be composed of very small particles that may be ice crystals about the size of smoke and that do have the property that they are more efficient "forward-scatterers" of light than they are reflectors. The actual thickness of the ring is quite uncertain. Voyager 1 put the upper limit to the thickness at 20 miles, while the better observations made by Voyager 2 reduced this to less than 4/10 of a mile!

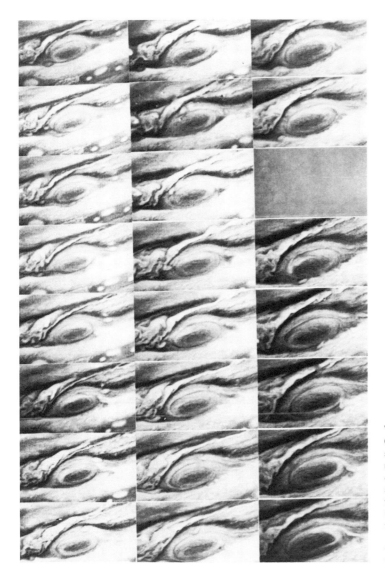

Voyager 1, time-lapse sequence of flow around the Great Red Spot (GRS). Every odd Jovian rotation is shown in blue light, starting with rotation 23 at the upper left, continuing down each column in turn, and ending with rotation 69 at the lower right. Rotation 59 was not imaged because of a spacecraft trajectory maneuver. Follow the two white spots initially in the dark band to the east (right) of the GRS. In the center frames, note also the relatively rare "eruption" of white material across the dark band to the north of the GRS, followed by rapid eastward flow out of the picture. (NASA.)

149

CHAPTER 31

VOYAGER PHOTOGRAPHS JUPITER'S FAMILY OF SATELLITES

Imagine, if you will, a small object deep in the remote recesses of the solar system, far from the sun and bitterly cold, yet pitted over its entire surface with scores of volcanoes, at least eight of which are active at any one time, all making the Mount Saint Helens eruption seem puny in comparison, and all ejecting sulphurous material over a hundred miles into space, where the clouds freeze and fall back to the surface as sulphurous snow that obliterates everything in its path.

A figment of the overheated imagination of a *Star Wars* scriptwriter? Not at all. These are precisely the conditions found by the Voyager spacecraft on Io, a satellite of Jupiter, that had the entire astronomical community agog and scrambing for explanations.

Io is only one of fourteen known satellites in orbit around Jupiter. A fifteenth is suspected, and there are probably more awaiting discovery. Four, including Io, are much larger and brighter than the others. These four were the first heavenly bodies ever discovered with a telescope, having been found by Galileo in January, 1610. Within a few weeks after his first observation, he determined their periods with surprising accuracy. They have been named the Galilean satellites in his honor, and are usually known as the first, second, third, and fourth in the order of closeness to Jupiter. They also have names: Io, Europa, Ganymede, and Callisto. Their periods of revolution are approximately 1¾, 3½, 7, and 16⅔ days. Thus the period ratio from one to the next is about 1:2. Their distances from the planet range from 262,000 to 1,169,000 miles. Io and Europa have diameters of around 2,000 miles (about the same size as our moon), while Ganymede and Callisto are each roughly 3,000 miles across, nearly the size of Mercury.

The four Galilean satellites revolve in nearly circular orbits and nearly in the planes of Jupiter's equator and of its orbit. Since their paths are turned almost edgewise to the earth, they appear to swing back and forth from one side of the planet to the other. When a satellite goes into the shadow of the planet, it becomes invisible. This is called an *eclipse*. It happens for each satellite at every revolution, with the exception of Callisto, which occasionally escapes eclipse. When a satellite is hidden by Jupiter itself, instead of by its shadow, the phenomenon is called an *occultation*.

When a satellite passes between the sun and Jupiter, its shadow can easily be seen in a telescope as a black dot moving across the planet's disk. This is known as a *transit* of the shadow. The Voyager spacecraft photographed transits of the satellites themselves, but such events are much more difficult to observe from the earth because the satellites look so much like the surface of the planet. The times of occurrence of eclipses, occultations, and transits of the satellites and their shadows are given in *The American Ephemeris and Nautical Almanac* each year. These phenomena provide very interesting observations with even a small telescope.

The four satellites figured prominently in Galileo's arguments in favor of the Copernican system, since it had been maintained that if the earth actually moved in an orbit around the sun, it would leave the moon behind. Here was a planet (Jupiter) that carried its family of satellites with it! The Galilean satellites later also made possible an important contribution to our knowledge of the velocity of light. A Danish astronomer named Roemer observed many eclipses of the satellites and discovered that the time intervals between eclipses were greater when the earth was receding from Jupiter than when it was approaching. In 1675 he correctly concluded that this must be because the light by which we see an eclipse occurring takes time to travel. It had previously been believed that

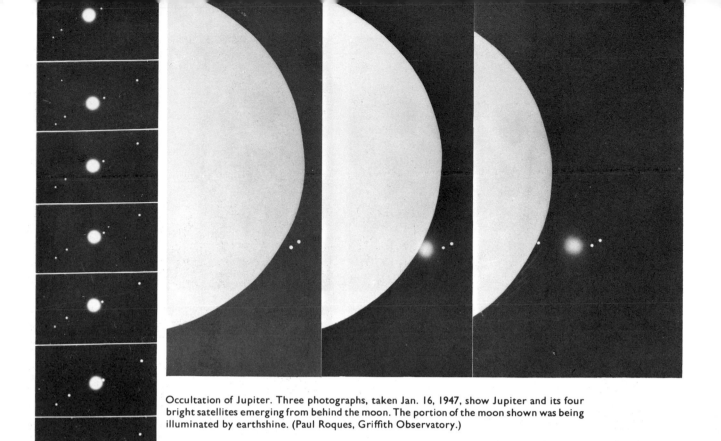

Occultation of Jupiter. Three photographs, taken Jan. 16, 1947, show Jupiter and its four bright satellites emerging from behind the moon. The portion of the moon shown was being illuminated by earthshine. (Paul Roques, Griffith Observatory.)

Jupiter and its four bright satellites. One-minute exposures were made at half-hour intervals on night of April 27-28, 1945. From left to right, satellites are IV, II, I, and III. Motion of I is clearly shown as it finally disappears behind Jupiter. (Paul Roques, with 12-inch refractor of Griffith Observatory.)

light did not take any time to travel. Roemer found that when the earth increases its distance from Jupiter, the light will be delayed a little in reaching us. When the earth is approaching Jupiter, we are coming to meet the light and the interval is shorter. Roemer's measurements showed that light takes 20 minutes to cross from one side of the earth's orbit to the other. This was not very accurate because of the experimental difficulties. It is now known that the time required to cross the earth's orbit is 16⅔ minutes, or 1,000 seconds. Since this distance is about 186,000,000 miles, the velocity of light is about 186,000 miles per second. Roemer's original suggestion of the finite velocity of light was neglected for more than 50 years, until long after his death, when an entirely different method was discovered for finding the velocity of light.

The period of Jupiter's satellite.

Roemer showed how the finite velocity of light would explain why the observed period of Jupiter's satellite was shorter when the earth was approaching Jupiter than when it was receding from it.

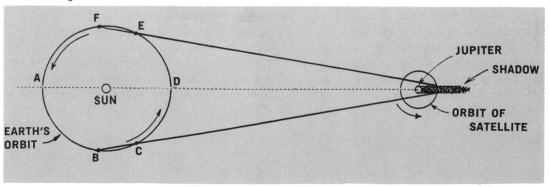

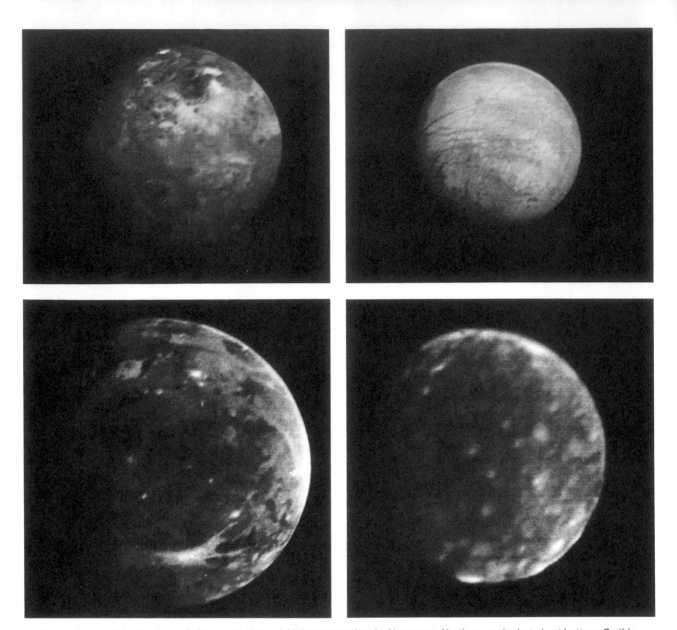

These photos of the four large Galilean satellites of Jupiter were taken by Voyager 1. North on each photo is at bottom. On this picture they are shown at their correct relative sizes; the two biggest, Ganymede bottom left and Callisto, are larger than the planet Mercury, while Io and Europa are about the size of our Moon. Io is thought to be covered with sulfur and salts, and Europa with water ice. Ganymede has both ice and rock exposed on its surface; while Callisto is primarily rocky. These surface properties contrast sharply with the interiors of the satellites: Io top left and Europa have rocky interiors, while Ganymede and Callisto contain large amounts of water or ice. The smallest markings on these images are about 50 kilometers (30 miles) across, except for Callisto, which has a resolution of 100 kilometers (60 mi). (NASA.)

The Voyager observations have, for the first time, made possible the detailed description of the surface conditions on these satellites. We have already said something about Io, the innermost of the four. The totally unexpected discovery of active volcanoes on Io was undoubtedly the most important result of the entire venture. Voyager 1 found eight volcanoes erupting simultaneously; six of the eight were still active four months later when Voyager 2 went by, and a seventh had started. As many as a hundred dark volcanic depressions have been counted on the surface, some as

Io, taken by Voyager 1 from a distance of about 862,000 kilometers (500,000 miles). The circular, donut-shaped feature in the center has been identified with a known erupting volcano; other, similar features can be seen across the face of the satellite. Io is the first body in the solar system (beyond earth) where active volcanism has been observed. (NASA.)

much as 120 miles in diameter. The active volcanoes proved to be extremely explosive; plumes observed at the limb extended more than 130 miles above the surface, with ejection velocities exceeding 2,200 miles per hour. Much of the material ejected must be composed of sulphur compounds, because infrared spectrometers on the satellites discovered that most of the surface material was sulphur dioxide that must have been spewed out from the volcanoes to obliterate any primeval features that may have been present. From the amount of material ejected in the observed eruptions, it has been estimated that the observed surface cannot be more than a million years old. It has been suggested that the darker vents may contain encrusted liquid sulphur.

The presence of active volcanoes on Io presents astronomers with a major problem: why do they exist? The heat from the distant sun could warm its solid surface only to −140°C; also, the satellite is too small to have much interior heat generated by radioactive processes. Two suggestions have come to the fore. One involves the satellites Europa and Ganymede. Like the moon, Io turns one face to Jupiter at all times, held permanently in that situation by the tidal forces of the planet. But Europa and Ganymede are massive enough to have gravitational influences on Io themselves, perturbing Io's orbit quite significantly. The result is that Jupiter's gravitational pull on Io changes considerably, producing changing stresses in Io's interior that may heat it to the point that it becomes molten. The second suggestion involves the very intense magnetic field that has been observed on Jupiter. The outer parts of this magnetic field extend well past the orbit of Io and are strongly perturbed by that satellite, a fact that produces marked changes in the radio static emanating from this field as charged particles in Jupiter's vicinity gyrate around in the field. It is suggested that there is an electric current between Io and the surface of Jupiter that may be strong enough to heat Io's interior sufficiently to produce the volcanic action. It is safe to say that the last word has not yet been said regarding the amazing behavior of this enigmatic satellite.

Next in order of distance from Jupiter comes Europa. The smallest of the Galilean satellites, it

153

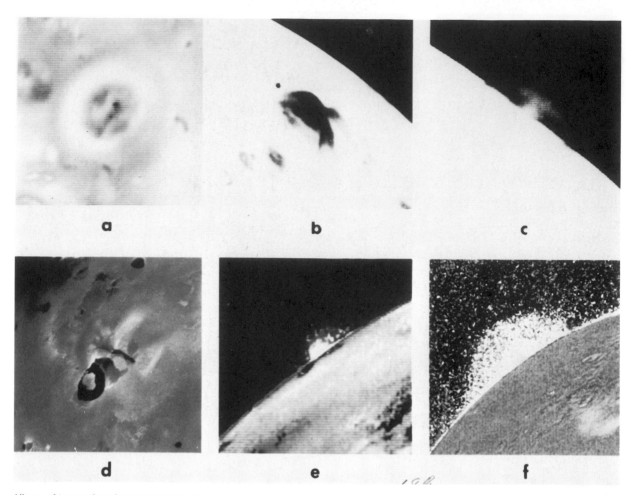

Views of two active plumes on Io from Voyager 1. (a to c) Images of the vent area (a) and eruptions over the disk (b) and bright limb (c). (d to f) Images showing the vent area (d) and two color-processed versions of the same image erupting over the limb. The first (e) is constructed from three visible filter images (violet, green, and orange) and shows the core of the plume is greenish relative to Io's surface. The second (f) is a ratio composite in which ultraviolet, violet, and orange images have been divided by green and used as blue, green, and red in the composite. It shows a second component, visible in ultraviolet light, extending far above the central core seen in visible light (e). (NASA.)

is at the same time the brightest, since it has a highly reflecting bright icy surface. The lack of numerous craters probably means that the surface is merely a thin ice crust that is very young, possibly overlying a warmer zone of softer ice or slush some 60 miles deep covering a rocky interior. A complicated array of dark streaks covers the surface. Many of the streaks are over 1,000 miles long and about 60 miles wide. They show that fracturing of the surface has taken place, without, however, any tectonic drifting of the various parts of the surface relative to each other. Again, tidal heating of the interior may be taking place, just as in the case of Io.

Ganymede is the largest satellite in the solar system, being slightly larger than Mercury. The fact that it is only half as dense as our moon shows

that only the innermost half of its bulk is rock; the rest must be ice. The surface shows a complex mix of a relatively dark, cratered surface pitted by bright craters that have exposed the underlying ice and from which material has been thrown out over the surface in systems of rays similar to the rayed craters on the moon. Voyager 2 photographed an enormous dark circular feature about 2,000 miles across, heavily cratered, that may be a piece of the primeval, cratered surface. Since the number of craters in this area is roughly similar to the number in a similar area on the moon, it has been suggested that this feature may be as much as 4 billion years old. Despite its complex surface appearance, Ganymede is relatively devoid of topographic relief, although there has been noticable movement of crustal blocks, producing fault lines,

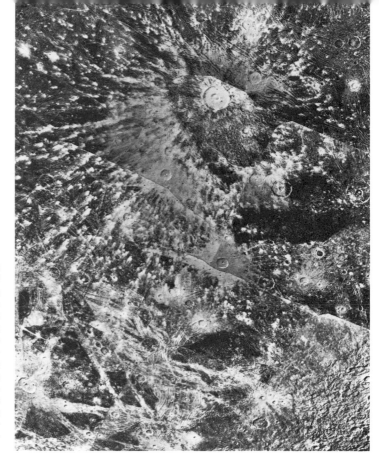

This photomosaic of Ganymede, Jupiter's largest satellite, was taken by Voyager 2 at a range of about 100,000 kilometers (62,000 miles). It shows numerous impact craters, many with bright ray systems. The rough, mountainous terrain at lower right is the outer portion of a large fresh impact basin which post-dates most of the other terrain. At bottom, portions of grooved terrain transect other portions, indicating they are younger. This may be the result of the intrusion of new icy material which comprises the crust of Ganymede. The dark patches of heavily cratered terrain (right center) are probably ancient icy material formed prior to the grooved terrain. The large rayed crater at upper center is about 150 kilometers (93 miles) in diameter. (Jet Propulsion Laboratory/NASA.)

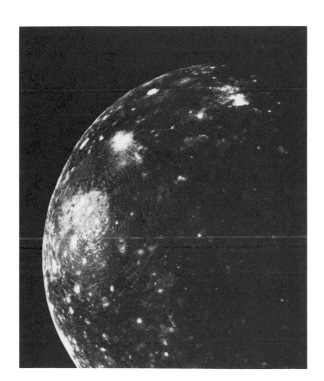

Computer-generated tour-frame color mosaic of Callisto, taken by Voyager 1. (NASA.)

and complex systems of closely spaced shallow grooves lying in broad, irregularly shaped stripes. It is likely that slow, glacial movement of the surface ice has flattened out any highlands, if any ever existed, and that, at the same time, enormous ice blocks have drifted relative to each other to produce the complex surface features that we see.

The most distant of the four Galilean satellites is Callisto. Slightly smaller than Ganymede, Callisto has the lowest internal density of all of the four satellites, showing that the greater part of its bulk must be ice, with a relatively small central core of rock. Its surface is the darkest of all, though area for area it still reflects sunlight twice as efficiently as does our moon. The most noticeable thing about the surface is the multitude of impact craters, many with central peaks. There are at least two large impact basins enclosed within concentric rings similar to the Caloris Basin on Mercury and Mare Orientale on the moon. The outermost ring on the larger of these basins is about 1,560 miles across, and it shows a bright central spot that is itself about 360 miles across. It has been theorized that these rings are the result of

155

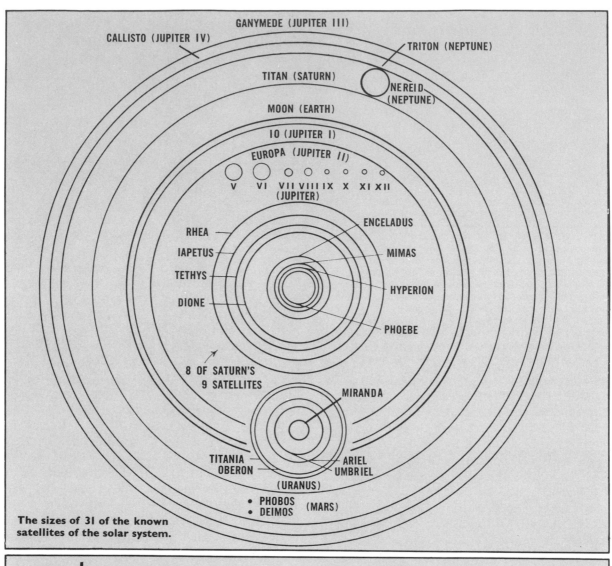

The sizes of 31 of the known satellites of the solar system.

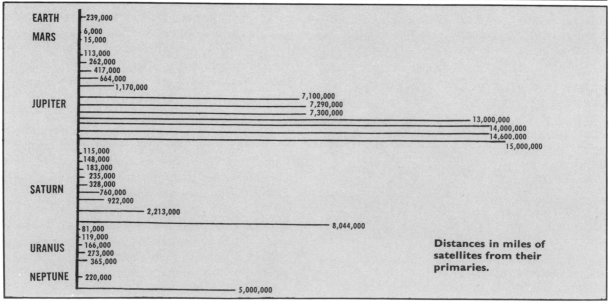

EARTH	—239,000
MARS	—6,000
	—15,000
JUPITER	—113,000
	—262,000
	—417,000
	—664,000
	—1,170,000
	—7,100,000
	—7,290,000
	—7,300,000
	—13,000,000
	—14,000,000
	—14,600,000
	—15,000,000
SATURN	—115,000
	—148,000
	—183,000
	—235,000
	—328,000
	—760,000
	—922,000
	—2,213,000
	—8,044,000
URANUS	—81,000
	—119,000
	—166,000
	—273,000
	—365,000
NEPTUNE	—220,000
	—5,000,000

Distances in miles of satellites from their primaries.

shock waves blasted out by the violent impact of a large object. The waves melted the surface slightly, but refreezing occurred so quickly that the pattern of waves has been preserved for billions of years. Certainly, the heavy cratering we see on the surface suggests that Callisto has the oldest surface of the four satellites, probably dating back 4 billion years.

In addition to the four Galilean satellites, the Voyager spacecraft photographed little Amalthea, Jupiter's innermost satellite. The photographs showed an irregularly shaped, elongated object, 165 miles long and 90 miles across. Tidal forces by Jupiter keep the long axis pointing toward Jupiter. An end-on view shows a surface pitted with numerous depressions that may be impact craters. The dark red color of the satellite shows that it is probably all rock.

The remaining satellites in orbit around Jupiter are completely insignificant little objects; indeed, most may be asteroids captured by Jupiter from the belt of asteroids lying between Mars and Jupiter. Typically, their orbits depart markedly from circles, with high inclinations to Jupiter's equatorial plane. Four of them are over 7 million miles from Jupiter, while another four are twice as far away. They are all extremely faint; Mount Wilson observers who discovered two of them estimated their brightnesses to be equivalent to that of a candle seen at the distance of 3,000 miles!

CHAPTER **32** **SATURN AND ITS RINGS**

It is probably fair to say that every time an amateur astronomer invites his friends to look through his telescope, he or she is asked to show them Saturn, and he is only too happy to oblige if the planet is above the horizon at the time. Even through a small telescope it is a striking object, with its golden globe and beautiful rings. When Galileo turned his crude telescope to the planet, he immediately saw that there was something very strange about it. It looked as though there were two appendages on either side, and he reported that the planet had two satellites. Fifty years later, in 1655, Christian Huygens, with a better telescope, discovered that these were actually the projecting parts of a ring that surrounded the equator of the planet but did not touch it. After another twenty years D. Cassini found a dark line that divides the ring into two rings. This so-called "Cassini's division" was thought to be completely devoid of material orbiting the planet. In the middle of the last century a much fainter ring was found within the others and was called the "crepe" ring. More recent observations have increased the number of recognized rings and zones of darkening to eight. In order of increasing distance from the planet, they are: D ring, C or crepe ring, B ring, Cassini's division, A ring, Encke's division, F ring, and G ring. It was thought that the reason why there was no material

in the 1,000-mile-wide Cassini's division was that an orbiting object at the center of the division would have a period of 11^h $17\frac{1}{2}^m$ as it orbited the planet under the influence of Saturn's gravitational attraction. This period turns out to be exactly half the period of the satellite Mimas, so that on every second circuit of this hypothetical object it would find Mimas at the same point. The gravitational attractions of the satellite on the object would, therefore, always come at the same place and act in the same direction, so that, with time, they would be cumulative and eventually would move that object out of that region of space, leaving a gap in the rings.

Galileo was fortunate to observe Saturn when he did, because every fifteen years the rings disappear almost completely. This is because the rings and Saturn's equator are tipped 27° to the orbit, so that twice during its 30-year period of revolution around the sun we see the rings edgewise. Despite their overall diameter of 171,000 miles, the rings must be very thin. During the nineteenth century the rapid development of physical theory proved that the rings must be composed of an almost infinite number of extremely small satellites, each one too small to be seen by itself. If the rings were solid, the forces acting on them would tear to pieces any material, unless it were thousands of

Saturn as seen by Voyager 1 from a distance of 106,250,000 kilometers (66 million miles). Considerable structure can be seen in the rings: Cassini's division, between the A ring and B ring. The three satellites visible are (left to right): Enceladus, Dione (just below planet), and Tethys. (NASA.)

times stronger than anything known to exist. Almost at the end of that century, observations by Keeler showed that the outer parts of the rings revolve more slowly than the inner, thus dispelling any doubt that might remain concerning their nature. The rings represent material that might have formed satellites if it had been farther from the planet. If a moon of fair size somehow had been placed at the distance of the rings from the planet, tidal forces of Saturn would have torn it to small pieces immediately and formed a ring from the fragments.

Even if Saturn had not possessed a ring system, it would still have been a spectacular object. It is second only to Jupiter among the planets in size, being over nine times the earth's diameter. It is 95 times the mass of the earth, which leads to the interesting result that on the average its material is only 7/10 as dense as water, so that if Saturn were dropped into a bathtub of water it would float. Obviously, the major portion of its volume must be made of gas.

Saturn is over 9½ times farther from the sun than is the earth, so a given surface area on Saturn receives just a bit more than one percent as much solar heat. It is, therefore, bound to be very cold. Measures show that the temperature of its gaseous surface is only −180°C, some 30° colder than Jupiter, but still warmer than it should be, so that there must be a residual source of internal energy, just as is the case with Jupiter. As a consequence of this lower temperature, we are looking down deeper into the atmosphere before reaching the opaque clouds. Spectroscopic evidence shows that the atmosphere, and undoubtedly most of its bulk, is composed of very much the same materials as found on Jupiter, that is, mainly hydrogen and helium with traces of other elements. Photographs from the earth have shown belts parallel to the equator, although they are not nearly as striking as those on Jupiter.

A list of the similarities between Saturn and Jupiter should also include their rotational periods. On Saturn the equator takes just over 10 hours to get around once while zones near the poles take longer, so that, on the average, the period is 10½ hours. As consequence of this rapid rotation, the planet has a pronounced equatorial bulge.

After the two Voyager spacecraft had completed their history-making visits to Jupiter in 1979, they both swept on to Saturn. Voyager 1 reached Saturn in November, 1980, and Voyager 2 in August, 1981. Tens of thousands of pictures were sent back to the earth of the planet's surface, rings, and satellites. These beautifully detailed pictures went far toward answering questions raised by earthbound observers, but, at the same time, they revealed a host of completely bizarre, outrageously unexpected phenomena that are leaving astronomers scrambling for explanations. Parenthetically, one might mention that this is frequently the situation in all of the physical sciences: An observation shedding light on one phenomenon raises further unexpected questions. This is what makes science so exciting for its practitioners. Far from being discouraged, astronomers are confident that the

The winds of Saturn. Some Voyager scientists interpret this ribbonlike structure as a large-scale atmospheric wave analogous to the jet streams on earth. Convection cells—analogous to Earth's high-pressure system—nestle in the hollows. (NASA.)

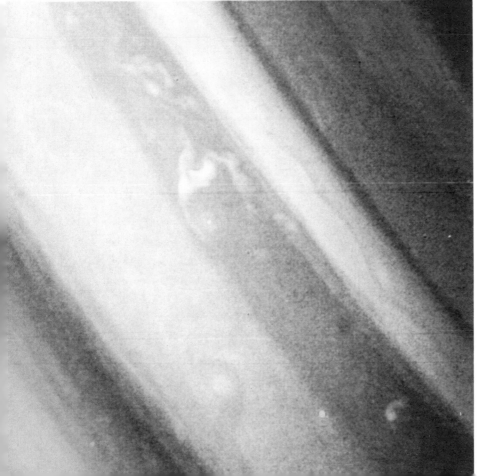

This enhanced image of the northern hemisphere of Saturn taken by Voyager 1 at a range of 9 million kilometers (5.5 million miles) shows a variety of features in Saturn's clouds: Small-scale convective cloud features are visible in the brown belt (middle); an isolated convective cloud with a dark ring is seen in the light brown zone (widest); and a longitudinal wave is visible in the light blue region (third form right). The smallest features visible in this photograph are 175 kilometers (108.7 miles) across. (NASA.)

Voyager 2 obtained this high-resolution picture of Saturn's rings when the spacecraft was 4 million kilometers (2.5 million miles) away. Evident here are the numerous "spoke" features in the B ring; their very sharp, narrow appearance suggests short formation times. Scientists think electromagnetic forces are responsible in some way for these features, but no detailed theory has been worked out. (Jet Propulsion Laboratory / NASA.)

Wide-angle view of Saturn's south polar region and mid-southern latitudes, from a distance of 442,000 km (265,000 miles). Note that the band and cloud features on Saturn continue to high latitudes. (NASA.)

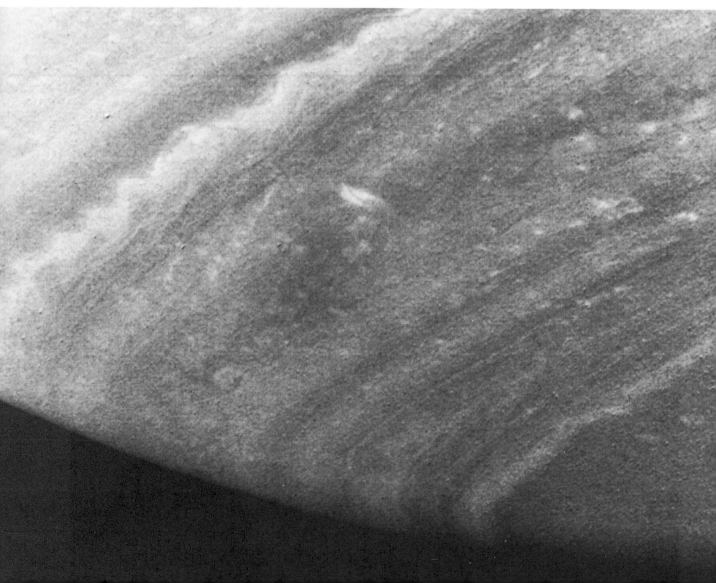

Voyager data for Jupiter and Saturn are going to go far toward providing a true picture of the origin and evolution of the solar system, including that small member, the earth.

Take the rings, for example. Instead of a mere six rings, Voyager revealed thousands. Voyager 2 even outdid this spectacular result. Advantage was taken of the fact that the light of the bright star Delta Scorpii passed through the rings as the satellite swept past the planet, twinkling on and off as ring after ring intercepted the light of the star. Across over 40,000 miles of ring material they counted hundreds of thousands of ringlets, many only 600 feet across. The edges of the complex that makes up the bright A ring are extremely sharp, giving way to darker zones in less than half a mile. The F ring turns out to be extremely thin, and made up of at least ten dense strands that are inexplicably braided around each other. After Voyager 1 had discovered this thinness and braiding, it was thought that two tiny, icy satellites discovered on either side were "shepherd" satellites that kept the thin ring between them. This mechanism for ring formation seemed to be supported by the discovery of a tiny "guardian" satellite just outside the bright A ring that might be responsible for its sharp outer edge. Thus it was proposed that the thousands of rings might be due to the presence of numerous small satellites strewn between them. Unfortunately for this theory, while Voyager 2 looked for these additional "shepherds," it found absolutely none at all. A further mystery

was the discovery of a faint, thin ring smack in the center of the dark Encke division, dubbed the "Encke doodle" because it showed a complex series of kinks and variations in brightness along its length, but with no "shepherd" satellite in sight.

Another strange result is that while most of the closely packed strands making up the C ring are blue and white, they are interspersed with sharp, isolated rings that are gold in color, similar to the color of the inner region of the B ring. Some mechanism must be operating to sort out ring material. The color of the B ring itself changes from gold at its inner edge to a beautiful turquoise at its outer edge.

When Voyager 1 passed behind Saturn, there was a period of time during which its radio signals to the earth passed through the rings. By comparing the variation in the strengths of the emerging radio signals at two different wavelengths, radio astronomers have been able to estimate the size distribution of the particles in the rings. They found virtually no objects greater than 35 feet in diameter, while between 120 and 200 per square kilometer (0.4 of a square mile) were found with sizes between 30 and 35 feet. The numbers increased steadily with decreasing size. Thus, between 1,700 and 6,000 per square kilometer were found with sizes between 10 and 15 feet. The radio data were incapable of determining how many smaller objects there might be; there might be millions, or there might be none.

Both Voyager 1 and Voyager 2 photographed

This computer-assembled mosaic, taken from a distance of 8 million km (5 million miles), shows the incredible complexity in the detailed structure of Saturn's ring system. Several ringlets are clearly visible inside Cassini's division, showing that it is far from empty. In the outermost region of the rings, the narrow F ring and satellite S14 are also visible. (NASA).

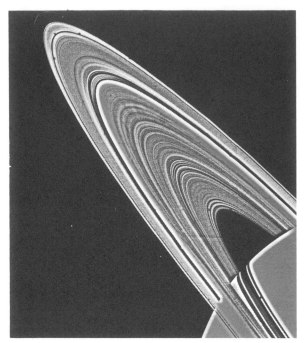

This photo shows the rings from the underside at a distance of 740,000 km (460,000 miles). The view from below shows the B ring as dark and Cassini's division as bright. The outer F ring is clearly visible. (NASA).

The narrow F ring photographed from the unilluminated side at a distance of 750,000 km. Several ring components appear to be "braided" around one another. (NASA.)

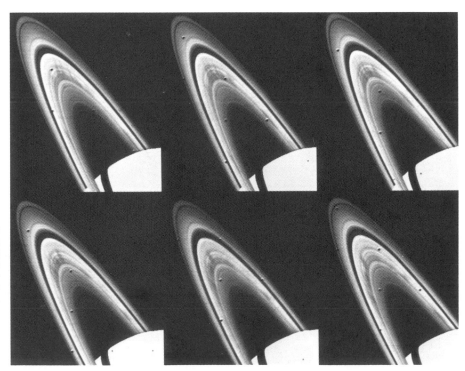

A sequence of photographs showing the spoke-like features Voyager I discovered in Saturn's B ring. In this series, taken about 15 minutes apart, the spokes can be seen to move around the ring and maintain their structure. (The dark round spots at regular intervals are calibration marks on the camera.) (NASA.)

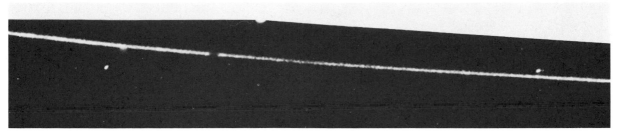

Two satellites, officially designated 1980 S26 *(left)* and 1980 S27 *(right)* "shepherd" the F ring, confining the ring particles in a narrow zone. (NASA.)

dark radial spokes that extended across the rings, particularly noticeable across the B ring. The orbiting particles in the rings seemed to carry the spokes with them. Each spoke did not last very long. Older spokes were seen to fade, to be replaced by fresh spokes that developed within the short span of 12 minutes. An initial suggestion was that the spokes were somehow related to the planet's magnetic field; particles in the ring may become ionized by sunlight and then drawn out of the ring by the magnetic field.

When we turn to the photographs of the surface of Saturn, we find them somewhat disappointingly devoid of the spectacular belts that made the images of Jupiter so interesting. Nevertheless, the photographs did show jet streams in an east-west pattern that extended much farther toward the poles than on Jupiter. In general, winds increased in speed with increasing distance from the equator, reaching speeds of 1,100 miles per hour. In some zones, narrow, undulating markings were quite distinct, looking like jet streams here on the earth. Elliptically shaped swirls within the loops of the jet streams may be convective cells, similar to high-pressure systems on the earth. Bright spots in some of the northern bands may be giant storm systems. One brown spot and a bright oval seemed to be semipermanent, since they lasted for several weeks. V-shaped trains of eddies could be discerned that were similar to the string of eddies found along Jupiter's equator and that may extend downwind from convective cells that are the seats of rising currents. In many ways, the atmospheric circulation may be quite similar to Jupiter's, the differences being attributable to the lower atmospheric temperature.

After a period of anxiety following the passage of Voyager 2 through the rings that resulted in the temporary jamming of the mechanism that pointed the cameras (fortunately eventually freed), the spacecraft was set on a course toward Uranus. It is hoped, after another billion-mile voyage when it reaches Uranus in January, 1986, it will still be operational, so that we can receive equally important photographs from that planet, revealing, most likely, additional mysteries to be solved.

Saturn's rings at different angles.

The rings keep the same direction in space while the planet revolves around the sun in about 30 years. They present their northern and southern sides to us alternately.

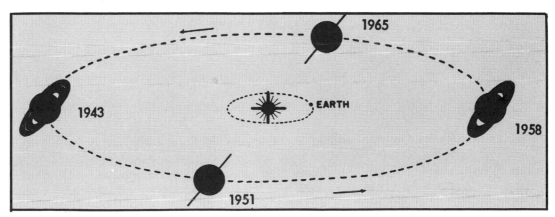

163

CHAPTER 33 THE SATELLITES OF SATURN

Saturn has a bumper crop of seventeen known satellites, more than any of the other planets, though the majority are insignificant little objects. Only five of them are more than 600 miles in diameter. From their bright, white colors, all of the smaller ones are little more than orbiting icebergs. Despite all this, this large family of satellites does display some interesting regularities, and presents us with a number of mysteries.

Let us first consider the regularities. Two of the orbiting icebergs are located at the so-called Trojan points associated with the 600-mile diameter Tethys, located a little less than 200,000 miles from Saturn. These Trojan points are points of gravitational stability located 60° ahead of, and behind, the orbiting Tethys, such that the triangle formed with Tethys and Saturn has three sides of equal length (an equilateral triangle). Any small object located at a Trojan point will remain locked at this 60° distance from Tethys. One of the two small satellites precedes Tethys by 60°, while the other follows Tethys by the same distance. A third small satellite is located at the Trojan point that precedes the satellite Dione, which is itself just slightly larger than Tethys.

In addition to these regularities, there are two so-called "co-orbiting satellites" located just above the rings, at about 100,000 miles from Saturn's center, that are chasing each other around the planet in exactly the same orbit, a fact that suggests that they may have been the result of the breakup of a single satellite in the past. Then there are the two "shepherd" satellites mentioned in the last chapter, located on either side of the F-ring

Saturn on November 23, 1943. (Photographed through 100-inch telescope of Mount Wilson and Palomar Observatories.)

The heavily cratered surface of Tethys was photographed from a distance of 1.2 million km (750,000 miles) by Voyager I. This face of Tethys looks toward Saturn and shows a large valley about 750 km long and 60 km wide (500 by 40 miles). The craters are probably the result of impacts and the valley appears to be a large fracture of unknown origin. The diameter of Tethys is about 100 km. (600 miles) or slightly less than one-third the size of our moon. (NASA.)

and believed to be responsible for its thinness. Finally, there is the "guardian" satellite just outside the A-ring and responsible for the ring's sharp outer edge.

The Voyager spacecraft passed close enough to a number of the larger satellites around Saturn to reveal a number of very unusual surface features. For example, a photograph of little Mimas, an icy sphere about 250 miles in diameter and slightly over 110,000 miles from Saturn shows an enormous impact crater on its surface more than 62 miles across; thus, the crater's width is about a quarter the diameter of the satellite. It has been calculated that if the impacting object had been only slightly larger than the object that actually made the crater, the shock would have split Mimas apart. This event must have happened fairly early in the history of Mimas, because the presence of numerous smaller craters suggests that the surface of Mimas is quite old.

Similarly, the surface of Dione, the satellite about 600 miles in diameter and 220,000 miles from the planet, is also pitted with numerous craters. Winding fault lines on Dione may have been the result of surface cracking. A little farther out, at 300,000 miles from Saturn, one of the larger

satellites, Rhea, with about half the diameter of our moon, has the most heavily cratered surface of all. On the other hand, little Enceladus, between Mimas and Dione in size and in distance from Saturn, not only has craters, but also wide surface areas wiped clean of craters, together with fissures and flow marks, suggesting massive crustal deformation and movement facilitated by internal heating. Possibly the same mechanism is operating on Enceladus that produced the fantastic volcanoes on Jupiter's Io, namely, that rhythmic shifts in Enceladus' orbit caused by close approaches to Dione and Tethys have forced the orbit sufficiently out of round that changes in the tidal forces by Jupiter have kneaded the interior like dough, heating the ices to the point that they could flow on the surface.

By far the largest satellite of Saturn is Titan, almost as big as Mercury and about ¾ million miles from Saturn. Ever since 1944 when Kuiper at the Yerkes Observatory found methane in the spectrum of Titan, this satellite has been known to be unique in the solar system in possessing an atmosphere of its own. Voyager 1 got a good look at Titan. It found an object so shrouded in orange-blue smog that the surface was invisible. The at-

Many impact craters—as a result of the collision of cosmic debris—are shown in this Voyager I photo of Dione, taken from a distance of 162,000 km (100,600 miles). The largest crater is less than 100 km (62 miles) in diameter and shows a well-developed central peak. Bright rays represent material ejected from other craters. Valleys probably formed by faults break the satellite's icy crust. (NASA.)

mosphere above the smog is mostly nitrogen (85 percent) and argon (12 percent), with the rest mainly methane. Hydrocarbons in the smog, such as ethane, acetylene, ethylene, and hydrogen cyanide are the result of either the effect of sunlight or charged particles in Saturn's radiation belts on the mix of methane and nitrogen. The smog layer is 120 miles thick. It takes a year for a smog particle to settle down to the surface. Since the surface is probably a chilly −180°, the hydrocarbon mole-

The satellite Mimas, as seen from a range of 208,000 km, shows extensive cratering. Two narrow troughs can be seen, one crossing from left to center, the other from the center to the right limb. (NASA.)

cules would freeze onto the surface, which might thus be covered with a layer of hydrocarbons hundreds of feet deep. As likely as this picture is, there are still some problems with it. For one thing, why is the smog orange when all of the hydrocarbons that have been identified within it are colorless? Another problem relates to the methane in the atmosphere, which should have completely disappeared in producing the hydrocarbons over the past 4½ billion years unless replenished from the surface. However, this replenishment would not have occurred if the hundreds of feet of hydrocarbons actually covered the surface. One intriguing suggestion made to circumvent this problem is that the surface of Titan may be covered with a deep ocean of liquid methane into which the hydrocarbons settle without a trace.

Another fascinating problem is provided by the little satellite Iapetus. Only some 850 miles in diameter, it is almost three times as far from Saturn as Titan. Tidal forces exerted by Saturn have resulted in one face of Iapetus always being turned toward Saturn, just as our moon always turns the same face to the earth. The Voyager photographs confirmed an observation made years ago by earthbound telescopes, namely, that the surface of Iapetus that faces forward along Iapetus' orbit is blacker than any other dark surface known in the solar system, while the surface that faces backward is as bright as new-fallen snow. In an attempt to explain this strange situation, it has been point-

166

This mosaic consists of high resolution images of the satellite Rhea, taken from a distance of 80,000 km (50,000 miles). Rhea's surface seems the most heavily cratered of all Saturn's satellites. (Astronomical Society of the Pacific.)

methane would have frozen out when the satellite was formed, resulting in an object that might be as much as 10 percent methane ice. The forward surface would have been extensively bombarded by impacting objects, producing craters and liquifying the subterranean methane, which welled up from within, filling the craters. Some process, probably the absorption of sunlight, might have broken up the methane molecules, releasing hydrogen into outer space and leaving carbon behind in the form of carbon black or even black pitch to blacken the surface.

Phoebe, the most distant of Saturn's family of satellites, at a distance of nearly 8 million miles, is probably a captured asteroid, because its orbit is not right over Saturn's equator, as is the case with the other 16 satellites, and also because it goes around Saturn the wrong way, from east to west rather than from west to east. Voyager 2 showed that Phoebe has twice the diameter (120 miles) that had been estimated previously and is darker than the others, which, in the main, are bright, icy objects. This dark surface is compatible with the idea that Phoebe is a rocky asteroid captured by the planet.

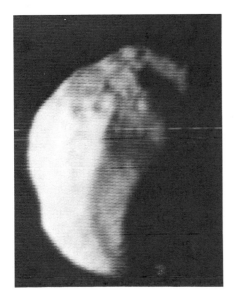 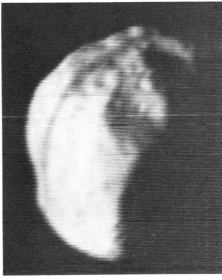

Two photographs of S 11 (Saturn's eleventh moon), the trailing member of the pair of co-orbital satellites discovered by Voyager. The two images, taken 13 minutes apart, reveal the narrow shadow of the newly discovered G ring across the face of the satellite. (NASA.)

167

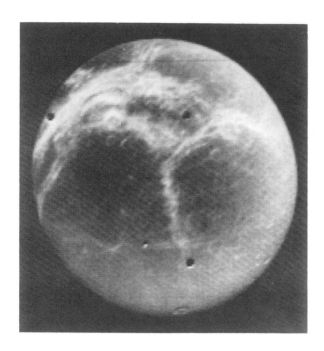

Circular impact craters up to 100 km (60 miles) in diameter are seen in Voyager I photo of Dione, Saturn's icy moon, from a range of 790,000 km (500,000 miles). Bright, wispy markings form complex patterns on the surface suggesting that they are frost deposits. (NASA.)

As we have already pointed out, the Voyager results for Jupiter, Saturn, and their families of satellites are going to have a profound effect on our theories to explain the origin and evolution of the solar system. Improvements in current theories are already being proposed. For example, there is a hypothetical scenario put forward by Eugene Shoemaker at the California Institute of Technology. It has been widely accepted that, early in the solar system's history, interplanetary space was filled with at least 100 billion cometary nuclei made up of balls of ice, pebbles, and rocks. By the end of the first billion years of the solar system's existence close encounters with the planets had ejected the cometary nuclei from the sun's vicinity, some destined to wander forever among the stars, but the majority formed a halo around the solar system with a diameter at least 40,000 times the

The cratered surface of Mimas is seen in this photograph taken by Voyager I from a range of 425,000 km (264,000 miles). The craters made by cosmic debris are shown; the largest is more than 100 km (62 miles) in diameter and displays a prominent central peak. (NASA.)

earth's distance from the sun. Occasionally a passing star will disturb the orbit of one of these little balls of ice, and it might come booming in toward the sun to be seen as a comet. Shoemaker suggests that originally Saturn may have had a number of large satellites similar to the system around Jupiter. During the first billion years, when the solar system was still crowded with cometary nuclei, the frequent collisions with the nuclei would have produced the craters we see on the satellites today. The great mass of Saturn would have attracted cometary nuclei toward it, resulting in so many collisions onto its inner satellites that they were broken up eventually to produce the debris of rings and little icy satellites we see today. Only Titan survived because it was outside the swarm attracted towards Saturn. Jupiter, being closer to the sun, was in a region where the swarms of cometary nuclei were less dense, so that its family of satellites escaped this bombardment and survived. Further study may show that this scenario needs modification, but it does explain a number of the current observations.

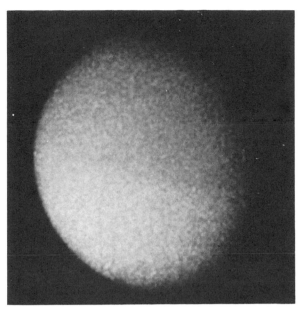

The clouds of Titan, photographed by Voyager I from a distance of 4.5 million km (2.8 million miles). Its southern hemisphere is the lighter region at the bottom half; note the well-defined boundary at the equator. The clouds in the northern hemisphere appear darker in contrast. (NASA.)

Iapetus, photographed from a distance of 3.2 million km (1.9 million miles). A large circular feature about 200 km (120 miles) across with a dark spot in its center is most probably a large impact structure. (NASA.)

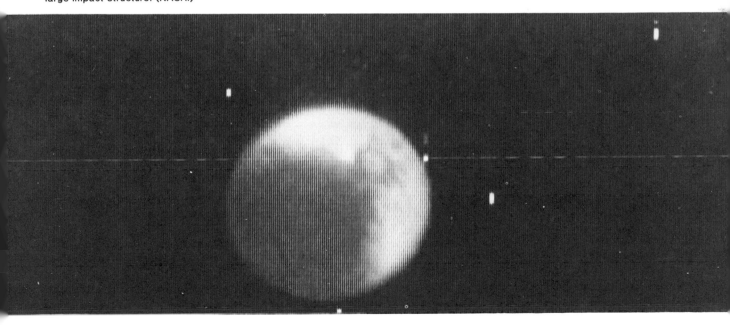

169

CHAPTER 34 THE DISCOVERIES OF URANUS, NEPTUNE, AND PLUTO

More than a century after the death of Isaac Newton, who lived from 1642 to 1727, his law of gravitation led to one of the most remarkable discoveries in the history of astronomy: the finding of Neptune in 1846. The position of this unknown world nearly three billion miles away was predicted by its gravitational effect on another world nearly two billion miles from the earth. This is a feat at which the imagination must stand in awe.

The story begins on March 13, 1781. William Herschel was examining with his seven-inch reflecting telescope some stars in the constellation of Gemini. He saw one that appeared larger than the rest and suspected it to be a comet. In his own words:

"The power I had on when I first saw the comet was 227. From experience I knew that the diameters of the fixed stars are not proportionately magnified with higher powers, as the planets are; therefore, I now put on the powers of 460 and 932, and found the diameter of the comet increased in proportion to the power, as it ought to be, on a supposition of its not being a fixed star, while the diameters of the stars to which I compared it were not increased in the same ratio. Moreover, the comet being magnified much beyond what its light would admit of, appeared hazy and ill-defined with these great powers, while the stars preserved that luster and distinctness which from many thousand observations I knew they would retain."

On March 19, Herschel found that the object was moving eastward among the stars at the rate of $2\frac{1}{4}$ seconds of arc per hour. On April 6 he reported that the object appeared perfectly sharp upon the edges, without any appearance of a tail.

After several months of observation and calculation, the orbit was found to be almost circular, very different from the long, narrow ellipse of the typical comet. At last he realized he had found a new planet—the first of the principal planets to be discovered with a telescope. Herschel suggested the name "Georgium Sidus" (the Georgian Star) in honor of his patron, King George the Third of England. The French astronomer Lalande proposed the name "Herschel," while the German astronomer Bode suggested the name of the Greek god who personified heaven, Uranus. This last name was finally adopted.

The keenness of Herschel's eye is shown by the fact that other observers had measured the position of Uranus about twenty times and had noticed nothing to distinguish it from a star. The first observation had been made in 1690. These inadvertent observations were used with later ones to determine the path followed by the planet. Its distance from the sun is nearly 20 times that of the earth, and it takes 84 years to go around the sun in a nearly circular orbit, which is inclined less than a degree to the plane of the earth's orbit.

The path followed by a planet would be an exact ellipse if it were not for the perturbations caused by the gravitational pull of the other planets. Jupiter and Saturn, the largest planets and the closest ones to Uranus, were the only known objects that could have had any appreciable effects. However, even after appropriate compensation Uranus did not travel exactly in the orbit calculated for it. By 1840 it was out of place by nearly two minutes of arc. Though this was a small amount, it was too large to be explained by errors of observation.

On July 3, 1841, John Couch Adams, an undergraduate student of St. John's College, Cambridge, wrote the following memorandum: "Formed a design, in the beginning of this week, of investigating, as soon as possible after taking my degree, the irregularities in the motion of Uranus, which are yet unaccounted for, in order to find whether they may be attributed to the action of an undiscovered planet beyond it, and, if possible, thence to determine approximately the elements of its orbit, etc., which would probably lead to its discovery."

170

The mathematical difficulties involved were tremendous, and, of course, cannot be gone into here. However, with the facts now known about the orbit of Neptune, we can understand how its pull on Uranus led to its discovery. Suppose we draw two concentric circles. One is two inches across and represents the orbit of Uranus. The other is three inches and represents the orbit of Neptune. A dot at the center stands for the sun. When Uranus was discovered in 1781, its position on the diagram was directly below the sun, corresponding to 6 on a clock. Neptune was at the right at the 3 o'clock position.

The planets go around the sun in a counterclockwise direction. Uranus on the inside track goes faster, and in 1822 caught up to Neptune at the 12 o'clock position. During about 40 years, Uranus went halfway around the sun and Neptune went a quarter of the way. During that period Neptune was constantly pulling forward on Uranus and putting it ahead of its predicted position. After 1822 Neptune was behind Uranus and pulled it back of its predicted position. Thus we can see how the general direction of the unknown disturber from the sun could be found. However, since its mass and distance were unknown, it was a difficult task to calculate its position in the sky as observed from the earth.

Adams attacked the problem in January, 1843, and obtained a solution, which he communicated in October, 1845, to Sir George Airy, the Astronomer Royal. If Airy had pointed his telescope to the spot indicated by Adams and had searched that immediate region, he would have seen the new planet within less than two degrees of the place assigned to it by Adams. However, it would not have been as easy as all that, because the planet looks very much like a star, and there was no good map of that region available to aid in the detection of a foreign object. Without a chart, the only method was to determine carefully the positions of many stars in that region and repeat the observations until the planet would betray itself by its motion.

Airy was busy and did not think that a young and unknown man could have successfully solved such a difficult problem. To test him, he asked him a question about a certain error in the orbit of Uranus. Adams could easily have answered the question, but he did not reply to Airy's letter. No further action was taken for about eight months.

In the meantime, Leverrier, a French mathematician, attacked this same problem without knowing of the work of Adams, who had published nothing of it. Leverrier published a paper in June, 1846, announcing his theoretical position for the planet. When Airy saw that this was within one degree of the place assigned by Adams, he asked Leverrier the same question he had put to Adams. The reply of Leverrier was prompt and satisfactory. At last the existence of the planet was officially believed in, as is shown by the remarks made by Sir John Herschel on September 10, 1846, at the meeting of the British Association: "We see it as Columbus saw America from the shores of Spain. Its movements have been felt trembling along the far-reaching line of our analysis with a certainty hardly inferior to ocular demonstration."

It was about time to begin to look for the planet. Airy thought a large telescope would be needed to reveal it, and there was no instrument at Greenwich adequate for the purpose. On July 9 he wrote to Professor Challis, director of the Cambridge Observatory, recommending a search with the large instrument there. Challis agreed to do so and soon started the slow and arduous task of plotting all the stars in that region down to the tenth or eleventh magnitude. On August 4 and 12 he actually saw the planet without knowing it. In his own words, "After four days of observing, the planet was in my grasp if only I had examined or mapped the observations." Had he done so, the first honors in the discovery, both theoretical and optical, would have fallen to the University of Cambridge. But this seemed to be another case of "too little and too late."

In the following month Leverrier wrote to Galle, a German astronomer at the Berlin Observatory. Possibly he was actuated by the knowledge that Germany was ahead of other nations in such matters as mapping the stars. On September 23, 1846, Galle received this letter, which asked him to direct his telescope at a certain point in the constellation of Aquarius. He had a new map by Bremiker of that region of the sky. That same evening he found the planet less than a degree away from Leverrier's predicted position. Its absence from the star map showed it to be no star. A close inspection revealed the existence of a disk. After several days, its position had changed in the predicted direction.

The controversy that ensued was not entered

into by either of the principal parties. Adams' papers had never been published, and the French were annoyed when a claim was set up on his behalf to a share in the discovery. The turn of events must have been heartbreaking for Adams, but he showed no resentment and endured his disappointment with quiet dignity. He also expressed his admiration for the genius of Leverrier. Both men now share equally the honor of having predicted the existence and position of Neptune, and Galle receives credit for being the first to observe it with the knowledge that it was a planet.

No surface markings have been observed on Neptune, and only faint belts have been seen on Uranus, but the planets are enveloped with atmospheres resembling those of Jupiter and Saturn. The spectroscope shows strong absorption of yellow and red light by methane (marsh gas), and the planets look green when observed directly. There is also a trace of ammonia vapor. Their surface temperatures are more than 300° below zero, Fahrenheit.

The rotation periods of the two planets have been determined by means of the spectroscope.

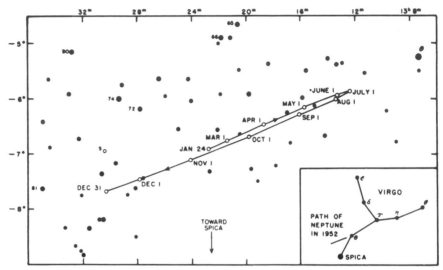

The path of Neptune.

The path of Neptune in 1952, retrograding, or moving westward, from January 24 to July 1, and moving eastward the rest of the year. The short, straight line in the inset shows the length of its 1952 path in the constellation of Virgo.

After the first approximate orbit of Neptune was computed, it was found that, like Uranus, it had been observed as a "star" many years before. Lalande, a French astronomer, observed it on the 8th and 10th of May, 1795, but finding that the two observations did not agree, he rejected the first as being in error. An investigation of this discrepancy would have led to the discovery of the planet 50 years earlier, but then we would not have had such an interesting story to tell.

Uranus and Neptune are practically identical twins. Their diameters are about four times that of the earth, Uranus being about 31,000 miles across and Neptune about 28,000 miles. The period of Neptune's revolution around the sun is 165 years, just about twice that of Uranus, which is 84 years. Neptune's distance from the sun is about 30 times that of the earth, while Uranus is only 19 times as far away.

The period of Uranus is 10.7 hours and that of Neptune is 15.8 hours. The equator of Uranus and the plane of revolution of its five satellites are tipped at an angle of 98° to its plane of revolution around the sun. Although Neptune rotates in the same direction that the earth does, one satellite goes around it in the opposite direction. This backward motion is also followed by the outer four satellites of Jupiter and the outermost satellite of Saturn.

Uranus and Neptune have diameters less than half those of Jupiter and Saturn, but they are considerably larger than the other planets. These four giants are really very much alike. They all rotate rapidly, have extensive atmospheres of methane and ammonia, and are made up of light materials highly concentrated toward the centers. Their densities range from 0.7 to 1.3 times that of water. (The four inner planets are small and

172

have higher densities, from 3.8 for Mercury to 5.5 for earth.)

The success in predicting the existence and position of Neptune led to efforts to find a still more distant planet. Even after the gravitational effect of Neptune was taken into account, the observed positions of Uranus did not agree exactly with the predicted ones. The differences were of the order of 2 seconds of arc, which is 1/60 the average size of the deviations that led to Neptune's discovery. At the beginning of the twentieth century several investigators attacked the problem. The most persistent and thorough of these was Percival Lowell, who had founded the Lowell Observatory at Flagstaff, Arizona, for the purpose of observing the planets, especially Mars. Lowell decided there were two solutions to the mathematical problem, the more probable of which put the trans-Neptunian planet in Gemini. The search was carried on from 1906 to 1916, when Lowell died.

After a number of years his brother, A. Lawrence Lowell, provided the funds for an excellent 13-inch photographic telescope, which was put in operation in 1929. It gives good images over almost the entire area of a plate, 14 by 17 inches, covering an area of the sky nearly 12° by 14°. An exposure of one hour records stars to the 17th magnitude. On an average plate the images of about 50,000 stars appear, but plates of the Milky Way region of western Gemini contain over 300,000 star images each.

The examination of so many images would have been impracticable without the blink-microscope, with which two plates of the same field can be viewed alternately in quick succession. Any object that has moved or changed in brightness becomes conspicuous. Many asteroids were found, but they could be distinguished by their relatively rapid motion. Finally, on February 18, 1930, after nearly a year of work, Clyde W. Tombaugh, a young assistant at the Lowell Observatory, found a faint object of the 15th magnitude whose shift in position between January 23 and 29 was about right for an object a billion miles beyond the orbit of Neptune. Subsequent observations proved that this was a trans-Neptunian planet, later named Pluto.

Pluto appears in the largest telescope almost as a point of light. Its mass and diameter were quite uncertain and controversial until James Christy discovered Charon, a satellite in orbit around Pluto. From the 6.4-day period of Charon, the calculated mass of Pluto turned out to be surprisingly small, only 0.2 percent that of the earth. Pluto is probably mainly ice, and about the same size as our moon. Its mass is so low that it is obvious that it could not have produced any appreciable part of the deviations of Uranus or Neptune, from which its position was predicted. In this case, Pluto's appearance so near its predicted position has to be regarded as a most remarkable coincidence.

The year 1977 also saw the surprising discovery, during an observation of an occultation of a star by Uranus, that that planet was endowed with at least five extremely faint, thin rings. Just as is the case with Saturn and Jupiter, the rings lie exactly over the planet's equator. The total amount of material in the rings must be extremely small, and is probably made up of millions of tiny satellites. Hopefully, Voyager 2 will shed further light on these rings when it passes the planet in a few years.

It has been suggested several times during the past 100 years that small residual deviations in the motion of Neptune are produced by an undiscovered tenth planet beyond Pluto. The most recent attempt to predict the location, motion, and mass of a tenth planet, popularly called Planet X, was carried out in 1972 by Joseph Brady of the Lawrence Livermore Laboratory, who investigated the possibility that Planet X was responsible for certain small irregularities in the returns of Halley's comet. The computation predicted a very remarkable object, with three times the mass of Saturn, moving in a retrograde direction in an orbit inclined 60° to the plane of the rest of the solar system. Such a massive object should be quite large and bright; Brady estimates it should be between the 13th and 14th magnitude, well within the range of moderate telescopes. Careful observations at the predicted position have revealed nothing. A large mass moving in such a highly-inclined orbit would have long-range effects on the solar system, shifting the orbits of outer planets out of the solar system's flat plane in about a million years. Most astronomers believe that a more likely explanation of a comet's irregular motion lies in the jet-action of ejected gases from a comet's nucleus as it comes close to the sun (see Chapter 36).

CHAPTER **35** **LITTLE PLANETS BY THE THOUSAND**

It may be a surprise to many people that, in addition to the nine planets (Mercury, Venus, earth, Mars, Jupiter, Saturn, Uranus, Neptune, and Pluto), there are thousands of little planets revolving around the sun. They are called asteroids (from the Greek word *aster*, meaning star) because in a telescope they look like stars. Each of the major planets except Pluto shows a disk when examined with a telescope, but the stars and almost all the asteroids appear as points of light. The asteroids are also known as minor planets or planetoids.

In spite of their number, the asteroids do not receive much attention, because they are faint. They are all invisible to the naked eye, with the occasional exception of one named Vesta. Their faintness is due to their small size, which ranges from a little less than 500 miles down to about one mile.

The discovery of the first asteroid makes an interesting chapter in the history of astronomy. In 1766, Titius of Wittenberg discovered a curious approximate relationship between the distances of the planets from the sun. It has come to be known as Bode's Law, because Bode, director of the Berlin Observatory, brought it prominently into notice in 1772. The relationship is obtained by first writing the following series of numbers: 0, 3, 6, 12, 24, and so on. Each number is found by doubling the preceding number. Now add 4 to each number, and divide the sums by 10. The resulting numbers are approximately equal to the mean distances of the planets from the sun, taking the earth's distance as 1. However, there was one gap. No planet had been found corresponding to the number 2.8.

In 1781 William Herschel found the planet Uranus, and its distance fitted in well with Bode's Law. However, the law breaks down completely for Neptune (found in 1846) and Pluto (found in 1930). Pluto is found nearly at the place predicted for Neptune. The following table shows the figures for the nine planets:

Bode's Law		Distance	Planet
0+4= 4	0.4	0.4	Mercury
3+4= 7	0.7	0.7	Venus
6+4= 10	1.0	1.0	Earth
12+4= 16	1.6	1.5	Mars
24+4= 28	2.8	?	?
48+4= 52	5.2	5.2	Jupiter
96+4=100	10.0	9.5	Saturn
192+4=196	19.6	19.2	Uranus
384+4=388	38.8	30.1	Neptune
768+4=772	77.2	39.5	Pluto

Before the discovery of Neptune and Pluto, Bode's Law was regarded as a fundamental relationship. So the position at 2.8 should have contained an undiscovered planet. A group of astronomical detectives was organized to find this fugitive subject of the sun. The heavens near the zodiac were divided into 24 zones, each of which was to be searched by one observer.

Meanwhile an astronomer named Piazzi was engaged in Sicily in making a catalog of stars. An error in a previous catalog had directed his attention to a certain region in the constellation of Taurus. On January 1, 1801, he noted the position of a star of the 8th magnitude. The next night he thought the star had slightly shifted its position to the west. He was sure of its motion by the 3rd, and believed that he had found a new kind of comet, without tail or nebulosity.

On January 14 this wandering body stopped its westward motion and started moving eastward among the stars. Such a reversal is characteristic of the apparent motion of a planet, as a result of the earth's revolution in its orbit. Piazzi wrote about his discovery to Bode, who did not receive the letter until March 20. He thought this might be the missing planet, but by that time it was too near the sun to be visible. In order to rediscover it after conjunction with the sun, a fairly accurate knowledge of its path was needed. No orbit of a planet had ever been calculated from observations extending over such a short interval. Illness had prevented Piazzi from observing it after February 11.

Astronomers tried to calculate an orbit, but were not successful. Then a young mathematician named Gauss came to the rescue. He had devised but not published several effective mathematical methods. By November he had calculated the orbit and predicted the place where the mysterious object could be found. After several weeks of cloudy weather, it was found on the last night of the year by Von Zach in the constellation of Virgo, near the place predicted by Gauss. It was also seen by Olbers the next night, exactly one year after Piazzi first found it. At Piazzi's request, it was named Ceres, after the protecting goddess of Sicily.

The distance of Ceres from the sun was found to be about 2.8 times the earth's distance, in agreement with Bode's Law. It was undoubtedly the missing planet, but it was very small (about 480 miles in diameter) and its orbit was inclined to the earth's orbit at an angle of 10°, more than the orbital inclination of any other planet known at that time.

Then a still more surprising discovery was made. In March, 1802, Olbers found another starlike object moving among the stars. Gauss found that it, too, moved between Mars and Jupiter at about the same distance from the sun as Ceres. It was named Pallas. A third one was found in 1804 and named Juno. A fourth was discovered in 1807 and named Vesta. Then no more were found until 1845. From then on, discoveries increased rapidly, and more than 300 asteroids had been found by 1890.

Photography has been of great use in finding asteroids. During a time exposure of several hours carefully guided on the stars, the asteroids change their apparent positions enough to make little trails on the photographic plate. The rate of discovery by this method has been as high as about one a day. Some are rediscoveries of old asteroids and some are single observations of asteroids that are not kept track of. At least three accurate observations separated by several weeks are needed to determine the orbit and predict the position of an asteroid for a year or more. Only a fraction of the newly discovered asteroids are sufficiently observed to allow the calculation of good orbits.

When a reliable orbit for a new asteroid has been computed, it is assigned a permanent number and the discoverer gives it a name. The earlier ones received mythological names, but these were soon used up. Asteroids have been named after flowers, cities, colleges, people, and even pet dogs and favorite desserts. It is customary to use the feminine form in the naming, except for a few having unusual orbits. The thousandth one was named after Piazzi, being called Piazzia. Number 1,001 was called Gaussia. Typical American names are Washingtonia, Hooveria, and Rockefellia.

More than 1,600 asteroids whose orbits have been calculated are revolving around the sun in the same direction in which the nine planets are going. However, their orbits do not lie so nearly in the same plane as those of most of the planets do. Six planets have orbits inclined less than 4° to the earth's orbit, Mercury's is inclined at 7°, and Pluto's at 17°. The average inclination for the asteroids is about 10°, but there are about 25 that are inclined more than 25°, the greatest (Hidalgo) being 43°.

The planets and asteroids all move in ellipses, but the orbits of most of the planets are more nearly circular than those of the asteroids. When the earth is nearest the sun on about January 3, its distance is 91,500,000 miles. When it is farthest away on about July 4, its distance is 94,500,000 miles. The mean distance is 93,000,000 miles, and each extreme differs from the mean by only 1,500,000 miles, which is about 1/60, or 0.017, of the mean distance. This fraction is called the eccentricity of the orbit. If it is very small the orbit is nearly circular. If it is large, the orbit is long and narrow.

The eccentricity is less than 0.1 for seven of the planets and is 0.2 for Mercury and 0.25 for Pluto. The average eccentricity for the asteroids is 0.15. The greatest is 0.83, which is that of Icarus. Its distance from the sun varies from about 17,000,000 miles to 184,000,000 miles. Thus it gets about 12,000,000 miles closer to the sun than Mercury does and goes about 30,000,000 miles farther from the sun than Mars does. Icarus is the only asteroid known to pass within the orbit of Mercury.

Hidalgo has the second largest eccentricity, 0.65. Its distance from the sun varies from twice the earth's distance (just beyond the orbit of Mars) to nearly 10 times the earth's distance (near the orbit of Saturn). Because of this great change in distance, its brightness varies during a period of about 14 years from the 10th magnitude, when it is readily observed, to the 19th magnitude.

Some asteroids with rather eccentric orbits come closer to the earth than any of the planets. The

Model of the solar system.
This drawing is from *Illustrated Astronomy* by Asa Smith, published about 1850. The orbits and sizes of the objects are not to scale. The orbits of five asteroids are shown between the orbits of Mars and Jupiter. Uranus is named Herschel here, after its discoverer. Neptune is named Leverrier here, after the mathematician whose prediction led to its discovery.

closest approach of a planet is that of Venus, which comes within 26,000,000 miles of the earth. The asteroid Eros, discovered in 1898, comes within about 14,000,000 miles. During the six years ending in 1937, four asteroids were found that came still nearer the earth. The distance of the closest one, Hermes, was uncertain, but probably was somewhat less than 1,000,000 miles. Although the news of its discovery was telegraphed around the world, no one other than the finder succeeded in photographing it, because it was many degrees ahead of its supposed position. At one time it moved, in one hour, five degrees, equal to the distance between the two pointers of the Big Dipper. In nine days it moved all the way across the sky. It was observed for such a short time that its orbit is very uncertain. It may never be observed again except by accident. The same is true of Apollo and

Adonis, which came within a few million miles of the earth.

The asteroid Eros has been used in determining the astronomer's yardstick, known as the astronomical unit. This method is described in Chapter 61, on measuring astronomical distances.

Eros is one of a small number of asteroids that vary in brightness. In 1901 it dropped from full brightness to about 20 percent in an hour and a quarter. The complete period of variation was found to be 5 hours and 16 minutes, with two maxima and two minima. The minima are not equally spaced and not equally dim. Sometimes the variation in light is not present, and often it can hardly be detected. But the period of variation has remained constant.

The remarkable changes in the range of variation can be easily explained by assuming that Eros is a long, thin, irregularly shaped body, like a rough brick, rotating about an axis nearly perpendicular to the greatest dimension. If its equator is highly inclined to the plane of its orbit, we may at times see only a pole of the asteroid. The same side would be turned toward us all the time and there would be no change in light. But more often, the earth lies in or near the plane of the asteroid's equator. Then all sides would be turned toward us in succession as it rotates, and we would get the full benefit of any changes in brightness arising either from irregularity of shape or variations in the diameter of the surface.

The guess as to the shape of Eros was confirmed by observations in 1931. It was seen to be elongated, and the position angle of this elongation was found to make a complete rotation in 5 hours and 16 minutes, agreeing exactly with that of the light variation. The approximate size of Eros is probably 15 or 20 miles long and 5 miles wide and thick. A number of other asteroids show similar variations in light and are undoubtedly irregular in shape.

We may wonder how these bodies can have an irregular shape. Large masses like the earth must be spherical or, at the most, flattened toward the poles by their rotation. The gravitational forces are so great that a body of irregular shape thousands of miles in diameter would slump under its own weight, like tar on a hot day, until its surface had become nearly uniform. However, the gravitational forces in an asteroid are so small that once an irregular shape has been assumed, it should endure.

The question still remains as to how the asteroids got their form in the first place. If they and the planets were formed by cooling and condensation from incandescent matter ejected from the sun, each body would have solidified in space under its own gravitational attraction into a sphere. The suggestion has been made that the asteroids are fragments of a planet. It is not known how the planet was broken up, but it could have resulted from a close approach to a large planet such as Jupiter. The gravitational attraction of the larger mass can become so intense that it will pull the nearer side of a smaller body away from the farther side and shatter it into many pieces. This may have been the origin of the asteroids.

The combined mass of all the known asteroids is uncertain, but has been estimated as about 1/3000 that of the earth. Ceres, nearly 500 miles in diameter, and Pallas, about 300 miles in diameter, are the two largest asteroids and account for about half the total mass. The undiscovered asteroids may run into many thousands, but they are probably so small that their combined mass is less than that of those already known.

The number of these bodies is so great that it is a serious problem to keep track of them. Their orbits are being disturbed, especially by the gravitational pull of Jupiter, and a great deal of labor has gone into the calculations of their positions. The American astronomer Watson found 22 asteroids and realized that they might easily become lost. In fact, one of them named Aethra was lost within a few weeks after its discovery in 1873 and was not found again until 1922, and then only by accident. To see that his asteroids were well looked after, Watson bequeathed a sum of money that has been used for that purpose.

Twelve of the known asteroids belong to an interesting group called the Trojans. All of these have about the same period of revolution as Jupiter. Each one is at Jupiter's distance from the sun, but 60° ahead of or behind the planet. Thus Jupiter, each asteroid, and the sun are at the corners of an equilateral triangle. Such an asteroid will remain in the same relative position, unless subjected to too great a gravitational pull from another planet.

There are conspicuous gaps in the orbits of asteroids corresponding to their periods of revolution, which are simple fractions of Jupiter's period —such as one-half, two-fifths, and one-third. It is believed that these gaps have been caused by the perturbing effect of the large mass of Jupiter. The gravitational pull of Jupiter on an asteroid may be very small at any one time, but the cumulative effect is great. We know how we can set a swing going in a large arc by small pushes at intervals equal to the time of swinging. Similarly Jupiter exerts a steady pull on an asteroid in the same part of its path, each time the asteroid passes Jupiter. This cumulative pull sets up changes that force the asteroid into a different orbit.

On an asteroid having a diameter less than about ten miles, the force of gravity is so small that it would be possible to throw a stone with enough velocity to send it off into space. Even the largest asteroids cannot hold an atmosphere which rules out the possibility of life like ours existing on these little worlds.

177

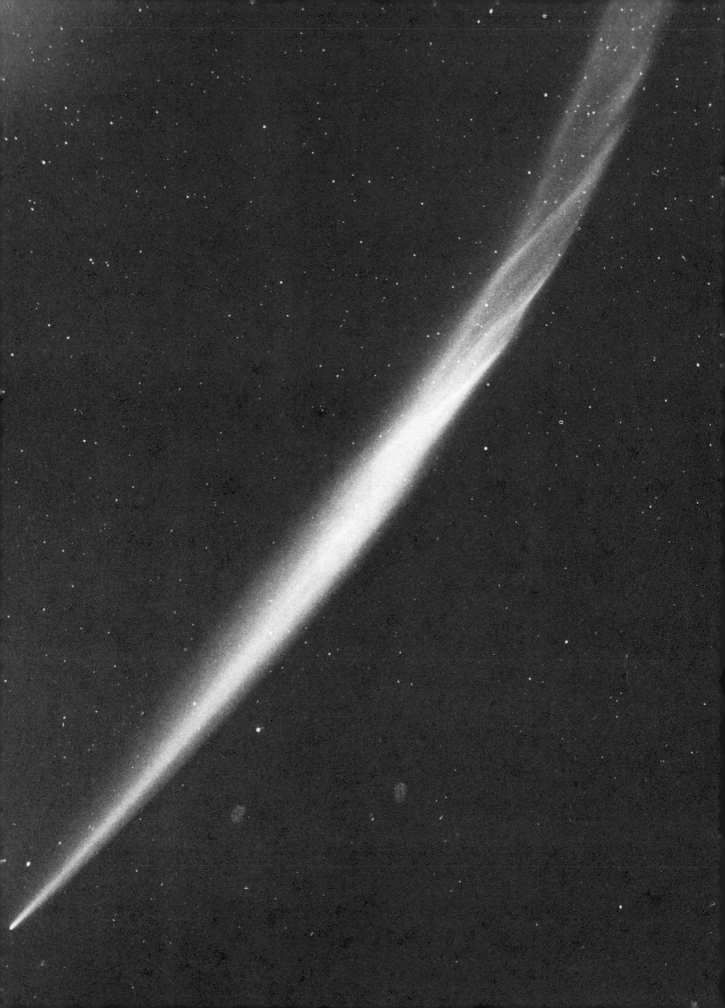

SECTION SIX

Comets and Meteors

CHAPTER **36** COMETS

An astronomer once defined a comet as the nearest thing to nothing that anything can be and still be something. An accurate description of a comet will prove this aphorism rather exact, especially for the tail of a comet. Comets have been observed with heads more than 100,000 miles in diameter, and yet when one such great comet moved in among the tiny moons of Jupiter, the gravitational force exerted by it was so small that the changes in their orbits were not apparent even after years had passed. Despite the volume of the comet, its mass must have been many thousands of times less than that of our earth. In 1910 one could look through 40,000 or more miles of the head of Halley's comet and see the stars beyond. The short dashes are the images of stars trailed out as the telescope followed the comet in its motion among the stars. In contrast, five miles of air the density of that we breathe would cut out half the light of such stars; and 100 miles of the same air would make even the brightest stars unobservable.

When one considers the extreme tenuity of the gases we breathe, the condition within the head of a comet is seen to resemble somewhat that within an X-ray tube. The tail of a comet must have a density many thousands of times less even than that of the head. In 1910 the earth passed through the tail of Halley's comet. Astronomers knew when the passage would take place and were watching

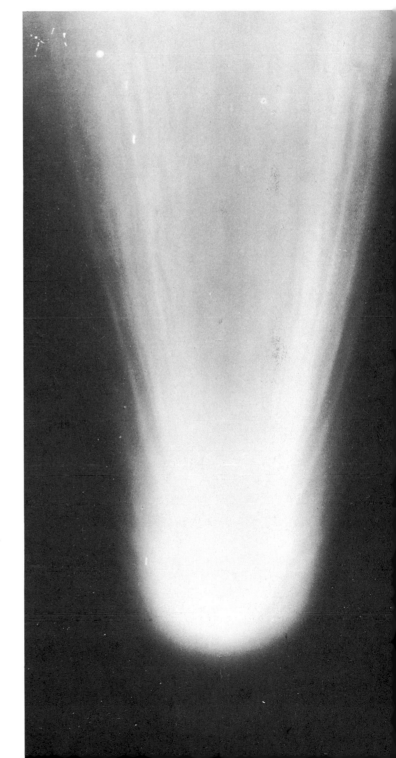

Head of Halley's comet, photographed May 8, 1910, through 60-inch telescope.

(Mount Wilson Observatory.)

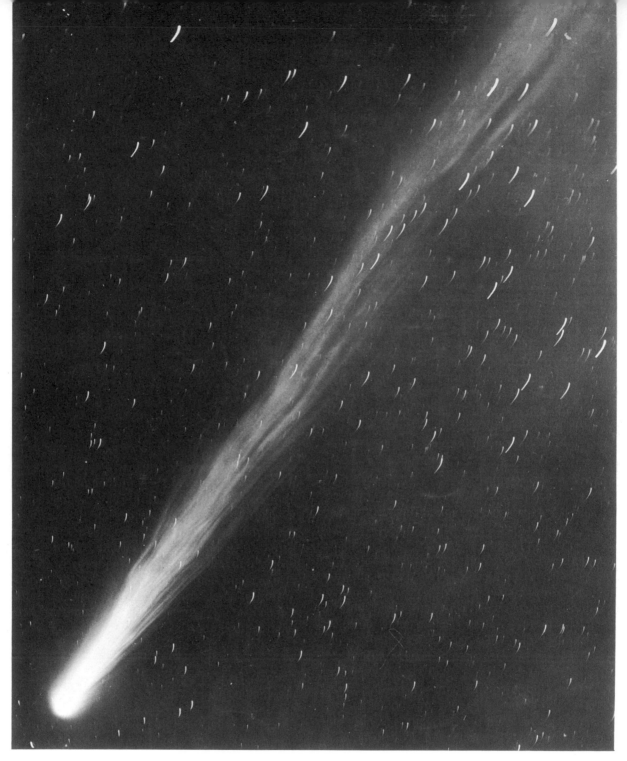

Comet Brooks, October 23, 1911. Since telescope followed comet during the exposure, stars are trailed and seen as short white lines. Notice beautiful filaments in comet's long narrow tail. (Yerkes Observatory.)

for it. Yet they did not see any evidence of it in our sky. Our air had brushed the tail aside as if it had not been there at all.

Barnard used to claim that by looking at a comet he could tell in advance of calculation whether it would show a short-period orbit or one of extremely long period—perhaps an orbit that permitted it to come only once around the sun. If the comet was sort of washed out and anemic, he would predict it to be a short-period

180

Comet Arend-Roland, April 24, 1957. Sharp spike, toward sun, is partly an illusion of projection. (Alan McClure.)

comet, but if it had sparkle, it was one that could not be observed again, at least for hundreds of years.

The phenomena that occur during the approach of a comet to the sun explain quite simply why these two types of comets differ so much in appearance. There is no bright short-period comet at all. Halley's comet, with its period of about three-quarters of a century, has the shortest period of all the really great comets. Most of those that come to *perihelion* (point of closest approach to the sun) every few years are not visible at all to the naked eye and form little, if any, tail. The tail of a comet is not formed until it gets rather close to the sun.

The spherical or ellipsoidal head of a comet is called the *coma:* it is almost entirely gaseous, although dust particles may also be present when the comet approaches quite close to the sun. The diameter of the coma depends markedly upon the distance of the comet from the sun. For instance, during the last appearance of Halley's comet, the coma was only 14,000 miles in diameter when the approaching comet was three times the earth's distance from the sun. At perihelion the comet was half the earth's distance from the sun, and the coma had increased to 120,000 miles in diameter. As the comet receded from the sun, the coma became even larger; at the earth's distance from the sun it had grown to a diameter of 320,000 miles. It subsequently shrank until it was only about 30,000 miles in diameter when lost to view.

Near the center of the coma is the *nucleus* of the comet. In some comets the nucleus can be seen as a starlike object. In other comets the coma is too bright for us to see the nucleus, or the nucleus is too small. Aside from some dust in the tail and coma, and a few meteoroids released from the nucleus, the nucleus is the only part of a comet that is solid. Recent estimates show that the nucleus is very small, possibly only a mile or two in diameter. It is composed of a heterogeneous mixture of meteoroids and ices. The "ices" include various kinds of frozen gases: frozen water, carbon dioxide, methane, etc. The evaporation of gases from the ices as the nucleus is warmed by solar radiation in the vicinity of the sun produces the coma and tail. At great distances from the sun this evaporation does not take place; hence comets near aphelion must consist solely of the little nucleus, with no coma or tail. As the ices of the nucleus evaporate, the meteoroids near its surface free themselves and gradually become dispersed into a meteor swarm.

In addition to the heat of the sun evaporating the ices of the nucleus, the radiations from the sun produce the cometary tails. Light presses on all substances that it touches. The sun, therefore, exerts two forces: one, a pushing away by means of its radiation; the other, a pulling toward it by its gravitational field. For bodies big enough to be seen even with a microscope, the sun's gravitational force is tremendous compared with its radiation pressure, but when one considers the far smaller atoms and molecules, the two forces compare quite well. Indeed, sometimes the radiation pressure is the greater, and tiny bodies will fall away from the sun instead of toward it.

This begins to occur when a comet has part of its material vaporized. The radiation pressure pushes the gases back from the sun, forming a tail, which in some cases is more than 100,000,000 miles in length. The tail of a comet, therefore, is always at least approximately away from the sun. This does not indicate the direction of motion of the comet. When the comet is moving toward the sun, its motion is in the direction that it is pointed, but as it leaves the sun, it moves toward its tail. Tails formed by radiation pressure are usually curved, with the convex side of the tail facing the direction in which the comet is moving in its orbital motion around the sun.

Sometimes comets are bombarded by charged particles from the sun. These are electrons, protons, and possibly other ions that are ejected from disturbed areas on the solar surface. Often sunspots are the visual evidence of the presence of such a disturbed area. When a comet is bombarded by such corpuscular emissions, ionized atoms and molecules in the coma are driven off very rapidly in a direction opposite to the direction of the sun. As a consequence of the high velocities of the ions ejected from the sun, the resulting tail of a comet is practically straight, in contrast to the markedly curved tails mentioned above. Sometimes when both processes—radiation pressure and particle bombardment—are operating simultaneously, a comet will possess two tails, one straight and the other curved.

Despite the very low density of the tail, the comet loses much material each time it comes near the sun. This material never can be recovered and is dissipated in space. Therefore, after each ap-

182

proach the comet becomes smaller, eventually reaching an insignificant size. It seems rather certain that this is the reason none of the short-period comets are spectacular. Thus Barnard's observation about comets has good physical background. Halley's comet is a partial exception to the rule.

speculate on this. One astronomer has thought that comets may have formed in a region surrounding the sun, a region far out from that body, but permanently connected with it, and that from this outermost part of our solar system the short-period comets have been trapped. If this spherical

DISCOVERY OF A COMET AT GREENWICH OBSERVATORY.

From Dec. 5, 1906, issue of *Punch*.

This comet has been observed for more than 2,000 years, and we have no means of knowing how much further in the past it may have made its visits to the sun. When one remembers that in all visits to the sun the tail of this comet has been extremely long, he is convinced that it cannot have existed with this short period indefinitely. It may be that it came from interstellar space accidentally toward the sun, and that the planet Neptune slowed down its motion sufficiently to trap it as a permanent member of the solar system. We do not know enough concerning the origin of comets to do more than

distribution of comets is part of the solar system and moves with it, then those comets that seem to come but once to the sun are almost equally likely to come from any direction. Observations support this. If the sun were plowing through a population of comets from outside the solar system, we would expect to find more of them approaching from the direction toward which the sun is moving.

Comets may be divided according to their orbits in *families* and in *groups*. A family of comets is defined as those comets whose aphelions (points of greatest distance from the sun) are near the orbit of a planet. The Jupiter family of comets is the only strongly marked one, though the other planets farther from the sun seem to have one or more comets that fulfill the requirement. A comet group refers to comets that have nearly identical orbits. The most striking of these groups is one that embraces four great comets—those of 1843, 1880, 1882, and 1887. The orbits are nearly identical, but in no case could any one of the comets possibly come back to the sun except after the lapse of centuries. We must regard these four as moving in one path but at different points along it. The implication is almost inescapable. All four must have had a common origin. Undoubtedly they broke apart many years ago, and each is but a fragment of a still greater comet. The disintegrating force may have come from the sun. Each of these comets approaches within about 500,000 miles of the sun's surface. At perihelion each actually passes through the outer atmosphere of the sun. They move three-quarters of the way around the sun in but a few hours and then take centuries to complete their paths.

We are not certain, but probably at some very distant date the original comet broke up as it swept through the solar corona. The comet of 1882, one of the members of this group, was observed to continue this breaking-up process into smaller comets. The picture of this comet on page 187 shows the formation of nuclei, which will become separate comets. We have observed other comets breaking, most notably Biela's comet in 1846. Other examples are on record. The tides that the sun raises in the head of a large comet passing close to it, as in the case of the four comets of this group, certainly must be disruptive, for their small mass permits only a slight gravitational force to hold the material together.

Each year several comets are discovered, but few of them become bright. Many are periodic comets that have been seen on previous approaches. Few great comets are seen each century. It has been estimated that the earth will strike the head of a comet about once in a hundred million years. This guess probably is extremely inaccurate, but nevertheless it does lead to speculation concerning the result. It seems probable that since the earliest rocks were formed, the earth has collided with a good many comets. Most of these probably did no more than give us a great shower of meteors, such as was viewed in 1833. Some, however, might have

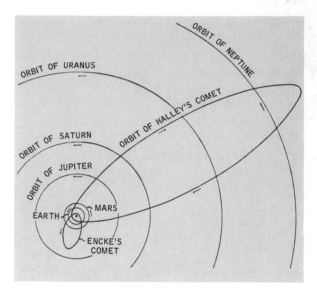

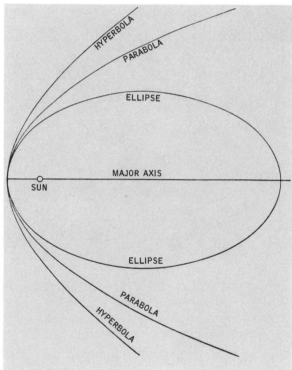

Cometary orbits.

done serious harm. Meteor Crater in Arizona may have been the result of a collision with a tiny asteroid or possibly with the nucleus of a comet. Such a collision probably killed all higher forms of life within a radius of many miles. If the point of a future collision were to be in a populous center, it might well result in the greatest disaster that the human race has known. Nevertheless, the guess is

These photos of Jupiter, taken almost four months apart by Voyagers 1 and 2, show that the planet's atmosphere undergoes constant change, presenting an ever-shifting face to observers. Note the white oval, located southwest *(below and left)* of the Great Red Spot, has drifted 60 degrees eastward *(right)* around the planet, allowing another feature, just west of the oval in the photo at far left, to move directly beneath the Red Spot. *NASA*.

The Galilean satellites of Jupiter as photographed by Voyager 1: *(clockwise from top left)* Io, Europa, Callisto, and Ganymede. *NASA*.

Closeup of Saturn's ring system. *(left)* This Voyager 1 photo shows the following rings in detail (outward from the planet) the C-ring, B-ring, Cassini's division, A ring (with Encke's division) and the F-ring. *(right)* This photo taken by Voyager 2 gives a closer look at the very dark and light "ringlets" in the planet's C-ring; the B-ring is also evident across the top. *NASA*.

This false-color image of Saturn taken by Voyager 1 brings out the details in the planet's atmosphere which are obscured by an upper-level haze. The southern hemisphere (below the rings) appears bluer than the northern hemisphere as a result of scattered sunlight. The bright spots in the northern hemisphere are a result of scattered sunlight. The bright spots in the northern bands of the planet may be giant storms. *Astronomical Society of the Pacific*.

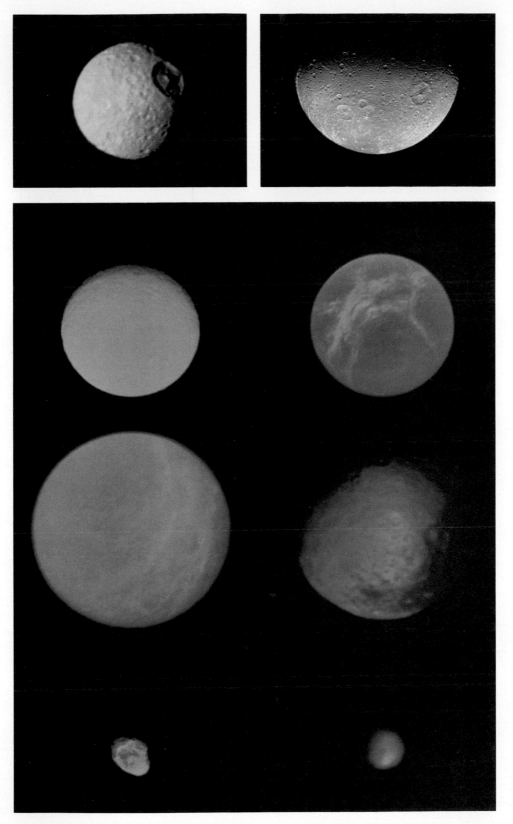

Global color images of eight satellites of Saturn: *(top to bottom, left)* Mimas, Enceladus, Tethys, Amalthea; *(top to bottom, right)* Dione, Rhea, Iapetus, and Titan. *Jet Propulsion Laboratory.*

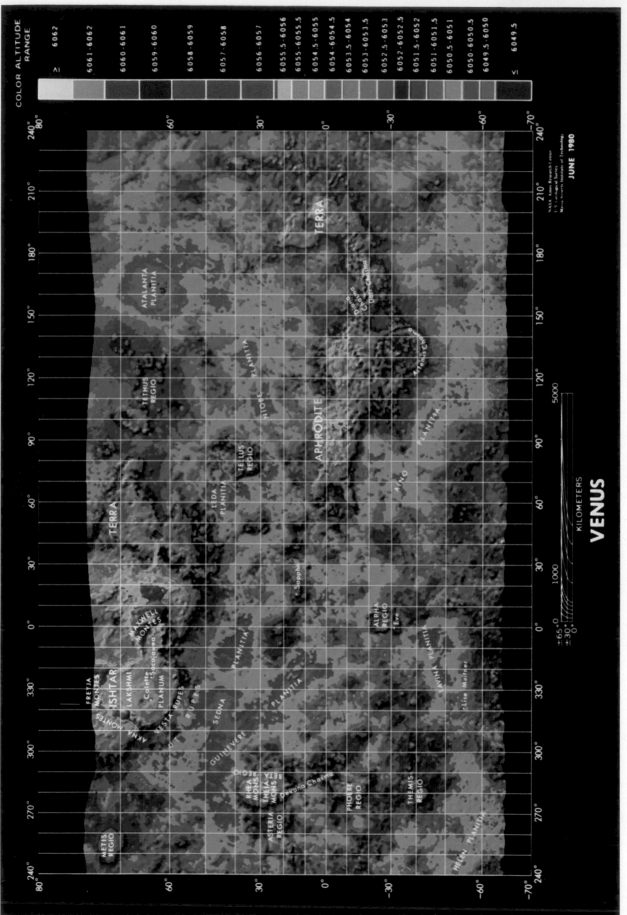

This color contour map of Venus, transmitted by the Pioneer-Venus orbiter, covers about 83 percent of the planet's surface. Venus's topography consists of highlands, lowlands, and a huge, planet-wrapping, rolling plain which covers 60 percent of the planet's surface. The highland areas are like continents themselves. *NASA.*

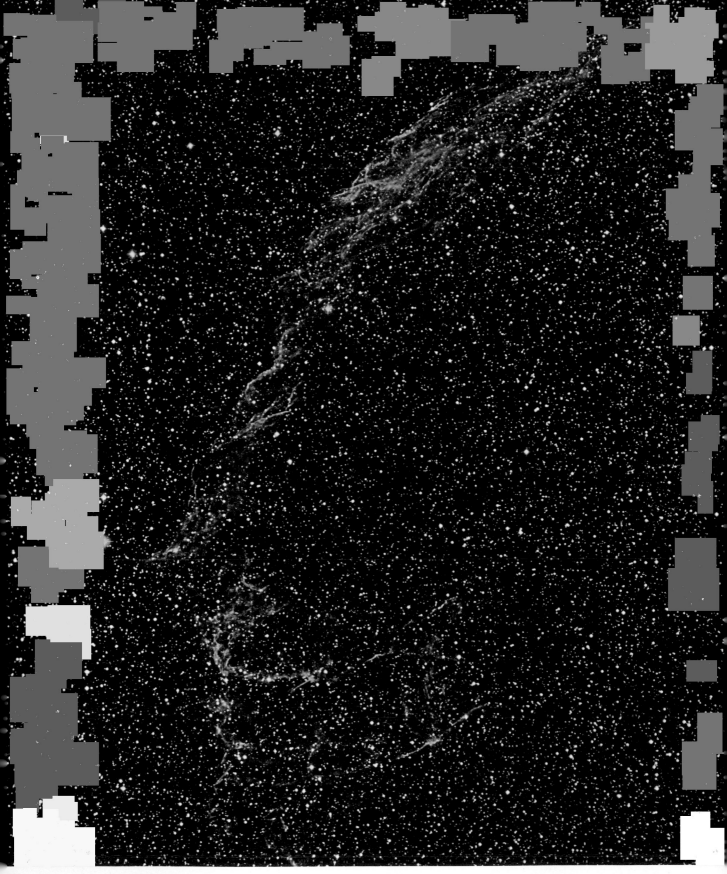

Veil Nebula in Cygnus photographed with the 48-inch Schmidt telescope at Palomar. Probably a remnant of a supernova that exploded in prehistoric times. © *California Institute of Technology and the Carnegie Institution of Washington.*

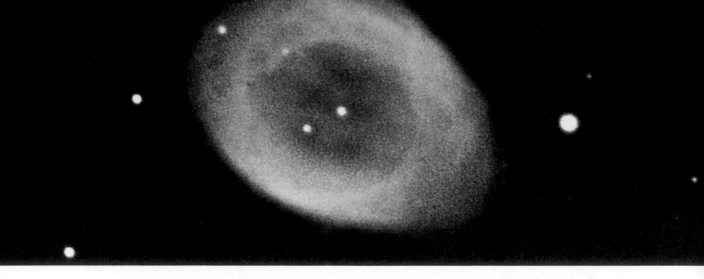

The Ring Nebula in Lyra, photographed with the 200-inch telescope at Palomar. The various colors in the ring show the effect of the particular physical processes taking place at different distances from the central star. Thus, the reddish outer periphery is a consequence of reddish hydrogen emissions at that distance. © *California Institute of Technology and the Carnegie Institution of Washington.*

The Crab Nebula in Taurus, photographed with the 200-inch telescope. Compare this color photograph with the four photographs on page 262 taken in the blue, yellow, red, and infrared. © *California Institute of Technology and the Carnegie Institution of Washington.*

The Andromeda Galaxy, photographed with the 48-inch Schmidt telescope at Palomar. The colors of the various regions of this great galaxy are typical of spiral galaxies of this type. The bluish color in the flat disk is the consequence of the presence of bright, young blue stars and diffuse nebulosities in that region, while the central bulge is yellow because the brighter stars in that region are much redder and older. © *California Institute of Technology and the Carnegie Institution of Washington.*

The Lagoon Nebula in Sagittarius. This is a very famous example of an emission nebula, similar in method of excitation to the Great Nebula in Orion. © *California Institute of Technology and the Carnegie Institution of Washington.*

The Great Nebula in Orion. This is the most conspicuous example of an emission nebula, easily visible through a small telescope. The greenish color is produced by extremely low density gas fluorescing when excited by ultraviolet light from hot, illuminating stars. © *California Institute of Technology and the Carnegie Institution of Washington*.

that few, if any, would entirely destroy our race. The principal danger from comets appears to be in the fright experienced by more or less ignorant people. In 1910 one man got rich selling anticomet pills.

COMETARY ORBITS

Comets which return to the vicinity of the sun move in elliptical orbits. These orbits are similar to those in which planets move around the sun, except that they are in general much more elongated. The first of the diagrams shown here illustrates the relative shapes of the orbits of two periodic comets and of the planets. Even the orbit of Encke's comet, which has one of the shortest of all periods, is much less circular than that of any planet. The second diagram on page 184 shows hyperbolic orbits as well as elliptical. The parabola may be considered as the boundary curve between ellipses and hyperbolas. A few comets seem to move in hyperbolas, but we are not certain whether, by the time they have moved farther from the sun than any of the planets, the orbits will continue to be such.

Some astronomers believe that in every case the orbits are elliptical in shape, though perhaps in some cases the comet may move many thousands of times as far from the sun at aphelion as at perihelion. This is quite a controversial point. As we have secured better and better observations of the comets, many more of them have been found to move in elliptical orbits, and the number believed to be moving in hyperbolic orbits has decreased. It is, however, a matter of custom to use the parabola instead of the ellipse until the observations definitely show the orbit to be a closed one. Part of the reason for the custom is historical. The other part is practical: it is much simpler to compute a parabolic than an elliptical orbit.

CHAPTER **37** GREAT COMETS OF THE PAST

Of all the astronomical objects that are observable on an average of once or more in a lifetime, a great comet is the most awe-inspiring. The only fairly great comets of this century have been the two comets of 1910, the far southern comet of December 1947. There has been no truly great one since 1882. Almost no living astronomer can remember any of the great spectacles of the last century. At their peak the greatest comets stretch almost across the sky and sometimes even are visible in the daytime. The conventional use of the term "great comet" does not apply to the size of the comet, but describes it as seen from the earth. Delavan's comet, which appeared shortly after Halley's, a generation ago, actually was a great comet, one of the very largest ever observed, but it did not approach at all close to the earth and excited almost no attention from the layman. A comet, to be "great," must not only be large and have a long tail, but it must approach closely enough to the earth to be spectacular. Some spectacular comets are designated as great by astronomers, yet attract very little attention from the layman. One such comet is the one that appeared in January, 1910, during the time that Halley's comet was observable but before it became conspicuous. This comet, although bright, set soon after the sun and as a result was not seen by many people.

Halley's comet. Halley's comet is the most famous of all, though not the most splendid. It is, however, the only great comet that returns at intervals of less than several hundred years. This comet is so important that a separate chapter will be devoted to its story.

Tycho Brahe's comet. The comet that made the greatest contribution to astronomy was the comet of 1577. The Danish astronomer Tycho Brahe was then at his prime and observed it with all the care characteristic of his genius. Before that appearance it had been believed that comets were vapors, probably poisonous, in the atmosphere of the earth; that they could not be a part of the immutable heavens. Tycho Brahe observed the comet at his observatory in Denmark and compared his observations with others made at Prague, about 400

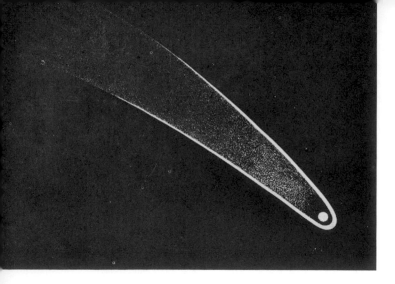

Great comet of 1811.

miles away. At any particular moment, the comet seemed to occupy the same place among the stars as viewed from each place. Since the moon's apparent position among the stars at any instant does change from place to place on the earth, his work proved the comet to be at least three times as far away as the moon and therefore truly a celestial body. These observations, with those of the nova of 1572, proved to any unprejudiced investigator that the ancient doctrine of immutable heavens was false. However, the doctrine bobbed up again and again with great power behind it. Despite absolute proofs of its falsity, it continued with some force even to the time of Newton (1642-1727).

Comet of 1811. Until the latter part of the nineteenth century there were no photographs of comets, so we must judge them by the descriptions of eyewitnesses who, especially before 1700, often were not trained in astronomical observation. After the beginning of that century there usually were astronomers to make observations, but even their

Great comet of 1843, as seen on Mar. 17 from Blackheath, Kent.

descriptions sometimes leave us in doubt as to which of two comets was the more spectacular. The accompanying drawing of the great comet of 1811 furnishes excellent evidence of how imagination must have played a part. The sharp line outlining the comet's tail must have been entirely unreal over at least most of its length. For later comets photography has played a very important part in separating phantasy from fact.

The comet of 1811 was discovered in France during March and was observed for 17 months, a record for those prephotographic days. It swung into the far northern part of the heavens during the autumn and was conspicuous all night long for many weeks. At its maximum the tail was longer than the distance from the earth to the sun and was about 15 million miles wide. Sir William Herschel measured the diameter of the small nucleus in the center of its head as 428 miles, but found that its boundaries were not sharp. A period of 3,065 years was assigned to the comet in the most accurate orbit then computed, but we must still regard the figure as uncertain.

A humorous bit of fame attached to the comet of 1811 through the fact that Portugal produced one of its most famous vintages of port while it was visible. Even as late as 1880, "comet wine" was advertised.

The comet of 1843. Almost a third of the century had passed before there was a rival for that first comet of the nineteenth century. Admiral Smythe, who observed both comets, spoke of the latter one as being far more splendid. Unfortunately the 1843 comet appeared first in the far southern sky, in which few astronomers could observe it. By the time it had moved north it had lost much of its splendor, and astronomers in northern observatories saw a much diminished spectacle. Nevertheless, in March of 1843 this great comet extended a quarter of the way across the sky.

The 1843 comet was notable in several ways. The tail was very long, narrow, and straight, and the comet had brilliance. However, its path rather than its general appearance excited the greatest attention from astronomers. At its perihelion it was only 500,000 miles from the center of the sun, actually passing through the solar atmosphere. The particles in its head must have been almost vaporized by the intense heat. In less than one day the comet moved three-quarters of the way around the sun, so great was its speed when closest to that body. But it will

Donati's comet, on Oct. 9, 1858.

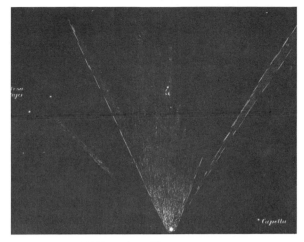

Great comet of 1861, on June 30.

Great comet of 1882, on Oct. 9.

take possibly thousands of years to complete the other quarter of the path. Of course, this last quarter will, so far as miles are concerned, be thousands of times the distance of the first three-quarters. It will carry the comet from a region hot enough to melt any substance out to a region that is almost at absolute zero. The great tail of this comet must be attributed principally to the fact that the tremendous heat near the sun generated far more gases than usually is the case.

Donati's comet. Donati's comet appeared in 1858 and, unlike the comet of 1843, was observed while still very faint and far from the sun. As a result, excellent preparations were made for its observation. This comet moved southward and, soon after it had passed its brightest phase, was not observable from Europe. While the comet was approach-

Great comet of 1882, showing four separate nuclei.

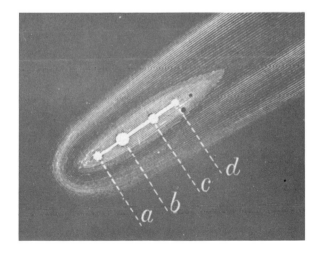

ing the sun, its chief characteristic was the unusual brightness of its nucleus. When nearest the earth, the comet's tail was almost perpendicular to the line of sight and was a magnificent spectacle.

The changes in the head of Donati's comet were remarkable, for envelopes of gas were seen to expand from the nucleus and then to be brushed backward to form the tail. Seven separate envelopes were observed.

Probably the orbit is elliptical, and a period of nearly 2,000 years has been assigned. If this is correct, the comet slows down to a speed of about 700 *feet* a second when farthest from the sun in contrast to about 30 *miles* a second when closest. Striking though these figures are, those for the comet of 1843 are even more so. When nearest the sun, that comet moved much faster than 30 miles a second, and at aphelion may move much more slowly than 700 feet a second. We do not know the period of the earlier comet but possibly it is much longer than 2,000 years.

Comet of 1861. Less than three years after Donati's comet disappeared, the fourth great comet of the century came suddenly into view. The great comet of 1861 passed through the northern part of the sky when at its best and was a splendid spectacle for England and other northern countries. The branched tail and the many envelopes which developed from the comet's bright nucleus distinguished this comet. At its maximum the tail extended about one-third of the way from the horizon to the zenith.

Comet of 1882. The last of the great comets of the nineteenth century was that of 1882. It was observable for three-quarters of a year and was conspicuous to the naked eye for several weeks. From this comet split off a mass that must be considered as being, for a time, a separate comet. Finally the separated part disintegrated too much to be visible. The nucleus of the comet was actually four nuclei moving together, any one of which would have been a fair comet alone. These separate condensations lead us to believe that one comet can break up into several comets. There is so much similarity between the paths of the 1882 comet and those of 1843, 1880, and 1887 that it is quite probable that originally all were one. If so, the multiple nuclei were in a phase in the disintegration process. Probably after hundreds of years the four nuclei will return as separate comets moving in similar paths but separated by many years. The fact that Biela's comet in 1846 was observed to break up strengthens such a hypothesis.

A world of speculation follows from these surmises, but most of it is too far from observation to be actual astronomy.

CHAPTER **38** **THE STORY OF HALLEY'S COMET**

In all ages the appearance of a bright comet has excited intense interest. A large comet is a magnificent object, with a bright, starlike nucleus and a nebulous head, which may appear as large as the moon, accompanied by a tail, which may be seen extending more than halfway from the horizon to the zenith. In ancient times and up through the Middle Ages, comets were generally regarded not as natural celestial bodies, but as supernatural phenomena. Rare and sudden in their occurrence, rapid in motion, and spectacular to see, comets have been blamed for all the misfortunes befalling the human race.

The work of Tycho Brahe, who showed that comets are true celestial bodies, led indirectly to knowledge of the paths of comets. Using Tycho's extensive observations of the planets, the German astronomer Johannes Kepler showed, after incredible labor, that the paths of the planets around the sun are slightly flattened ellipses. Then it was natural to suggest that comets might move in elongated ellipses or parabolas. It remained for Isaac Newton to prove that, according to the law of gravitation, a body could move around the sun, situated at the curve's focus, in an ellipse, a parabola, or a hyperbola. This work was continued in Newton's *Principia*, which was published through the efforts of Edmund Halley, who assumed all the responsibility and expense of the publication.

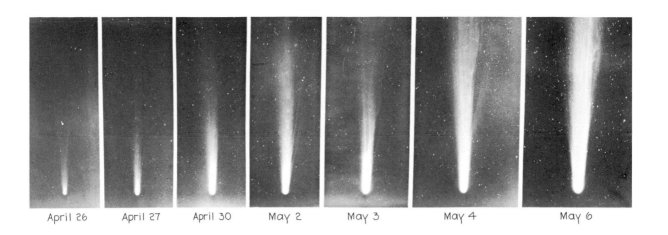

| April 26 | April 27 | April 30 | May 2 | May 3 | May 4 | May 6 |

Halley's Comet
in 1910

| May 15 | May 23 | May 28 | June 3 | June 6 | June 9 | June 11 |

Fourteen observations of Halley's comet. (Photographed at Honolulu by Ellerman of Mount Wilson Observatory.)

Newton applied the law of gravitation to the comet of 1680, and Halley carried on this study of the paths of comets. He collected all the observations of the positions of comets he could find. They extended over several centuries and included twenty-four comets. Halley found that three of these comets, those of 1531, 1607, and 1682, were moving in practically identical paths. He thought that this might be the same comet returning at intervals of about 75 years and predicted that it would return about the end of 1758.

Halley died in 1742, and as the year 1758 drew near, a French astronomer, Clairaut, assisted by Lalande and Madame Lepaute, undertook the laborious calculations of the disturbing effects of Jupiter and Saturn on the comet. In November, 1758, Clairaut predicted that the comet would pass its perihelion (the point of its orbit nearest the sun) about April 13, 1759, but said that the date might be wrong by a month either way, because of small quantities neglected in the calculations.

On Christmas night, 1758, the comet was discovered with a small telescope by an amateur astronomer named Palitzsch, who lived near Dresden. The comet passed perihelion on March 12, 1759, one month before the date set by Clairaut. Thus Halley's prediction, which was made more than fifty years before, was triumphantly vindica-

189

ted, and it is fitting that the comet bears his name.

The next return of Halley's comet was due in 1835. In the meantime, theoretical astronomy had made advances and the planet Uranus had been discovered, adding another body whose disturbing effect had to be considered. The comet was found in August and reached perihelion in November, within two days of the predicted time. The last return was in 1910, and the next one will be in 1986.

The motion of Halley's comet in the past has been very thoroughly studied by the English astronomers, Cowell and Crommelin. They have

*240, *163, *87, *12, 66, 141, 218, 295, 374, 451, 530, 607, 684, 760, 837, 912, 989, 1066, 1145, 1222, 1301, 1378, 1456, 1531, 1607, 1682, 1759, 1835, 1910.

Because of perturbations—that is, the disturbing effects of the planets—the period of Halley's comet has varied by nearly five years. The longest period, 79 years and 4 months, was between 451 and 530. The shortest period, 74 years and 5 months, was between 1835 and 1910. The average of the above 28 intervals is 76 years and 9 months.

The return of the comet in 12 B.C. is of interest

Halley's comet in 1066 as represented in Bayeux Tapestry.

determined the time of every perihelion passage of the comet back to 240 B.C., and have found records of the appearance of a comet at the proper season and in the right part of the sky for every one of these twenty-nine returns. Many of these observations were made by the Chinese, and some of them are given in enough detail to prove that the comet not only appeared at the right time, but followed an orbit similar to that of Halley's comet.

Because of the appearances of Halley's comet at or near the times of several important events in history, it will be of interest to include a complete list of the dates of its return to perihelion, starting in 240 B.C. The first four starred dates are B.C.:

because of its occurrence within a few years of the birth of Christ. Attempts have been made to identify it with the Star of Bethlehem, but authorities are agreed that the date of the Nativity could not be so early. The next return in A.D. 66 is probably the "sword of fire" mentioned by the Jewish historian Josephus as suspended over Jerusalem shortly before its siege.

In 218, the comet was described as presenting the appearance of "a fearful flaming star," which preceded the death of Emperor Macrinus. The return of 451 coincided with the defeat of the Huns under Attila at the Battle of Chalons.

The appearance of Halley's comet in the spring

190

of 1066, about five months before the Battle of Hastings, was regarded in Europe as a warning of the conquest of England by William, Duke of Normandy. The comet is given a place in the famous Bayeux Tapestry, in one section of which people are shown gazing in wonder at it. In the adjoining section a messenger announces the news of the comet to Harold, who appears to totter on his throne. The tapestry is preserved in the museum at Bayeux, Normandy. It consists of a band of linen, 231 feet long and 20 inches wide, on which have been worked with a needle, in worsteds of eight colors, scenes representing the conquest of England by the Normans.

In 1453, Constantinople was captured by the Turks. All of Europe appeared during the following years in grave danger of being conquered. Therefore, the return of Halley's comet in 1456 caused the greatest terror. The statement that the Pope issued a bull excommunicating the comet has been proved false, but the Pope evidently thought it prudent to invoke the Deity against the probable ills to follow the comet's appearance.

At its latest return, Halley's comet was detected photographically on September 11, 1909, and was followed photographically until July 1, 1911, when it was 520,000,000 miles from the sun. It was due to pass across the sun's disk on May 18, 1910, but this was during the night for Europe and America. Therefore, an expedition was sent to Hawaii for the special purpose of observing the transit. Ellerman used a 6-inch telescope and was able to see small sunspots clearly, but he could detect no trace of the comet against the sun's disk. This fact, combined with other observations, tells us that the mean density of a comet is very low. Its volume is enormous, but the amount of material in it is very small. The brightness of Halley's comet when first photographed in September, 1909, could be explained by supposing that in every cubic mile of its volume there were a dozen bodies as big as small marbles, and nothing else.

Though Comet Kohoutek, which swung around the sun on December 29, 1973, never lived up to its spectacular billing, astronomers learned much from it: Water vapor was detected in its tail; an "anti-tail," probably composed of heavier particles, was discovered pointing toward the sun, and —perhaps most exciting—traces of complex hydrocarbon molecules were observed in the comet, similar to those that have been detected by radio emissions from clouds of dust and gas in outer space (see page 312).

CHAPTER **39** **METEORS AND METEORITES**

Meteors are familiar to all of us as starlike objects that occasionally shoot across the night sky. The fainter ones, which are commonly called shooting stars, are much more frequent than the brighter ones, which are known as fireballs. Occasionally they are bright enough to light up the landscape and cast shadows for several seconds.

These streaks of light are caused by small, swiftly moving celestial bodies, most of which are probably no larger than grains of sand. Ordinarily these particles are invisible, but those that happen to enter the earth's atmosphere are heated enough by friction to become visible at a height of about 75 miles above the earth's surface. Fortunately for us, the average particle is destroyed by its passage through the air. If this were not so, the earth would

Great fireball of Mar. 24, 1933.
(Photographed by Charles M. Brown.)

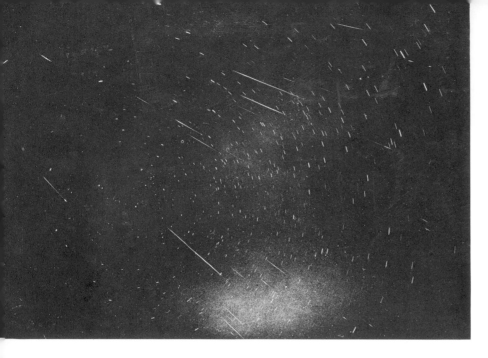

be uninhabitable, for millions of these projectiles enter our atmosphere daily at an average speed of about 25 miles per second. Even a grain of sand moving at such a velocity would be dangerous.

On any clear night a meteor can be seen about every ten minutes. At certain times of the year meteors are more numerous, and it is noticed that many of the luminous trails diverge from the same

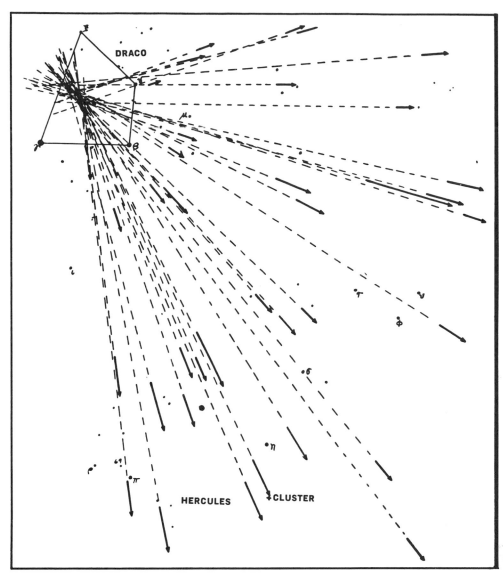

The meteor shower of October 9, 1946 plotted from the preceding photograph and others.

spot in the sky, though they usually do not start there. When this happens, we say that a meteoric shower is in progress, which means the earth has encountered a swarm of *meteoroids*. This is the term usually used for the particles when they are still in space. The term *meteor* refers to the streak of light produced by such a particle in going through our atmosphere. And when the solid object reaches the earth it is called a *meteorite*.

Examples of meteoric showers are the Perseids and Leonids. The Perseids may be seen for several weeks, reaching a maximum on August 11, when an observer may count about one a minute. For the average year the Leonids, which appear in the middle of November, are not as numerous as the Perseids, but at intervals of 33 years brilliant showers have occurred.

The finest meteor shower seen in North America since 1885 occurred on the evening of October 9, 1946. At the time of maximum occurrence a single observer could see meteors appearing at a rate of about one every second. This was true in spite of the brilliance of a full moon.

The bright moonlight interfered with the photographing of many meteors. At the Griffith Observatory two cameras were used. One was stationary and was pointed near the zenith for some views and high in the southwest for others. The other camera was fastened to the tube of the 12-inch refracting telescope, which was used for guiding. The head of Draco was kept in the upper left corner of the field. The exposure times were between 10 and 15 minutes apiece. Because of the rotation of the earth, stars appear as curved trails on pictures taken with a fixed camera.

It will be noticed in the accompanying photographs that the meteors radiate from one part of the sky. This is well brought out in the diagram, which shows the positions of 36 meteors photographed on six different negatives. Each negative was placed in an enlarger and its image was projected on white drawing paper. The principal stars were marked, and the meteors were drawn in heavy lines with arrows marking their direction of flight. Dashed lines were drawn back far enough so that they all intersected each other. The dashed lines do not intersect in one point, but in a small area inside the head of the constellation Draco. The center of this area has a position of R.A. 17^h 47^m and Decl. $+54°$. This is known as the radiant. A shower of meteors is named for the constellation in which the radiant is situated. So these recent meteors are known as the Draconids.

Appearance of meteors.

Meteors appear to radiate from one spot in the sky for the same reason that the parallel lines in this sketch diverge from one point.

The reason that the meteor trails, extended backward, seem to diverge from one small area can be seen from the drawing of the familiar effect of perspective that we get when looking down a straight road or line of telegraph poles. The lines along the road are really parallel, but they seem to diverge from a point in the distance. Similarly the paths of the meteors in the earth's atmosphere are very nearly parallel, but they seem to diverge from the radiant. Occasionally a stationary meteor is observed. This is one that appears as a point of light at the radiant and is produced by a particle coming directly toward the observer.

The radiant is important because a straight line from it to the observer, corrected for the earth's motion, gives the direction from which the meteors are coming. If the velocity of the meteors is also known, their orbit around the sun can be computed. The Draconid meteors are following the same path as a comet discovered on December 20, 1900, by Giacobini at Nice. It has a period of about $6\frac{1}{2}$ years, but it was not seen at its expected return in 1907. It was rediscovered in 1913 by Zinner, and so it is referred to as comet Giacobini-Zinner.

The diagram on page 195, not drawn to scale, may help to make clear what happened on October 9. On that date the comet was about 30 million miles away from the earth. Its tail does not necessarily travel along behind its head, but it always points away from the sun, which is in the direction of the lower part of the diagram. A comet seems to be composed of solid particles and gas. Because of slight differences in the gravitational attraction of the planets, mainly Jupiter, on the different parts of the comet, many of its particles become separated from the comet. They follow the comet's path, but may run ahead of it or lag behind it and may be scattered to some distance on all sides of the path.

In 1933 the earth was 500,000 miles away from the comet's orbit when it encountered large numbers of meteors. In 1946 the earth was only about 130,000 miles from the orbit during the shower. We may compare the comet to a locomotive that leaves a lot of smoke and cinders behind it. The earth may be compared to an auto traveling on a road passing near the railroad track. It runs into smoke and cinders left by the train that has just passed by. The earth arrived at the nearest point to the comet's orbit eight days after the comet had passed by. The comet was visible as a faint object in the telescope for several months. The particles left behind it produced the display of October 9.

The average meteor appears at a height of about 75 miles and disappears about 50 miles above the earth's surface. Bright ones will appear a little

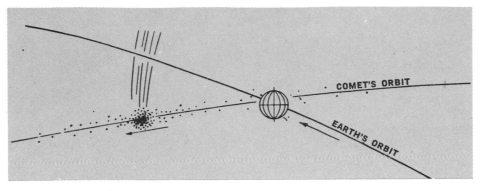

A comet showers the earth.
How the earth, passed near the orbit of a comet and was hit by many particles traveling along the comet's orbit. Each dot shown must be thought of as representing many millions of particles. Comet's tail points away from direction of sun.

higher and disappear a little lower. The particles producing a shower of meteors are so small that they are entirely consumed by the heat developed from friction with the earth's atmosphere. There is no case of a meteorite reaching the earth's surface during a shower of meteors and known to be a member of the swarm of particles that produced the shower.

Meteorites may be divided into three broad classes:

1. The irons, called siderites, are composed of nickel-iron alloy, containing 5 to 15% nickel and 85 to 95% iron.

2. The stones, called aerolites, are similar to rocks of the earth, but usually contain metallic iron particles scattered throughout the mass.

3. The stony irons, called siderolites, are composed of a spongelike mass of iron with the spaces filled with rock materials.

Iron meteorites are rough, angular, and irregular in shape, and have a pitted outside surface. Those of recent falls are black, but those of old falls are rust brown. They show beautiful markings in cross section, due to the crystalline structure of nickel-iron alloys. These Widmanstetten figures, as they are called, are brought out by etching a polished surface, as shown in the photograph.

Stony meteorites have a burnt, pebbled crust, which is black on recent falls and rustbrown on old falls. A good test is to hold the stone in question against a revolving emery wheel for a few seconds. If iron particles are revealed to the unaided eye or under a magnifying glass, the stone is fairly certain to be a meteorite.

The stones are heavier than ordinary rock due to the iron, which runs from about 10 to 15%. They are always solid masses, never porous or like cinders. The specimen from Holbrook, Arizona,

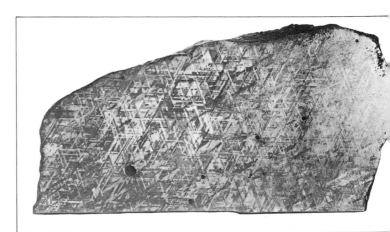

Polished and etched section of an iron meteorite (siderite) from St. Genevieve County, Mo., showing Widmanstetten crystallization figures typical of most iron meteorites. (Griffith Observatory.)

Stone meteorites (aerolites) from Forest City, Iowa, and Holbrook, Ariz., showing the black or brownish-black fusion crust typical of freshly fallen stones. Both of these meteorites are fragments of large masses that burst when they struck the earth's upper atmosphere. The shower at Forest City occurred May 2, 1890, and that at Holbrook July 19, 1912. (Griffith Observatory.)

195

Sections of stone meteorites (aerolites) from Gilgoin Station, New South Wales, Australia, and Richardton, N. D., showing the metallic grains found in many stone meteorites. (Griffith Observatory.)

shown in the photograph, fell on July 19, 1912, and is one of about 15,000 stones which spread over an area three miles long and half a mile wide. This is a good example of a shower of meteorites, which results from the bursting of a large mass during its passage through our atmosphere. The total weight of the pieces recovered from this shower is about 500 pounds, and the largest individual weighs 15 pounds. On May 2, 1890, 500 stones weighing 268 pounds fell near Forest City, Iowa, and on June 30, 1918, several stones totaling about 200 pounds fell near Richardton, North Dakota. These showers are not to be confused with meteor showers described earlier, which are associated with comets.

The surface of a stony iron is pitted and is black on recent falls and yellow brown on old falls. One of the commonest types is pallasite. A cross section shows a spongelike matrix of iron with the spaces filled with rock. Due to their structure, these

Polished sections of ironstone meteorites (siderolites) from Brenham Township, Kiowa County, Kan., and Springwater, Saskatchewan, Canada, show the metallic matrix and enclosed stony material characteristic of this type of siderolite (pallasite). (Griffith Observatory.)

meteorites are more easily broken up. Several thousand masses were recovered in 1885 from a fall (date unknown) near Brenham, Kansas.

Until 1969 there were only about 2,000 meteorites in the world's collections. As a result of an accidental discovery in that year, thousands of meteoritic fragments have been found in Antarctica. They are not all small pieces. For example, one rock weighing about 900 pounds was found in 33 pieces scattered on top of the ice over about two acres.

Very few of the Antarctic meteorites are iron ones, whereas in museum collections they are quite numerous. The reason is that stony ones are likely to be passed over as ordinary rocks by casual observers. However, in Antarctica, where there are very old ice surfaces with little or no snow accumulation and where there are no ordinary rocks, all types of meteorites are equally likely to be discovered.

From the number of meteorites which have been observed to fall it has long been known that the stony ones are more numerous than the irons. This is confirmed by the unbiased relative frequencies of the Antarctic specimens. The high concentration of meteorites has been found in only a few areas in Antarctica, where the ice has apparently been worn away by evaporation, sublimation (changing from solid to gas), and wind action.

The past thirty years have seen great advances in techniques brought to bear on the problem of the structure and possible origin of meteorites. Electron microscopes and mass spectrographs have played a key role. Meteorites show roughly the same mix of chemical elements we find in the earth, though iron and nickel are much more abundant. The Widmanstetten crystallization figures shown by the irons must have been formed when iron cooled very slowly under considerable pressure. Many meteorites show internal evidence of fracture, or shock, as though they were fragments of a larger body that broke up, probably in a collision with another object. Microscopic diamonds found in some are probably further evidence of sudden shock producing momentary great increases in pressure. Estimates of the size of the original body based on their rates of cooling and the pressures needed show that it could not have been much larger than 150 miles in diameter. Radioactive dating techniques show that meteorites are about the same age as

Meteor Crater in Arizona is nearly a mile across. (Fairchild Aerial Surveys, Inc.)

the earth—almost 5 billion years old. The irons may have originated from deeper zones near the central core of the parent body, and the stony meteorites nearer its surface. Some of the latter have gases trapped within their weaker structures, and several have an overabundance of carbon compounds filling pits and hollows. These *carbonaceous chondrites* have attracted much scientific and popular attention when long molecular chains were found in a couple of samples that were reminiscent of similar complex molecules found in living organisms.

Many meteorites have not been seen to fall, but have been recognized from their resemblance to others that *have* been observed. The largest meteorite now known is an iron one that lies where it fell at some unknown date at Grootfontein, South Africa. Its weight has been estimated as at least 60 tons. The largest meteorite in any museum weighs 34 tons and is also iron. It was brought from Greenland by the Arctic explorer Robert Peary in 1897 and is in the Hayden Planetarium of the American Museum of Natural History in New York.

Approximately thirty tons of iron meteorites have been found in the vicinity of Meteor Crater, Arizona. The largest specimen weighs about 1,500 pounds and is in the Colorado Museum of Natural

History in Denver. The Griffith Observatory has a piece weighing 395 pounds and several smaller ones. They are referred to as the Canyon Diablo meteorites, named after a gorge located about three miles west of the crater, which is about 20 miles west of Winslow and 35 miles east of Flagstaff. The crater is nearly a mile in diameter and nearly 600 feet deep. The rim of the crater rises about 150 feet above the plain and is covered with "rock flour" and rocks as large as a house.

The largest fall in modern times hit Siberia in 1908, killed a herd of 1,500 reindeer, and felled trees over an area of several hundred square miles. A man fifty miles away from the place of the fall saw a great light in the sky and felt so hot that he thought his clothes would catch fire. The explosion that followed knocked him unconscious and almost completely demolished his house. Imagine the devastation if this meteorite had landed on some major city such as New York or London!

No impact crater and no fragments of the 1908 meteorite have been found. This can be explained if it was of the stony type, which would have exploded before impact and scattered particles over a wide region. In Europe and Great Britain, brilliant sky glows were seen on the night of the fall and on several succeeding nights. The sky was so

197

bright that it was possible to read newsprint at midnight without any other light. Many observers thought that they were seeing northern lights, but there were no arcs or streamers and no flickering. Also, the spectroscope did not show the characteristic lines produced by the aurora.

There seems to be no doubt that the night glows were caused by sunlight shining on dust very high in the earth's atmosphere. This dust must have been part of the tail of a small comet, the head of which was the meteorite which exploded in the atmosphere over central Siberia.

In 1947 there was a great shower of iron meteorites in eastern Siberia, as a result of a large meteoritic body that exploded in the atmosphere. The larger fragments, weighing several tons each, split into smaller pieces on impact and produced craters as wide as 25 yards. In 1976 a two-ton stone fell in northeastern China, and more than 100 additional pieces were recovered from a wide area.

Two large meteorites which were observed to fall in the United States and later were recovered were the 800-pound stone near Paragould, Arkansas, in 1930 and the one-ton stone in Furnas County, Nebraska, in 1948. The latter was not located until six months after the fall. The other one required one month to be found, in spite of the fact that it landed about one-fifth of a mile from a farmhouse, made a hole eight feet deep, and scattered clods of clay for more than 50 yards. This illustrates how easy it is for even a freshly fallen meteorite to escape discovery.

An extraordinary number of meteorites has been found in Kansas, but there is no reason to believe that this state has received more than a normal quota of falls. There are two reasons that may explain it. The Kansas soil is almost free of terrestrial rock, and so meteorites are more conspicuous than in rocky country. Kansas has been meteorite-minded ever since 1890. For five years a farmer's wife had been collecting heavy black "rocks," because they resembled a meteorite she had been shown in childhood by a teacher. Her husband and the neighbors laughed at her, and she received no encouragement from scientists to

whom she wrote. Finally a geologist arrived and was surprised to find about a ton of stony iron meteorites. He bought the choicest ones and paid several hundred dollars, which was almost enough to buy another farm in those days. The story spread until today hardly a farmer turns up a stone with his plow without thinking of meteorites and the prices paid for them.

Though meteorites have devastated large areas and killed many animals, there is no proven case of a man having been killed by one. The reason is simple. Only a small fraction of the earth's surface is occupied by people. Thus, by the laws of chance, for every meteorite falling near a habitation, thousands will fall in the oceans and in uninhabited land areas. Yet there have been a number of narrow escapes. The best known is that of July 14, 1847, at Brannau, Bohemia, where a forty-pound meteorite crashed through the roof of a house into a room in which three children were sleeping, covering them with dust and debris but leaving them unharmed. An interesting fall occurred in Benld, Illinois, on the morning of September 29, 1938. A woman was working in her yard, when she heard a roar and a crash, which sounded as if an airplane had fallen nearby. Later in the day it was learned that a four-pound meteorite had penetrated the roof of a frame garage about 50 feet from where the woman had been standing. It went through the top of an automobile in the garage, made a hole in the cushion to the right of the driver's seat, broke the floor board, and dented the muffler. It did not reach the ground, but became entangled in the springs of the cushion, being snapped back up into the cushion by the recoil of the springs. This is the first authentic case of a meteorite striking an automobile, and the first where the end course of the meteorite could be accurately measured from three points penetrated in its fall.

The first authentic case of a person being struck by a meteorite occurred in 1954. A woman in Sylacauga, Alabama, lay napping on her living room sofa when a ten-pound stone crashed through the roof and struck her a bruising blow on her left hand and side.

198

SECTION SEVEN

Stars and Nebulae

CHAPTER **40** THE VARIETY OF THE STARS

The stars are essentially like the sun. They differ in appearance from the sun because of their greater distance from us. The nearest star is 270,000 times the sun's distance from the earth. If we could view the sun from the nearest star, it would appear as a point of light. Viewed from the distance of most stars, the sun would not even be visible to the naked eye.

The stars differ from one another in apparent brightness for two reasons. The first is that the stars are at different distances from us. The farther away a light is, the fainter it appears. If its distance from us is made ten times greater it appears 100 times fainter. This is in accordance with the well-known law that the brightness of a point-source varies as the inverse square of the distance. As soon as we know the distance of a star, we can find out how bright it would appear at some standard distance from us. When this is done for many stars, it is found that they would differ enormously in apparent brightness even if they were all at the same distance from us. There are as great differences in the intrinsic brightness, or luminosity, of stars as in the candle power of lights on the earth. A candle, an ordinary electric light, and a powerful searchlight differ no more in relative brightness than do the stars. This is the second reason that the stars do not all have the same apparent brightness.

Sirius appears as the brightest star in the whole sky as a result of a favorable combination of both of the above factors. Its distance of about 50 million million miles, or about 550,000 times the sun's distance, makes it the second nearest of all the stars visible to the naked eye. It is also very bright in itself, having a luminosity of about 30 times that of the sun. On the other hand, there is a star close to Sirius which appears about 10,000 times fainter than Sirius.

The intrinsic luminosity of a star depends on two factors, the size of the star and the amount of radiation it emits from each square mile of its surface. There are great difficulties in measuring the sizes of the stars directly. If we could get near enough to them, they would look like disks, just as the sun does. However, they are so distant that they appear as points of light even in the most powerful telescopes. In a photograph, the images of bright stars appear larger than those of faint stars, but this is merely because the excess of light spreads in the emulsion on the photographic plate. The diameters of a few of the largest stars have been measured by an intricate instrument known as an interferometer, but most of our knowledge of the sizes of stars has come indirectly, as will be explained later.

The apparent brightnesses of stars are expressed in terms of what are called apparent magnitudes, an archaic system originating about 2,000 years ago when about 20 of the brightest stars were

Effect of exposure time on brightness of stars is shown in these photographs of the globular star cluster (Messier 13) in Hercules. Four exposures, of 6, 15, 37, and 94 minutes, show an increase of one magnitude (an increase in brightness of about 2.5 times) on each succeeding exposure. The brighter stars in these photographs show what are known as diffraction patterns, which are due to the effects of light passing the supports of the secondary mirror at the upper end of the tube of a reflecting telescope. (Mount Wilson and Palomar Observatories.)

called stars of the first magnitude. The faintest stars visible to the naked eye were classified as of the sixth magnitude. It was found that an average star of the first magnitude is just about 100 times brighter than one of the sixth magnitude. The scale has been fixed on this basis of a difference of five magnitudes corresponding to a difference of 100 times in brightness. The light-ratio from one magnitude to the next is 2.512, the number that may be multiplied by itself five times to give a result of 100. A star of the first magnitude is about 2.5 times brighter than one of the second magnitude, and about 2.5 × 2.5, or 6.3, times brighter than one of the third magnitude. Fractional magnitudes are also used, as is illustrated in the following table:

Constellation of Orion, photographed with an exposure of 10 hours. Red star, Betelgeuse (upper left), when compared with blue star, Rigel (lower right), appears much fainter here than it does in sky, because photographic plate is not as sensitive to red as eye is. (Mount Wilson and Palomar Observatories.)

DIFFERENCE OF MAGNITUDE	RATIO OF BRIGHTNESS
0.25	1.26
0.5	1.58
0.75	2.00
1	2.51
2	6.31
3	15.85
4	39.8
5	100
6	251
7	631
8	1,585
9	3,981
10	10,000
15	1,000,000
20	100,000,000
25	10,000,000,000

Some celestial objects are brighter than the first magnitude and therefore have magnitudes less than one. For instance, the apparent magnitude of Sirius is −1.6, and that of the sun is −26.6. Thus the Sun appears 25 magnitudes or 10,000,000,000 (10 billion) times brighter than Sirius. This tremendous difference in brightness is obviously a consequence of the closeness of the sun. The relative *intrinsic* brightnesses of the two objects could be derived by calculating their magnitudes if they were placed at the same (standard) distance. By international agreement, this standard distance is taken to be 10 parsecs, or 32.6 light-years (the *parsec* as a unit of distance is described in Chapter 61). The calculated magnitudes at the standard distance are called *absolute* magnitudes. The sun's absolute magnitude is 5, while that of Sirius is 1.3, or, intrinsically, 30 times brighter than the sun.

STELLAR DATA

FIVE NEAREST STARS

Star	Magni-tude	Paral-lax	Distance, Light-Years
1. Alpha Centauri	−0.01, +1.4	0″.760	4.3
2. Barnard's Star	+9.54	.545	6.0
3. Wolf 359	+13.66	.421	7.7
4. Lalande 21185	+7.47	.410	8.1
5. Sirius	−1.42, +8.7	.398	8.7

NOTE: Alpha Centauri and Sirius are double stars. Two degrees from Alpha Centauri is an 11th-magnitude star called Proxima, meaning ''nearest.'' It was so named when it was believed to be the nearest star of all, but its distance is the same as Alpha Centauri's and it is considered a part of that star's system.

FIVE BRIGHTEST APPEARING STARS

Star	Apparent Magnitude	Distance, Light-Years	Absolute Magnitude	Luminosity (Sun − 1)
1. Sirius	−1.42, +8.7	8.7	+1.4, +11.5	20
2. Canopus	−0.72	98	−3.1	1320
3. Alpha Centauri	−0.01, +1.4	4.3	+4.4, +5.8	1.25
4. Arcturus	−0.06	36	−0.3	100
5. Vega	+0.04	26	+0.5	48

FIVE MOST LUMINOUS STARS IN OUR GALAXY

Star	Apparent Magnitude	Distance, Light-Years	Absolute Magnitude	Luminosity (Sun − 1)
1. Rigel	0.3	540	−5.8	21,000
2. Deneb	1.3	400	−4.2	4,800
3. Betelgeuse	0.9 *	300	−3.9 *	3,600 *
4. Canopus	−0.9	100	−3.2	1,900
5. Antares	1.2	250	−3.2	1,900

*Varies by about one magnitude.

NOTE: Since the distances of these stars are not accurately known, the absolute magnitudes and luminosities are somewhat uncertain. These stars are in our galaxy. Supernovae in other galaxies have reached for a short time an absolute magnitude of −14. Except for these exploding stars, the most luminous star known is S Doradus. It is located in the Large Magellanic Cloud at a distance of about 150,000 light-years. Its estimated absolute magnitude is about −9.5, which would make it 500,000 times brighter than the sun. There is evidence that this star is double. Even if this is so, each component is far more luminous than any other known star.

TWO LEAST LUMINOUS STARS

Star	Apparent Magnitude	Distance, Light-Years	Absolute Magnitude	Luminosity (Sun −1)
1. Companion of BD + 4° 4048	18	19	19.2	1/500,000
2. Ross 614B	14.8	13	16.8	1/70,000

SEVEN LARGEST STARS
Diameter (Sun = 1)

1. Epsilon Aurigae	2,700
2. VV Cephei	1,200
3. Alpha Herculis	800
4. Mira	500
5. Betelgeuse	400
6. Antares	300
7. Zeta Aurigae	300

NOTE: Because of the uncertainties in the distances of these stars, these figures are not exact, but are of the right order of magnitude.

SMALLEST STARS
Diameter

LP 357–186 ....less than 1700 miles
Neutron Stars ....12 miles for neutron star of one solar mass

EXTREMES OF MASSES OF STARS
Mass (Sun = 1)

HD 698 113
Ross 614B 1/12

EXTREMES OF DENSITIES OF STARS
Density (Water = 1)

Epsilon Aurigae..10^{-8}
LP 357–186 ....10^{8}
Neutron Stars...................10^{15} for neutron star of one solar mass

EXTREMES OF TEMPERATURES

The hottest stars belong to Class O. An example is Zeta Puppis, which has a temperature of more than 30,000° C. They radiate so much ultraviolet energy that they appear relatively faint in the visual region of the spectrum. The central stars of planetary nebulae are of this type, and some are estimated to have temperatures of more than 50,000° C.

The lowest temperature is about 1,300° C., which is reached by a long-period variable star, Chi Cygni, at minimum brightness.

STARS OF GREATEST RADIAL VELOCITY

	km/sec	mi/sec
1. CD —29° 2277	547	340
2. HD 161817	360	224
3. Be B 1366	339	211
4. Helsingfors 956	325	202
5. O As 14320	307	191

STARS OF GREATEST PROPER MOTION
(Seconds Per Year)

	Proper Motion	Distance, Light-Years
1. Barnard's Star	10″.3	6.0
2. Kapteyn's Star	8 .8	12.6
3. Groombridge 18.	7 .0	25.5
4. Lacaille 9352	6 .9	12.0
5. Cordoba 32416	6 .1	14.4

STARS OF GREATEST SPACE VELOCITY

	km/sec	mi/sec
1. O As 14320	660	410
2. CD —29° 2277	612	380
3. Be B 1366	425	264
4. HD 161817	*	*
5. Helsingfors 956	359	223
6. Groombridge 1830	348	216

*Since this star's distance has not been determined, its space velocity is not known exactly, but is at least 360 km/sec, or 224 mi/sec.

PERIODS OF VARIABLE STARS

Shortest—DQ Herculis	1.18 min
Longest—Harvard Variable 10446	1,380 days

PERIODS OF ECLIPSING BINARIES

Shortest—Nova DQ Herculis	4 hr 39 min
Longest—Epsilon Aurigae	27 years

(CPD—60° 3278 was reported in 1951 as having an eclipse lasting more than 17 years and a period probably of the order of 200 years.)

PERIODS OF SPECTROSCOPIC BINARIES

Shortest—Gamma Ursae Minoris	2 hr 38 min

(The spectroscopic line shifts, which indicate a radial velocity variation, may be due to a pulsation of a single star.)

Longest—Epsilon Hydrae	15.3 years

PERIODS OF VISUAL BINARIES

Shortest—BD—8° 4352	1.7 years

(Capella, having a period of 104 days, has been resolved into two stars with the interferometer. It is also a spectroscopic binary.)

Longest—Castor	380 years

(Greater periods must exist for binaries that have not been observed long enough.)

THREE BRIGHTEST GALACTIC SUPER-NOVAE

Constellation	Year	Maximum Magnitude
1. Taurus (Center of Crab Nebula)	1054	—4
2. Cassiopeia (Seen by Tycho)	1572	—4
3. Ophiuchus (Seen by Kepler)	1604	—2

THREE BRIGHTEST GALACTIC NOVAE

Constellation	Year	Maximum Magnitude
1. Aquila	1918	—1
2. Perseus	1901	0
3. Puppis	1942	0.4

GLOBULAR CLUSTERS

	Distance, Light-Years
Nearest—Omega Centauri	20,000
47 Tucanae	20,000
Farthest—NGC 2419	185,000

NEAREST GALAXIES

	Distance, Light-Years
1. Large Magellanic Cloud	160,000
2. Small Magellanic Cloud	180,000
3. Ursa Minor System	220,000
4. Sculptor System	270,000
5. Draco System	330,000
6. Fornax System	600,000
7. Leo II System	750,000
8. Leo I System	900,000
9. NGC 6822	1,500,000
10. NGC 147	1,900,000
11. NGC 185	1,900,000
12. NGC 205	2,200,000
13. M 32	2,200,000
14. IC 1613	2,200,000
15. M 31 (Andromeda)	2,200,000
16. M 33	2,300,000

When we calculate how bright the stars would appear if they were all placed at the standard distance of 10 parsecs, we find great differences in their luminosities. The star having the brightest absolute magnitude (−9.5) is located in the Large Magellanic Cloud and is called S Doradus. The faintest known star has an absolute magnitude of +19.2. This difference of 28.7 magnitudes represents a difference in real brightness of over 300 billion times.

A glance at the two brightest stars in the constellation of Orion reveals that they are of different colors. If you are facing south, and Orion is near the meridian, Betelgeuse in the upper left-hand corner is red, while Rigel in the lower right-hand corner is blue. This is because the stars have different temperatures. When a piece of iron is heated, its color changes from red to yellow, and finally to what we call white hot. Its temperature can be measured by an examination of the light emitted. In a similar way the temperature of the stars may be found. A camera is useful in distinguishing the colors of stars, because the ordinary photographic plate is very sensitive to blue light and not very sensitive to red light, as compared with the eye. A photograph of Orion shows Betelgeuse somewhat fainter than Rigel, whereas they appear about the same brightness to the eye.

Thus we see what a great variety there is among the stars. They are all suns having masses comparable to that of our sun, but differing enormously in luminosity, temperature, size, and density. In each of these respects, our sun is somewhere about average. Of all of these stellar properties, the two that are the most directly observable for the nearer stars are the absolute magnitude and the temperature, or spectral type. When these two properties of each of the nearer stars are plotted, one against the other, as in the illustration on page 207, a very remarkable diagram results. The points representing the majority of the stars fall on a diagonal band, known as the *main sequence*. It shows that hot blue stars on the main sequence are also very luminous, and that as one goes to cooler and cooler stars the luminosity progressively decreases. There are, however, some stars that do not lie on the main sequence. One group of red stars lies well above the main sequence, with much greater luminosities than main sequence red stars of the same surface temperature. They have been dubbed, very appropriately, *red giant* stars, because their greater luminosity must be a consequence of much greater size. A second group lies in the lower left corner of the diagram, with very high temperatures but very low luminosity. These are the white dwarf stars that were discussed above. Thus, despite the wide variety found among the stars, there *are* these regularities, revealed in the illustration on page 207, that will have to be explained in any valid theory of the structure and evolution of stars. The successes and deficiencies of modern theories will be taken up in a later chapter.

To understand how these different colors originate, it is necessary to look at the structure of the atom. Our mental picture of an atom is merely schematic. It accounts for the behavior of atoms.

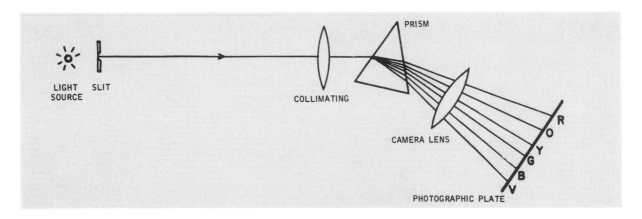

A diagrammatic representation of a spectograph, for photographing spectra. If the photographic plate and camera lens are replaced by an eye lens and telescope objective, the instrument becomes a spectroscope.

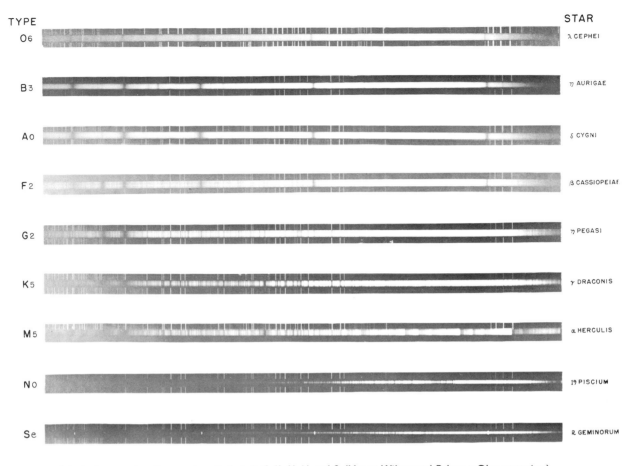

TYPE		STAR
O6		λ CEPHEI
B3		η AURIGAE
A0		δ CYGNI
F2		β CASSIOPEIAE
G2		η PEGASI
K5		γ DRACONIS
M5		α HERCULIS
N0		19 PISCIUM
Se		R GEMINORUM

Principal types of stellar spectra: *O, B, A, F, G, K, M, N,* and *S.* (Mount Wilson and Palomar Observatories.)

All that we can know for certain is that they are built up from smaller particles, which may have positive or negative electrical charges or even no charge at all. The same sorts of particles go into the formation of all the elements, but their number and arrangement differ from one element to another. Most of the material (mass) is concentrated in the central part of the atom and is called the nucleus. In our mental picture tiny particles with negative charges revolve around the positively charged nucleus somewhat as the planets of the solar system move around the sun. These particles are called electrons. The total positive and negative charges cancel out each other in the normal condition of an atom. We say such an atom is neutral. If it loses one or more of its orbital electrons it is ionized. Neutral hydrogen has one orbital electron, helium has two, and so on to uranium, which has 92.

Under many sorts of conditions an atom radiates energy into the space around it. We receive this radiation as light, as radio waves, or as similar manifestations. Our instruments interpret this radiation as waves that travel through space with a speed of about 300,000 kilometers (186,000 miles) per second. Perhaps radiation actually does travel as waves; but whether it does or not, the wave concept gives us a convenient and useful picture of what does happen. We speak of the number of waves received per second as the frequency. The ratio between the frequency and the speed is the wave length.

An atom that is not too closely crowded by neighbors does not give out all frequencies of radiation but only a certain set of them. No other element has the same set. This set is called the emission spectrum of the element. The spectrum changes with temperature of the source, with its pressure at the source, and with any electrical or magnetic fields that may envelop it at that time.

If a very large number of atoms are too closely crowded—in other words, if the pressure on a gas

205

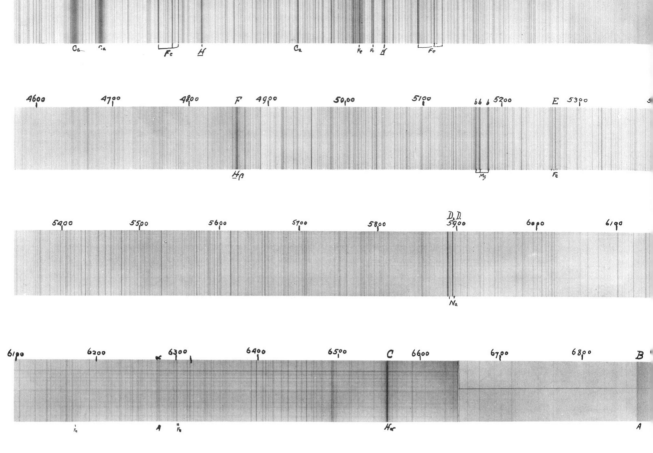

Solar spectrum, photographed with 13-foot spectrograph. (Mount Wilson and Palomar Observatories.)

is too high—they interfere with each other's radiation and they cannot give their characteristic frequencies. There will also be numerous ionizations, followed by recombinations of ions with electrons. As a result, we get all frequencies from the assembly. Physicists call such radiation a *continuum*. Increase in temperature causes the emitted light to appear bluer to us.

If such a continuum is passed through a gas, the gas absorbs from the light the same frequencies that the gas itself emits when incandescent. Indeed, a gas may at the same time absorb and emit these frequencies. A continuum from which the frequencies peculiar to one or more elements have been subtracted is called an *absorption* spectrum.

When light passes from one transparent substance into another of a different density, it generally changes direction. This refraction enables us to use suitably curved pieces of glass to form images in our cameras and telescopes. A complica-

tion arises from the fact that the various frequencies change direction by different amounts. As a result, a lens made from a single piece of glass will produce an image spoiled by a colored border. Most of the trouble can be eliminated by using a lens made from two or more different kinds of glass.

This difficulty, however, proves to be a blessing in disguise. If we pass sunlight through a glass prism we get a band of colored light. The path of the violet in the "white" light has been changed most in direction, the blue next, then green, yellow, and orange, with red least. We can combine the prism with a very narrow slit through which the light must pass before it reaches the prism. To this we can add a system of lenses to make an image of the slit and either a screen or an eyepiece to observe the image. The prism does not spoil the image made by the lenses but it does send the different frequencies to different parts of the screen.

206

We therefore get an image of the slit for each of the frequencies that the gas has emitted. We see these images as narrow, bright, colored lines on the screen. The name spectroscope has been given to the combination of slit, prism, lenses, and eyepiece or screen. If the screen is a photographic plate we call the instrument a spectrograph. The accompanying diagram shows the essential parts of a simple spectrograph. By means of this instrument a nebula in a distant galaxy tells us of what elements it is composed and gives us other important information.

A star consists of extremely hot gases. In its central part the pressure is high enough for the gases to emit all frequencies (a continuum). Surrounding this central part is a sphere of thinner gases through which the continuum must pass on its way to us. These gases absorb the frequencies that are natural to the gases. As a result, the images of such frequencies are missing from the continuum, and we see the spectrum as crossed by dark lines. The photograph on page 206 shows the spectrum of our sun, a rather typical star.

If a source is moving away from us the frequencies appear to be less than normal—in other words, the lines of the spectrum are shifted toward the red. For ordinary velocities the shift is small. If the source is approaching us, the lines are shifted toward the violet. This gives us the opportunity to determine how fast any object in space is moving toward or away from us, as long as we have enough light to examine the spectrum. The shift is called the Doppler effect and is used continually in the study of even distant galaxies.

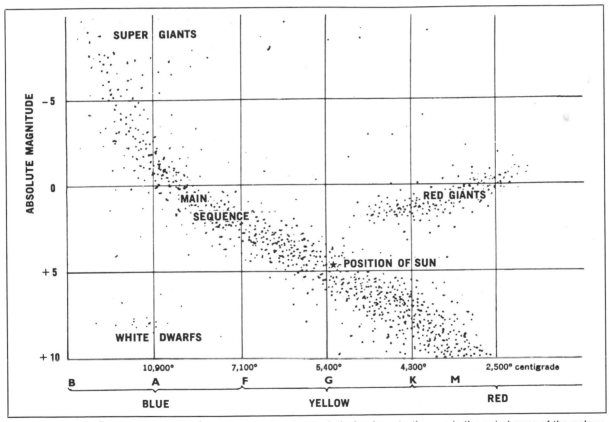

The absolute-magnitude-versus-temperature diagram for stars relatively close to the sun in the spiral arms of the galaxy.

This is known as the Hertzsprung-Russell diagram. The letters in the horizontal scale refer to the kind of spectrum the star has and are arranged in order of decreasing temperature. The numbers on the vertical scale refer to the absolute magnitude, the smaller numbers indicating the more luminous stars. Stars with surface temperatures near 10,900° C are in spectral class A, and those near 7,100° C in class F. The sun is a G star, and the coolest stars, with surface temperature below 3,500° C, are in class M. In the lower left corner are more than two hundred peculiar stars called white dwarfs. These unbelievably dense stars are hotter than the sun, but all of them give much less light than the sun does.

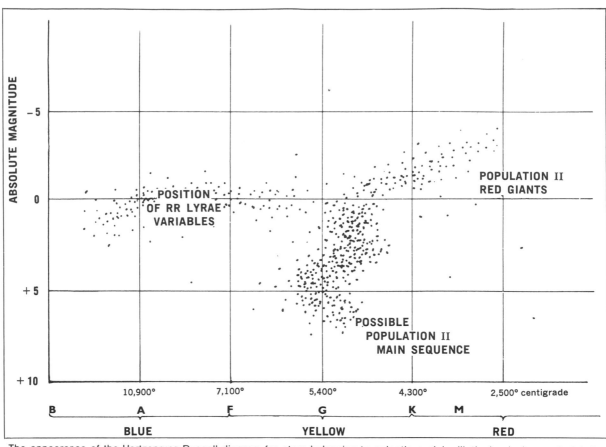

The appearance of the Hertzsprung-Russell diagram for stars belonging to galactic nuclei, elliptical galaxies, and globular clusters (stars of population II).

Diagram showing schematically the Hertzsprung-Russell arrays for both stellar populations. Also indicated on the diagram are the positions of two kinds of variable stars.

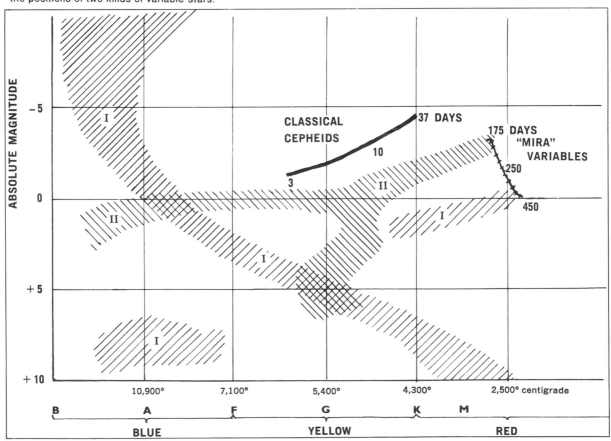

The number and arrangement of lines in a spectrum vary with the color of the star. The pattern of lines in the spectrum of the blue star Rigel is much simpler than that in the spectrum of the red star Betelgeuse. When the spectra of the stars are arranged in order of increasing redness of the stars themselves, their patterns of lines form, except for slight differences in detail, a single continuous series. Going from one end of the series to the other, we pass through a continuous range of nine stellar temperatures. The principal classes of spectra are designated by the letters O, B, A, F, G, K, M, N, and S. Rigel belongs to Class B and has a temperature of about 15,000° Centigrade. Betelgeuse belongs to Class M and has a temperature of about 3,000°. Our sun is a G star with a temperature of about 6,000° Centigrade (10,000° Fahrenheit).

The hotter a star is, the more light it gives per unit area. When the temperature of a star has been determined from its color and its spectrum, we can calculate how much radiation is coming from each square mile of its surface. We also can find the star's total radiation, from its apparent brightness and its distance. If we divide the amount of radiation per square mile into the total radiation of the star, we find how many square miles of surface the star has. Thus the size of a star is determined.

The results of such calculations are sensational. They show that stars range in diameter from about two thousand miles to more than two billion miles. If the sun were placed at the center of Betelgeuse, the earth could revolve at its present distance of 93 million miles from the sun and still remain inside the surface of Betelgeuse.

If we imagine all the stars placed in a row in order of size, we find that to a large extent they also are arranged in order of color. The largest are red and comparatively cool. They give out a small amount of radiation per unit area, and so they need a large surface to work off their heat. As we pass to somewhat smaller stars we find the color becoming less red. We finally come to a blue star, which may have only about a thousandth part as much surface as a red one, but its luminosity may be about the same, because it is giving out one thousand times as much light from each square mile of its surface. So we have a group of stars of high luminosity, including large red ones and

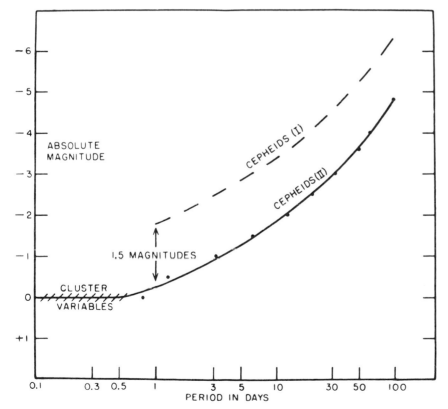

Period-luminosity relations for cluster variables and the two types of Cepheids.

209

smaller blue ones, which are called "giant" stars.

As we leave the hot blue stars and go to smaller stars, we find them cooler and again redder. They have not only smaller surfaces, but also give out less light from each square mile. Therefore they are fainter stars than the giants and are called "dwarfs." About halfway down this sequence we come to our sun, a yellow star. Its luminosity is about 1/10,000 that of a red giant, but it is about 10,000 times that of a red dwarf. Thus there is a range of about 100,000,000 times in brightness between a red giant and a red dwarf. With the temperatures of the two about the same, the difference in luminosity is due to the surface area of the giant being about 100,000,000 times greater. That would make its diameter about 10,000 times that of the dwarf.

Astronomers have discovered a few stars that are even smaller than the red dwarfs, approaching the earth in size. They are called "white dwarfs," because they are mostly white in color, showing spectra that usually correspond to temperatures of

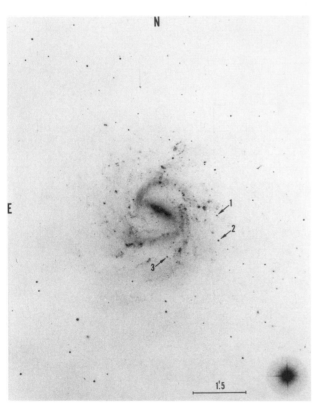

A photograph of the spiral galaxy NGC 1073 taken by Alan Sandage. The locations of three quasars discovered by Halton Arp and Jack Sulentic are indicated. According to Arp and Sulentic, it is statistically very likely that these three quasars are intimately associated with the galaxy, and are not very remote background quasars that just happen to be in the same direction. This, and other examples, cast some doubt on the widely accepted view that quasars are the most distant objects in the universe.

10,000° Centigrade or more. They are noted for their very high densities. A white dwarf has a large amount of material in a small volume of space. A well-known example is the companion of Sirius. It forms with Sirius a binary system, or a double star. The two stars revolve around each other somewhat as the earth revolves around the sun, but there is a difference. The sun's mass is over 300,000 times the earth's so that the sun is hardly disturbed by the small gravitational pull of the earth. But the two components of a double star are much more nearly equal in mass. They revolve about some point between them. If that point is halfway between them, their masses are equal. The position of the point can be determined by observing the motions of the stars about each other, and hence the ratio of their masses can be found. If the dimensions of the system can be measured, the *actual* masses can be determined. Thus the stars are weighed by noticing how much each star pulls on the other. The process is much like weighing an object on the earth, which means that we measure the gravitational pull between it and the earth.

The results of weighing the stars are interesting. Taken as a whole, the stars show only a small range in mass. The great majority have masses between one-fifth and five times the sun's mass. This means that there is a great diversity in the densities of the stars. The density tells us how closely the substance of a star is packed, and is found by dividing the mass of the star by its volume. The mass of the companion of Sirius is about the same as that of the sun, but its volume is perhaps only 1/30,000 as great. Since the sun's density is about $1\frac{1}{2}$ times that of water, the density of the companion of Sirius is nearly 50,000 times that of water. A pint of water weighs a pound. A pint of the material of this white dwarf star would weigh almost 25 tons. W. J. Luyten has recently discovered on plates taken by the 48-inch Palomar telescope a remarkable white dwarf which probably is less than 1,700 miles in diameter. The mean density of the material of this object is probably at least 1,600 tons per cubic inch.

Such a statement sounds incredible until we look again at the nature of the atoms of which matter is composed. We have seen that the atom is like a miniature solar system, consisting of a central nucleus around which revolve electrons. The planets and the sun are very small compared with the space occupied by the whole solar system. Similarly the electrons and the nucleus of an atom are exceedingly small in comparison with the space included within the orbits of the electrons. It is assumed that atoms do not come close enough together so that the orbits of their electrons overlap. However, at the center of a white dwarf star it is believed that the temperature is so high that the atoms are broken up into their constituent nuclei and electrons, which may be packed closely together by the intense pressure of all the rest of the star. We may compare an atom to an empty box, consisting of a small amount of material occupying a large volume of space. If the box is knocked to pieces, its material may be packed in a much smaller volume of space. And so it is with the atoms in a white dwarf.

At the other extreme of density are the red giant stars, which in some cases have only about 1/1,000 the density of ordinary air. Such a star has been described as a red-hot vacuum, since in the physical laboratory a volume of gas at this density would be regarded as a vacuum.

For a while, after the diagram shown on page 207 had been established, astronomers believed that it represented the characteristics of all the stars in the galaxy. However, during World War II, Dr. Walter Baade of the Mount Wilson Observatory investigated the colors of stars in the Andromeda Galaxy and showed that not all stars fit the pattern.

In 1923 the outer spiral arms of the Andromeda Galaxy were resolved into individual stars with the 100-inch telescope of the Mount Wilson Observatory. However, there was no resolution of the galaxy's central part, which appeared simply as a uniform cloudy mass on even the best photographs. These had been taken on blue-sensitive plates, which were the only kind then available.

During World War II the observing conditions on Mount Wilson were improved by the dimout in the nearby cities, whose lights produce a glare in the sky. Also, a new type of red-sensitive plate had been developed. When Dr. Baade used this, his photographs showed that the central part of the Andromeda Galaxy could be resolved into stars. The reason that the nucleus could not be resolved before was that its brightest stars are red, whereas the brightest stars in the spiral arms are blue.

CHAPTER **41** HOW TO RECOGNIZE THE STARS

The first step in getting acquainted with the night sky is to find the North Star with the aid of the Big Dipper. The North Star is also known as Polaris, the Pole Star, because it appears almost directly over the North Pole of the earth. If the earth's axis were extended out into space, it would pierce the imaginary sphere of the sky in a point called the north celestial pole. This is one degree away from the direction of Polaris. As the earth turns on its axis, Polaris describes about the pole a circle with a radius of one degree. This is such a small circle that the star always appears in about the same position with respect to the horizon.

Since the direction of north is toward the North Pole of the earth, the north celestial pole is directly above the north point of the horizon. Its angular distance above the horizon (known as altitude) is equal to the latitude. At Chicago, which is 42° north of the equator, the north celestial pole is 42° high. Therefore, the direction of the North Star is always within one degree of true north and its altitude is always within one degree of the observer's latitude.

To find the North Star, draw a line through the two stars forming the side of the bowl farthest from the handle of the Dipper. These two stars are named Dubhe and Merak and are called the Pointers. They are 5° apart, and the North Star is 29° from Dubhe. The North Star is at the end of the handle of the Little Dipper, a part of Ursa Minor, the Little Bear. Four of the stars in the Little Dipper are so faint that the group is hard to find in our large cities, whose many lights make the sky too bright to see the fainter stars.

The turning of the earth on its axis causes the Big Dipper to revolve around the pole in a counterclockwise direction. If the earth turned exactly once in a day, the Dipper would be in the same position at a certain hour every night. Observation shows that this is not so, and a little explanation is necessary. While the earth is turning once on its axis, it also moves about 1/365 of its way around

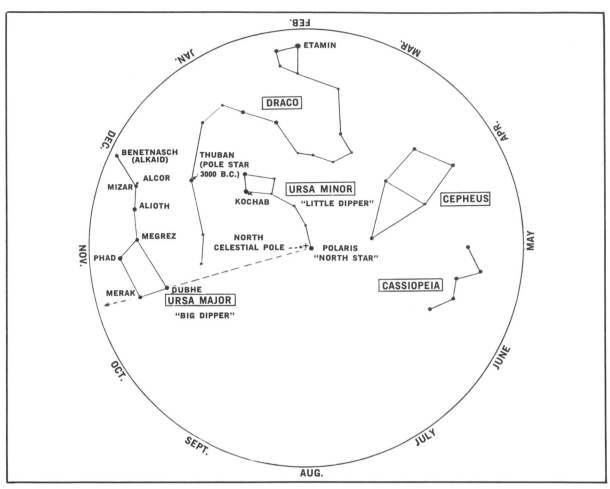

Star map.
When held toward the north with the correct month at the bottom, this map shows the northern stars in the evening.

the sun. Since there are 360° in a circle, a line from the earth to the sun sweeps through an angle of about one degree in a day. This changes the sun's direction in space as observed from the earth. If the stars, whose directions are practically fixed, could be seen in the daytime, the sun would appear to move eastward with respect to them about one degree per day. Thus the earth has to turn through about 361° to complete a rotation with respect to the sun. Such an interval is called a solar day and is about four minutes longer than the true period of the earth's rotation, which is known as a sidereal day.

If the two Pointers, Merak and Dubhe, are on a horizontal line to the left of the pole at 2130 (9:30 P.M.) Standard Time on July 15, they will reach that position four minutes earlier the next night. They make a complete revolution around the

pole in 23 hours and 56 minutes of solar time. Thus if one looks at the sky each night at the same hour, he will see that the Dipper slowly shifts from night to night in a counterclockwise direction. During 24 hours a line from the Pointers to the pole sweeps through about 361°. At the end of a month the Pointers at a certain hour will be 30° beyond their former position at that hour. After six months they will be 180° away from it. After a year of 365 days they will be back at their first position, having completed 366 trips around the pole.

The accompanying circular chart shows the principal stars within about 40° of the north celestial pole. In its present position it shows the appearance of the northern sky at 2030 (8:30 P.M.) Standard Time on August 15. One hour must be added to Standard Time to get Daylight Saving Time. To see how the northern sky will look at the

212

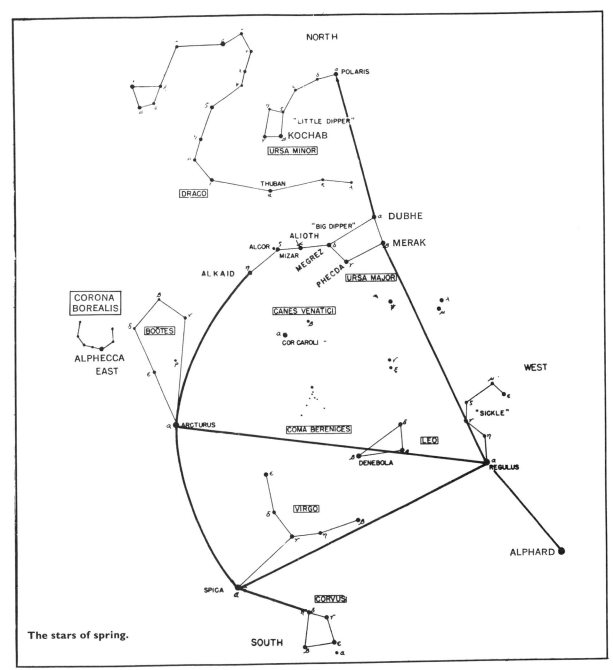

The stars of spring.

same hour in some other month, turn the map around until the correct month is at the bottom. For instance, in November the Big Dipper will be directly below the pole in the evening, and in May it will be above the pole.

On the opposite side of the pole from the Big Dipper is Cassiopeia, in the shape of a W. When the Big Dipper is below the pole, Cassiopeia is above it. If the line from the Pointers, Merak and Dubhe, is extended beyond Polaris, it leads to the

northern star of Cepheus. This constellation looks like a house with a high, steep roof. Winding around between the two Dippers is Draco, the Dragon. The star Thuban in Draco was the pole star around 3,000 B.C. It is about halfway between the middle of the handle of the Big Dipper and the bowl of the Little Dipper. The brightest star in the head of Draco is Etamin, and the brightest one in the bowl of the Little Dipper is Kochab.

We now turn to the accompanying map, which

213

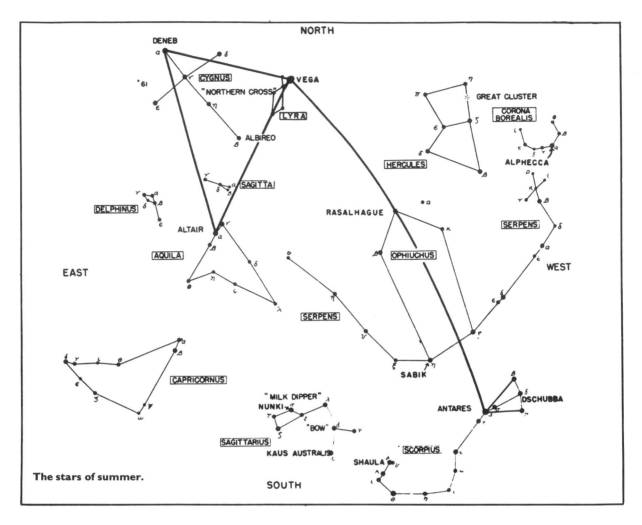

The stars of summer.

shows how to find the stars that are prominent south of the Big Dipper on a spring evening, such as May 15, at about 2030 (8:30 P.M.). At this time the Big Dipper is above the pole as we face the north. In fact, it is nearly overhead. This becomes more apparent if the map is held above the head and viewed from below. Viewing the map from this position also shows why the directions of east and west on a map of the sky are the reverse of their positions on a map of the earth. This is because the sky and the earth are facing each other. Of course, it is understood that north is always at the top of either map.

The star at the end of the handle of the Big Dipper is Alkaid. It is also called Benetnasch. Imagine a curve passing through the handle and prolonged about as far as the total length of the Dipper. There it will meet Arcturus, a very bright, orange star. Continue the curve to the south about an equal distance and it comes to Spica. This curve

from Alkaid to Spica is shown as a heavy line on the map. Arcturus belongs to Bootes, which looks like an ice cream cone. Just to the east of Bootes is Corona Borealis, the Northern Crown. It is a semicircle of faint stars that looks like a crown. The brightest gem in the crown has the name Gemma, but is usually called Alphecca.

Spica is the brightest star in Virgo, the Virgin whose stars form a Y. Below and to the right of Spica is Corvus, the Crow. It resembles a ship's spanker sail, and so it is often called Spica's Spanker. The two stars at the top of the sail point to Spica. Directly south from Spica's Spanker is the Southern Cross, but it is always below the horizon for observers in the United States.

There are at least two good ways of finding Regulus, the brightest star in Leo, the Lion. It is to the west of Arcturus and Spica, and forms a large triangle with them. The angle at Spica is nearly a right angle. Also a line drawn southward from the

214

two Pointers of the Big Dipper leads to Leo. If the line is swung slightly to the west, it passes through Regulus. To the southwest of Regulus is Alphard, the brightest star in Hydra. Leo consists of a small right triangle and a sickle. The star at the left corner of this triangle is Denebola. In the old mythological figure of the Lion, it marked the Lion's tail. That is the origin of the name, "deneb" meaning "tail" in Arabic. The sickle opens to the west and has Regulus at the bottom of its handle.

A connecting link between the maps of spring and summer is Corona Borealis, on the eastern edge of the spring map and on the western edge of the summer map. This constellation lies one-third of the way from Arcturus to Vega, the brightest star of the summer sky. Vega belongs to Lyra, the Lyre, consisting of a tiny equilateral triangle joined to a parallelogram.

A large right triangle is formed by Vega, Deneb, and Altair. Deneb marks the tail of Cygnus, the Swan, which is better known as the Northern Cross. This is tilted to the east, with Deneb at the top and Albireo at the bottom. Vega and Altair can also be identified by bisecting the right angles formed by the crossarms of the Northern Cross with the southern arm. Vega is the northern star of the two and is nearer the cross. Altair is in the rather faint constellation of Aquila, the Eagle, and can be identified by its two little guard stars pointing in a line to Vega.

A slightly curved line has been drawn on the map from Vega to the southwest, passing through Rasalhague and ending at Antares. Rasalhague marks the head of Ophiuchus, the Serpent-Bearer. The head of Serpens, the Serpent, is marked by an X, just south of Corona Borealis. Antares, like Altair, has a faint star on either side of it, but the three stars are not quite in a straight line. Lines have been drawn from Antares to three stars forming the head of Scorpius, the Scorpion. These suggest the suspension lines of a parachute. The star at the top of the middle line is Dschubba. The Scorpion looks more like its name than do most constellations. Its tail curls up over its back like a real scorpion. There are two stars close together at the end of the tail. The brighter one is Shaula.

East of Scorpius is Sagittarius, the Archer. All that we find of the Archer is his bow, with an arrow aimed at Scorpius. The southern star in the bow is Kaus Australis, "Kaus" meaning "bow" and "Australis" meaning "southern." The rest of the

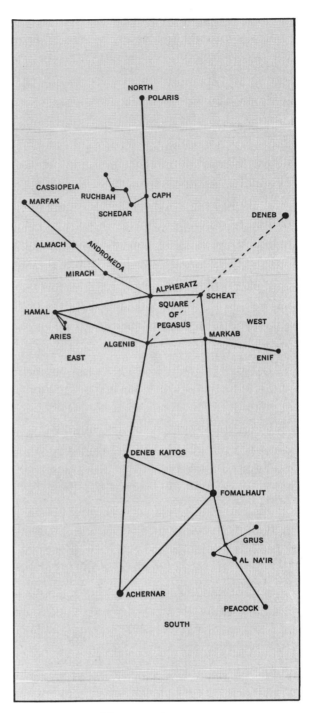

The stars of autumn.

constellation forms a dipper. It is called the Milk Dipper, because it lies in the Milky Way. The brightest and most northerly star in the bowl of this dipper is Nunki.

To the east of Sagittarius is Capricornus, the

215

Sea Goat, in the shape of an inverted cocked hat. It is a faint constellation, having no stars brighter than the third magnitude. A line to the south from Altair and its two guard stars points toward it. Also near Altair are two tiny constellations. Sagitta is just north of Altair and looks like its name, the Arrow. To the east of Altair is Delphinus, the Dolphin. Four of its stars make a diamond, and the brightest two of these point to a fifth star to the south. Delphinus is also called Job's Coffin, but the origin of this name is not known.

A line from Vega through the center star of the Northern Cross and extended about twice the distance between those two will come to the Square of Pegasus. This is the most important group of autumn stars (see map). The two eastern stars of the Square begin with a letter "A." The southern one is Algenib, ending in a "b," and the northern one is Alpheratz, ending in a "z." A line through the eastern side of the square points straight north to Polaris. The distance from Alpheratz to Polaris is about 60°. Halfway between them is Caph, at the end of the W that marks Cassiopeia. The stars in the troughs of the W are Schedar and Ruchbah. An aid to memory is to note that the consonants in the word Caesar, C, S, and R, are the first letters of the three navigational stars in Cassiopeia. To emphasize that there is an unnamed star between two named stars, one navigator suggested learning the four western stars in Cassiopeia as Caph, Schedar, "Skip One," and Ruchbah.

The western side of the Square of Pegasus is made up of Scheat at the top and Markab at the bottom. A diagonal of the Square from Algenib to Scheat points in the direction of Deneb. Using Scheat and Markab as pointers to the south, we come to Fomalhaut. This line then curves a little to the westward and passes through Al Na'ir in Grus, the Crane. This looks like a hook or a bent pin. To the south and west of this is Peacock. Al Na'ir and Peacock are not very bright, but are mentioned because they are used by navigators. Peacock is too far south to be visible from latitudes north of about 30° N.

A line through Alpheratz and Algenib points southward to Deneb Kaitos, which marks the tail of Cetus, the Sea Monster. Thus the Arabic word for tail has given us the names of three stars, Deneb, Denebola, and Deneb Kaitos. This line

can be continued to Achernar, a very bright star, but like Peacock it is too far south for us to see at our latitudes.

To the east of the Square is Hamal, which forms an isosceles triangle with Alpheratz and Algenib. Hamal has two faint stars close to it, which make a tiny triangle with it. This is the distinguishing feature of the constellation of Aries. About 2,000 years ago the vernal equinox (where the sun's apparent path crosses the celestial equator in March) was in Aries, but the slow wobbling of the earth called precession has shifted the equinox to the west by about 30°. The equinox is now nearly in line with Alpheratz and Algenib and about 15° south of the latter. Although the vernal equinox is no longer in Aries, it is still called the First Point of Aries.

Starting at Alpheratz and extending to the northeast is the constellation of Andromeda. Two stars in it, Mirach and Almach, point to Marfak, the brightest star in Perseus. This constellation is not shown on these maps, since it is rather faint and happens to lie on the dividing line between two of them.

Marfak is also marked on the winter map in the upper right corner. A curved line has been drawn from Marfak passing through Capella, Castor, Pollux, Procyon, Sirius, Rigel, and Aldebaran, and coming back to Capella again. In order to distinguish the twin stars, Castor and Pollux, it may help to remember that Castor is nearer to Capella (both beginning with a C) and that Pollux is nearer to Procyon (both beginning with a P).

The Big Dipper is the best known constellation in the northern hemisphere, and the Southern Cross in the southern hemisphere. However, the finest constellation in the whole sky is Orion. It contains more bright stars than any other. Fortunately it can be seen from all the inhabited parts of the world. It lies on the celestial equator and passes overhead at the earth's equator. Even at the poles half of Orion is visible, the northern half at the North Pole and the southern half at the South Pole.

Orion is nearly a rectangular figure. The eastern and western sides are almost parallel, but the eastern side is a little longer. The two brightest stars in it are Betelgeuse at the northeast corner and Rigel at the southwest corner. Betelgeuse has also been spelled Betelgeux. Just to the west of it is Bellatrix, which also begins with a B and ends with an X. In the center of Orion is his belt, marked by

216

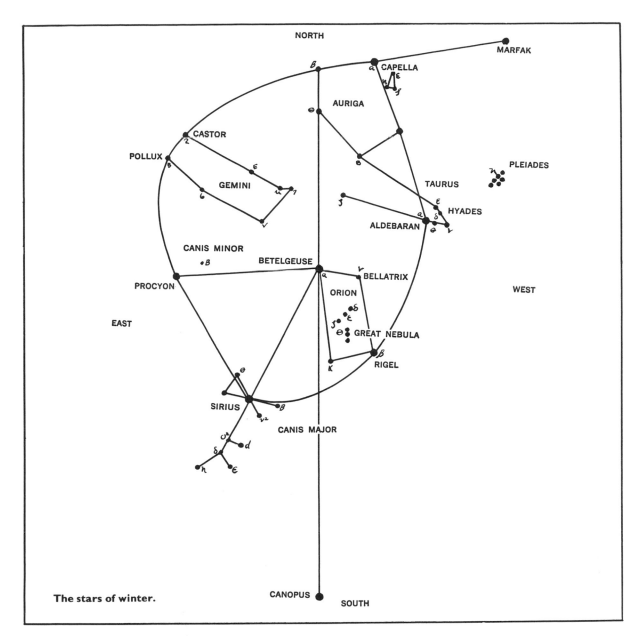

NORTH

MARFAK

CAPELLA

AURIGA

CASTOR

POLLUX

GEMINI

TAURUS

PLEIADES

HYADES

ALDEBARAN

CANIS MINOR

BETELGEUSE

BELLATRIX

PROCYON

ORION

WEST

EAST

GREAT NEBULA

RIGEL

SIRIUS

CANIS MAJOR

The stars of winter.

CANOPUS

SOUTH

three stars nearly in a straight line. The middle star is Alnilam, a navigational star, and the northern one lies almost exactly on the celestial equator. Below the belt are three fainter stars in a vertical line. They represent the sword of Orion. The middle one of these three marks the position of the Great Nebula in Orion, a tremendous mass of tenuous gas and dust particles lighted up by the stars embedded in it. Optical aid is needed to see it.

The belt of Orion points up to the northwest in the direction of Aldebaran, the brightest star in Taurus, the Bull. The head of the Bull is marked by

a star cluster in the shape of a V, called the Hyades. Aldebaran is at one end of the V. Its name means "the follower," coming from the fact that it always follows the Pleiades across the sky in their apparent westward motion each day. The Pleiades are well known as the Seven Sisters, but only six of them are usually visible to the average unaided eye.

A line from the belt of Orion points down to the southeast toward Sirius, the brightest star in the sky. It is also called the Big Dog Star, belonging to Canis Major, the Big Dog. An equilateral triangle can be formed by joining Betelgeuse, Sirius,

217

and Procyon, the last being in Canis Minor, the Little Dog.

The second brightest star in the sky is Canopus, far to the south of the eastern side of Orion. It is so far south that it gets only 3° above the horizon at the latitude of Los Angeles. The best time to see it in the evening is in February. It is not visible at all at latitudes north of 37° N.

Since the planets are constantly moving, their positions cannot be shown on a map used year after year. They are always found on or close to the ecliptic (the apparent annual path of the sun among the stars), which passes near Aldebaran,

Regulus, Spica, and Antares. Two of the planets, Venus and Jupiter, appear brighter than all the stars. Mars is red and occasionally is brighter than any star. Saturn moves so slowly that once it is located, it is easy to keep track of it.

There is a sense of satisfaction in being able to recognize a star or constellation, just as there is in calling by name a bird or flower. There is no season of the year and no place on the earth where the interest in astronomy ceases. There is no time of life when we are too old to enjoy it. It is hoped that these few suggestions will help you make friends with the stars.

CHAPTER **42** THE SKY FROM POLE TO POLE

The accompanying charts show the principal stars and constellations in the sky. Each one extends from pole to pole, with the celestial equator running through the center. Declination, which corresponds to latitude on the earth, is marked at each side at 10° intervals. It is measured from the equator northward to $+90°$ and southward to $-90°$.

Right ascension in the sky is similar to longitude on the earth. It is measured eastward along the celestial equator from the vernal equinox, which is also known as the First Point of Aries. It is expressed in hours, each hour being equal to 15°. Fractions of an hour are expressed in minutes. Rigel has a right ascension of 5 hours 12 minutes, and is one-fifth of the way between the numbers 5 and 6. Note that these numbers increase to the left, which is to the east in the sky. The declination of Rigel is $-8°$, which means that it is 8° south of the equator.

Inclined to the equator at an angle of $23\frac{1}{2}°$ is the ecliptic, which is the apparent annual path of the sun among the stars. It crosses the equator at the equinoxes. The vernal equinox has a right ascension of 0, and the autumnal equinox has one of 12 hours. The moon and the planets are always found on or close to the ecliptic, but they are not shown because their positions change.

The months at the bottom of each chart show when the stars on that chart reach their highest

positions above the horizon during the middle of the evening. At such times they are found on or near the celestial meridian, which is an imaginary circle running overhead in a north and south direction.

The stars visible from any place depend on its latitude. From the equator all the stars can be seen. From the North Pole only the stars north of the celestial equator are visible, and from the South Pole only the stars south of the celestial equator. If one goes northward 40°, the stars within 40° of the South Pole remain always below his horizon. His zenith is always 40° north of the celestial equator. Similar statements are true for other latitudes, north and south.

The accompanying list includes the principal stars and constellations of general interest, especially to the navigator. Faint groups of stars have been omitted. The pronunciations are given in parentheses and without the use of any symbols. With that limitation, an attempt has been made to follow the recommendations of a committee of the American Astronomical Society, as published in the August, 1942, issue of "Popular Astronomy."

Right ascension (R. A.) and declination (Dec.) are given for each star. The approximate declination of the middle of a constellation and the month when it is on the meridian during the evening are given for each constellation on page 224.

218

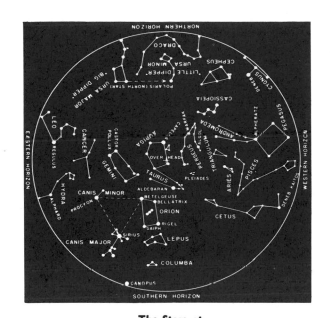

**The Stars at
11 P.M., Dec. 15;
9 P.M., Jan. 15;
7 P.M., Feb. 15.**

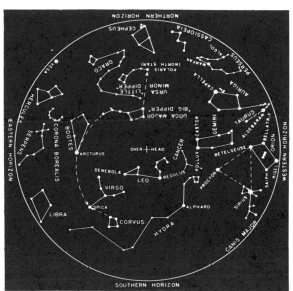

**The Stars at
11 P.M., March 15;
9 P.M., April 15;
7 P.M., May 15.**

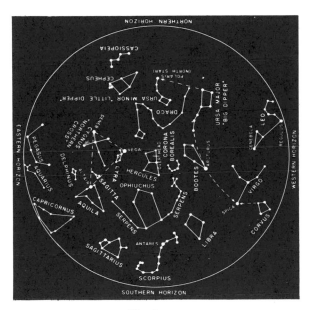

**The Stars at
11 P.M., June 15;
9 P.M., July 15;
7 P.M., Aug. 15.**

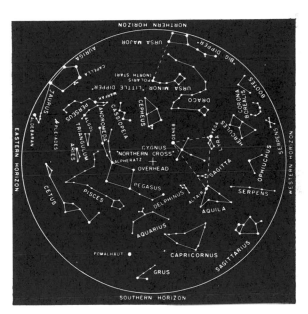

**The Stars at
11 P.M., Sept. 15;
9 P.M., Oct. 15;
7 P.M., Nov. 15.**

The night sky throughout the year.

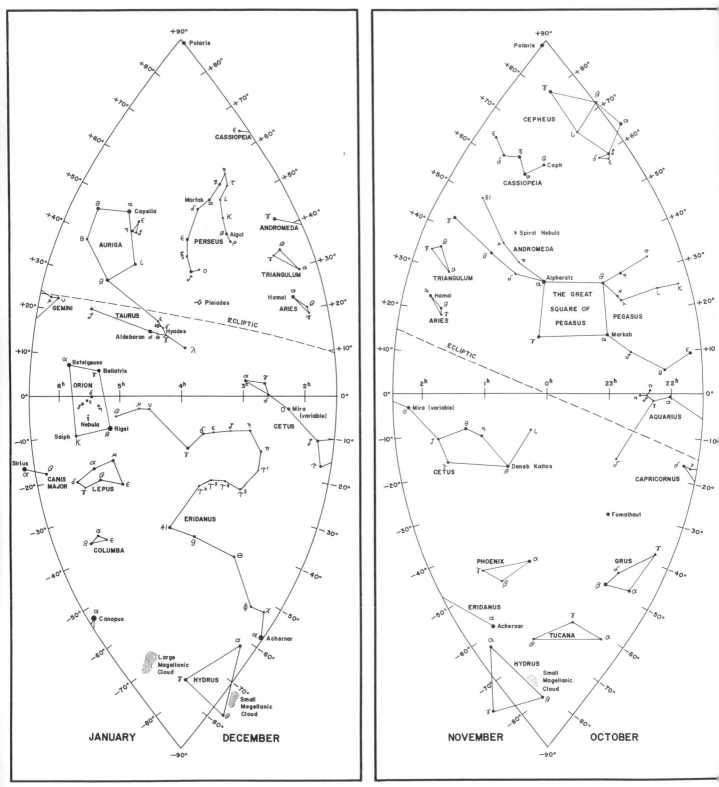

The sky from pole to pole, October to January.

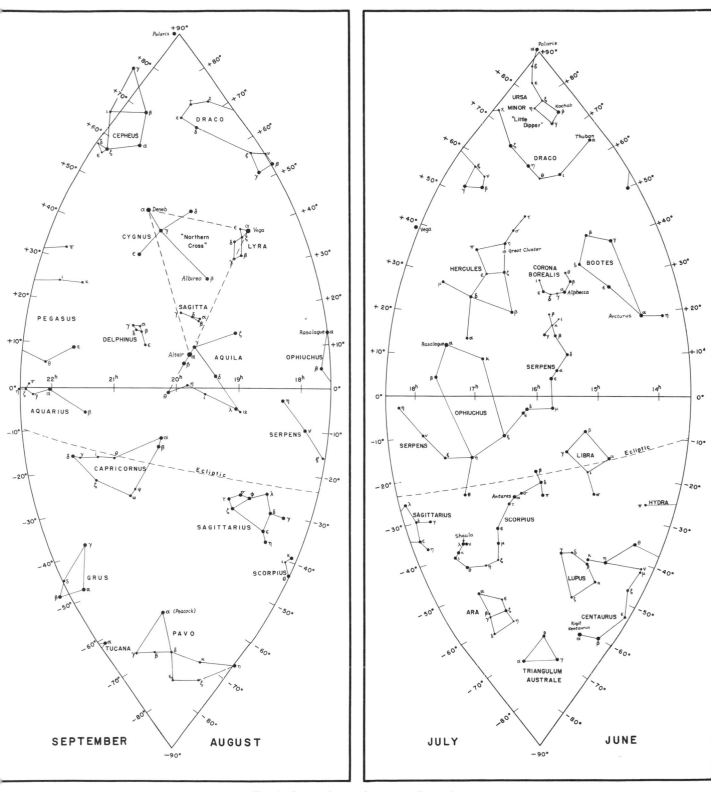

The sky from pole to pole, June to September.

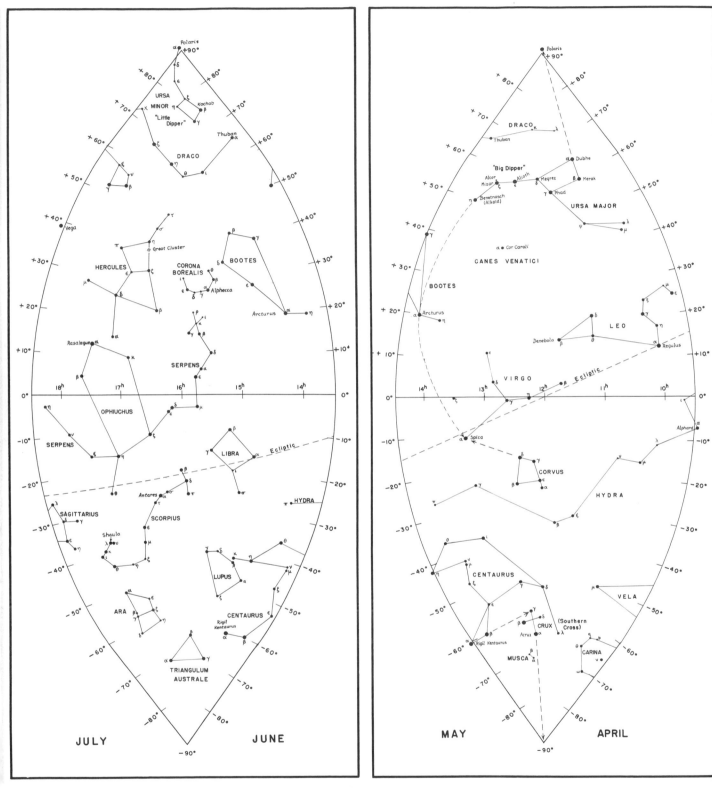

The sky from pole to pole, April to July.

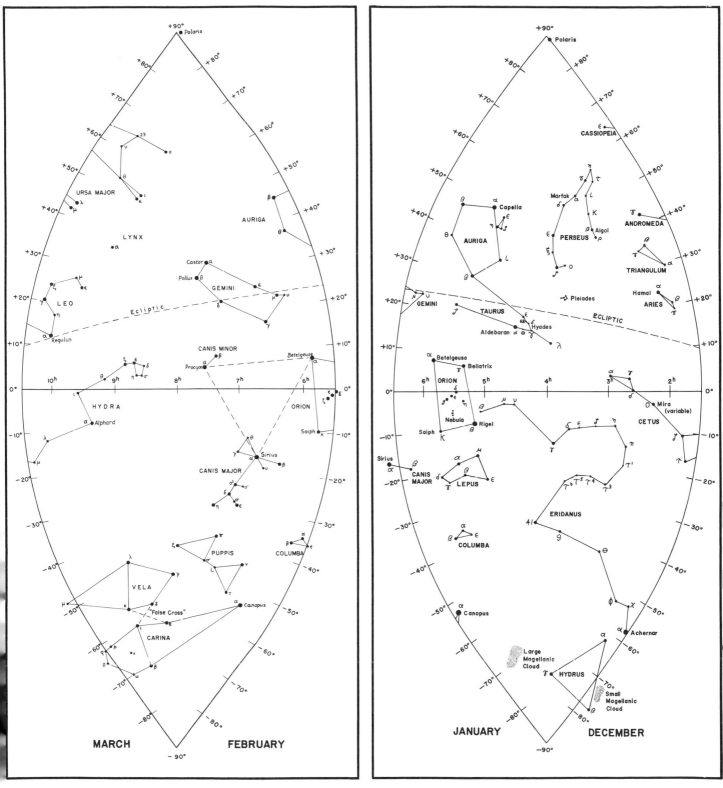

The sky from pole to pole, December to March.

223

STARS

	R. A. h m	Dec. °
Achernar (ay′ker-nar)	1:36	−58
Acrux (ay′kruks)	12:23	−63
Adhara (ad-hah′ra)	6:56	−29
Albireo (al-bir′ee-oe)	19:28	+28
Alcor (al′kor)	13:22	+55
Aldebaran (al-deb′a-ran)	4:33	+16
Algol (al′gol)	3:04	+41
Alioth (al′i-oth)	12:51	+56
Alkaid (al-kayd′)	13:45	+50
Alphard (al′fard)	9:25	− 8
Alphecca (al-fek′a)	15:32	+27
Alpheratz (al-fee′rats)	0:05	+29
Altair (al-tair′)	19:48	+ 9
Antares (an-tay′reez)	16:26	−26
Arcturus (ark-too′rus)	14:13	+19
Bellatrix (be-lay′triks)	5:22	+ 6
Benetnasch (be-net′nahsh)	13:45	+50
Betelgeuse (bet′el-jooz)	5:52	+ 7
Canopus (ka-noe′pus)	6:23	−53
Capella (ka-pell′a)	5:12	+46
Caph (kaff)	0:06	+59
Castor (kas′ter)	7:31	+32
Cor Caroli (kor kayr′o-lee)	12:53	+39
Deneb (den′eb)	20:39	+45
Deneb Kaitos (den′eb kay′tos)	0:41	−18
Denebola (de-neb′o-la)	11:46	+15
Dubhe (doob′ee)	11:00	+62
Fomalhaut (foe′mal-awt)	22:54	−30
Hamal (ham′al)	2:04	+23
Kaus Australis (kaws aws-tray′lis)	18:20	−34
Kochab (koe′kab)	14:51	+74
Marfak (mar′fak)	3:20	+50
Markab (mar′kab)	23:02	+15
Megrez (mee′grez)	12:12	+57
Merak (mee′rak)	10:58	+57
Mira (my′ra)	2:16	− 3
Mizar (my′zahr)	13:22	+55
Phad (fahd)	11:51	+54
Polaris (poe-lay′ris)	1:43	+89
Pollux (pol′uks)	7:42	+28
Procyon (proe′si-on)	7:36	+ 5
Rasalhague (ras-al-hay′gwee)	17:36	+13
Regulus (reg′yoo-lus)	10:05	+12
Rigel (rye′jel)	5:12	− 8
Rigel Kentaurus (rye′jel ken-taw′rus)	14:36	−61
Saiph (sah′if)	5:45	−10
Shaula (shou′la)	17:30	−37
Sirius (seer′ee-us)	6:43	−17
Spica (spy′ka)	13:22	−11
Thuban (thoo′ban)	14:02	+65
Vega (vee′ga)	18:35	+39

CONSTELLATIONS

	Month	Dec. °
Andromeda (an-drom′e-da)	Nov.	+35
Aquarius (a-kwair′i-us)	Oct.	− 5
Aquila (ak′wi-la)	Aug.	+ 5
Ara (ay′ra)	July	−55
Aries (ay′ri-eez)	Dec.	+20
Auriga (aw-rye′ga)	Jan.	+40
Bootes (bo-oe′teez)	June	+30
Canes Venatici (kay′neez ve-nat′i-sye)	May	+40
Canis Major (kay′nis may′jer)	Feb.	−20
Canis Minor (kay′nis my′ner)	Feb.	+ 5
Capricornus (kap-ri-kor′nus)	Sept.	−20
Carina (ka-rye′na)	Feb.	−65
Cassiopeia (kass-i-o-pee′ya)	Nov.	+60
Centaurus (sen-taw′rus)	May	−50
Cepheus (see′foos)	Oct.	+70
Cetus (see′tus)	Nov.	−10
Columba (koe-lum′ba)	Jan.	−35
Corona Borealis (koe-roe′na boe-ree-ay′lis)	June	+30
Corvus (kor′vus)	May	−20
Crux (kruks)	May	−60
Cygnus (sig′nus)	Sept.	+30
Delphinus (del-fye′nus)	Sept.	+15
Draco (dray′koe)	July	+60
Eridanus (ee-rid′a-nus)	Dec.	−30
Gemini (jem′i-nye)	Feb.	+25
Grus (grus)	Oct.	−45
Hercules (her′kue-leez)	July	+30
Hydra (hye′dra)	Apr.	−15
Hydrus (hye′drus)	Dec.	−70
Leo (lee′oe)	Apr.	+20
Lepus (lee′pus)	Jan.	−20
Libra (lye′bra)	June	−15
Lupus (loo′pus)	June	−45
Lyra (lye′ra)	Aug.	+35
Musca (mus′ka)	May	−70
Ophiuchus (off-i-oo′kus)	July	0
Orion (oe-rye′on)	Jan.	0
Pavo (pay′voe)	Aug.	−70
Pegasus (peg′a-sus)	Oct.	+20
Perseus (per′soos)	Dec.	+40
Phoenix (fee′niks)	Nov.	−45
Puppis (pup′is)	Feb.	−45
Sagitta (sa-jit′a)	Aug.	+15
Sagittarius (saj-i-tay′ri-us)	Aug.	−30
Scorpius (skor′ pi-us)	July	−30
Serpens (ser′penz)	July	0
Taurus (taw′rus)	Jan.	+20
Triangulum (try-ang′gyoo-lum)	Dec.	+30
Triangulum Australe (—aws-tray′lee)	July	−65
Tucana (too-kay′na)	Oct.	−60
Ursa Major (er′sa may′jer)	Apr.	+55
Ursa Minor (er′sa my′ner)	June	+80
Vela (vee′la)	Mar.	−50
Virgo (ver′goe)	May	0

CHAPTER 43 DOUBLE STARS

The star in the middle of the handle of the Big Dipper has played an important part in the history of double stars. Its name is Mizar, and close to it is a fainter star named Alcor, which is visible to the unaided eye, though the casual observer might well overlook it. Many centuries ago the Arabs are said to have regarded Alcor as a test of good eyesight. They referred to these two stars as "the Horse and the Rider." In 1650, about 40 years after Galileo first turned a telescope to the sky, G. B. Riccioli, another Italian astronomer, looked at Mizar with his telescope. He was surprised to find that Mizar itself appeared as two stars. It was the first double star to be discovered.

Mizar was also the first double star to be observed photographically. In 1857, G. P. Bond at the Harvard College Observatory obtained measurable images of its two components. Then in 1889, E. C. Pickering found by means of the spectroscope that the brighter member of the pair consists of two stars revolving around each other. This was the first so-called "spectroscopic binary" to be discovered. In 1908, Edwin B. Frost at the Yerkes Observatory found that the fainter component of Mizar is also a double star, or binary. At the same time he discovered that Alcor is a pair of stars, too. Thus there are six stars at the middle of the handle of the Big Dipper.

After Mizar was seen to be double in the telescope, other stars, which appeared single to the naked eye, proved to be double when viewed with optical aid. However, their significance was not understood at the time, and little attention was given them until William Herschel made a systematic search for them near the end of the eighteenth century. He looked for pairs of stars that differed considerably in brightness. It was then believed that two stars appearing close together were really far separated in space and just happened to be viewed in nearly the same direction from the earth. Presumably the brighter star of such a pair was nearer. Herschel hoped to detect a slight shift of the brighter star with respect to the fainter one, due to the revolution of the earth around the sun. If one holds a pencil at arm's length and looks at it with one eye and then with the other, he sees that the pencil shifts its position with respect to more distant objects. Similarly a nearby star should shift its position with respect to a more distant star when viewed from opposite sides of the earth's orbit at intervals of six months.

After many years of observations, Herschel was unable to find the displacements he expected, but he made a discovery of absorbing interest. He found in a number of cases that the direction of the imaginary line joining a pair of stars was slowly changing in such a way as to leave no doubt that the two stars were revolving about each other. Instead of the two stars being far apart, they were close enough to revolve about each other in accordance with the law of gravitation, which was thus shown to prevail outside the solar system.

When the 36-inch refracting telescope of the Lick Observatory has been used under the best observing conditions, stars have been found to be double when the separation between the two components was only one-tenth of a second of arc. That is about the same as the separation of the two headlights of an auto, if they could be seen from a distance of one thousand miles. Such resolving power is possible only when the two stars are of about the same brightness. Sirius, the brightest star in the sky, has a very faint companion, which has not been seen in any telescope when the separation approached its minimum value of two seconds of arc. The story of the discovery of the companion of Sirius is so interesting that it is worth repeating here.

The stars are commonly described as being fixed in the sky, because to the unaided eye they do not appear to change their positions with respect to each other, as do the planets. However, observations with a telescope show that the stars are in motion. They are moving through space with speeds expressed in miles per second, but their great distances make their angular displacements in the sky extremely small. The velocity of the planet Jupiter around the sun is less than the velocity of the star Sirius through space. Yet during one year Jupiter moves about 30 degrees in the sky, because it is so much closer to us than Sirius, which moves only 1.3 seconds of arc in a year. This proper motion of Sirius, as it is called, is one of the largest, because Sirius is one of the nearest stars. Suppose that an ant moved so slowly that it took one year to crawl across the diameter of a nickel. If the nickel were observed through a telescope from a distance of two miles, the ant would appear to move with the same speed with which Sirius appears to change its position among the other stars.

In 1834 the German astronomer F. W. Bessel found that Sirius was not moving uniformly in the

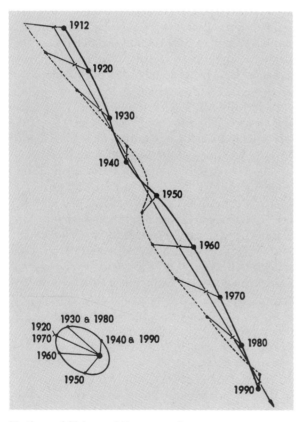

Motions of Sirius and its companion.
The oval shows the orbit in which the companion moves relative to Sirius in a period of fifty years. The long straight line is the path of the center of gravity of the system. The heavy curved line shows the absolute positions of Sirius. The broken line shows the absolute positions of the companion from 1912 to 1990.

heavens, but in a wavy line. He decided that it must be revolving around the common center of gravity between itself and an invisible companion in a period of about 50 years. In 1862, Alvan G. Clark, an American telescope-maker, was testing a new 18-inch lens. There seemed to be a defect in the lens when it was used to view Sirius, but the defect disappeared when other stars were examined. Clark then realized that he had observed the companion of Sirius, whose existence had been predicted by Bessel. Subsequent observations have shown that Sirius and its companion revolve about each other in the predicted period of 50 years.

It is probable that the majority of the 25,000 or more close, visual double stars consist of pairs of stars revolving about each other, but in most cases the orbital motion is so slow that it cannot be detected until after many years. The separation of the stars of a binary in most of those systems whose orbits are known is about equal to the distances of such planets as Jupiter, Saturn, and Uranus from the sun. In shape the orbits of visual double stars are much more eccentric than those of planets. For example, Sirius is four times as far from one end of the orbit of its companion as from the other. Also, unlike the orbits of the planets, which are nearly in the same plane, the inclination of the orbits of double stars is entirely at random. If an orbit were edgewise to the earth, the stars would seem to go back and forth in a straight line. The orbit of Sirius and its companion lies at about a half slant (43°) to the plane of the sky.

So far we have considered only the case where the two stars of a binary are far enough apart to be seen separately in a telescope. If the largest telescope could just barely separate two stars a billion miles apart, it alone would not reveal two stars only a million miles apart. Fortunately, the spectroscope tells us about such close pairs.

One component of Mizar was referred to earlier as the first spectroscopic binary to be discovered. At one time its spectrum may consist of single lines. That is when one of the two stars is nearly behind the other. Each star produces a set of lines, but they coincide exactly, because the stars are neither approaching nor receding from us. As succeeding photographs of the spectrum are obtained, each line is found to split up into two lines. One star is moving away from us, and its lines are shifted slightly toward the red end of the spectrum. The other star is approaching us, and its lines are

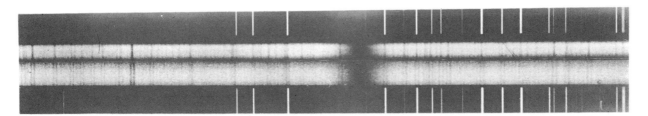

Spectra of Mizar. Appearing to the unaided eye as a single star at middle of handle of Big Dipper, Mizar actually consists of two pairs of stars. Two light bands above show dark lines of the spectra of one pair at two different times. When upper spectrum was taken, neither member of pair was approaching or receding, in relation to the other. Spectra of the two stars are superimposed and appear as one. In lower spectrum, lines are doubled. One star is approaching, and its dark lines are shifted slightly to violet end of spectrum. Other star is receding, and its dark lines are shifted slightly to red end of spectrum. Bright lines at top and bottom of photograph form a comparison spectrum and were produced by a source of light attached to telescope through which photograph was taken. (Mount Wilson Observatory.)

shifted toward the violet end. From the amount of the shift, the velocity of the star can be determined. If the orbit is circular and edgewise to the earth, the maximum velocity of approach or recession gives the true speed of the star in its orbit. When this is multiplied by the time taken for a star to make one revolution, the size of the orbit is found. In case one of the component stars is several times brighter than the other, only the lines of the brighter one can be seen.

If the plane of the binary's orbit is nearly edgewise to the earth, the stars eclipse each other. Eclipses of spectroscopic binaries are much more probable than eclipses of visual binaries. If stars are far enough apart to be seen separately, their orbits must lie almost exactly edgewise to us to permit one star to hide the other. The members of such a visual binary move so slowly about each other that the shifting of the spectral lines is not usually noticeable and the pair is not a spectroscopic binary. In a typical binary of the spectroscopic type, the two members are close to each other and move rapidly. In some cases the two stars are so close that they almost roll around upon each other's surfaces. Stars of such a pair will partially eclipse each other when seen from many directions.

The most famous eclipsing binary is Algol, which is on a line between the Pleiades and Cassiopeia. In 1669, G. Montonari noted that it changed in brightness, but the general character of its light variation was not established until 1783 by John Goodricke. This young man is one of the most remarkable characters in the history of science. He was a deaf-mute from birth, but was very proficient in his studies, especially in mathematics. By persistent observation of Algol and the use of nearby

stars as standards, he followed its changes in light and determined its period as 2 days, 20 hours, 49 minutes. At minimum the star appears one-third as bright as at maximum. Goodricke suggested that the variation was caused either by spots on the surface of the star or by eclipse by a dark companion. About a century later the latter explanation was confirmed when Algol was found to be a

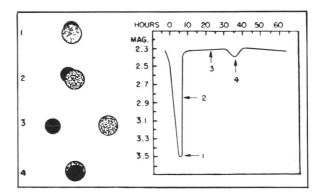

Algol and its light curve.

Although the brighter star is smaller than its companion, it emits over 90 percent of the light of the system, because it is much hotter. The light curve shows a difference of 1.2 magnitudes, which means that the system appears one-third as bright at position *1*, when the primary star is 70 percent eclipsed, as at position *3*, when no light is cut off. A secondary eclipse occurs at position *4*, when part of the light from the faint star is obstructed, but this change is too small to be detected with the eye.

spectroscopic binary, with a period equal to that of its light variation.

For nearly 2½ days Algol remains so nearly constant in brightness that only the most accurate observations can detect any change. In the following five hours it is gradually dimmed to one-third of its usual brightness. In five more hours it rises to normal again. During the ten hours of conspicuously varying light, the bright star is being partially eclipsed by the fainter one. The latter is not a dark body, as Goodricke supposed, but is a self-luminous star several times larger than the sun. This is known from the fact that half a period after the eclipse of the brighter star, there is a very small decrease in light, which can be detected only with accurate instruments. At that time, the brighter star is in front of the fainter one and cuts off some of its light.

By methods that cannot be gone into here, much can be learned about the two members of such a binary star as Algol. Here are a few facts about that system. The distance between the centers of the two stars is $6\frac{1}{2}$ million miles, less than one-fifth the radius of Mercury's orbit. The diameter of the brighter star is about three times that of the sun, and its mass is nearly five times as great. The fainter star has about the same mass as the sun, but its diameter is 3.7 times greater. The brighter of the pair is smaller in size, but it emits over 90 percent of the light of the system, because it is much hotter than its companion.

This information has all been obtained from observations of a star that appears as a single point of light even in the most powerful telescope. The star is so distant that the light we see from it has been traveling to us at the rate of 186,300 miles a second for over 100 years. As we look at the sky, we wonder not only at the marvels revealed there, but also at the ingenuity of man, which enables him to decode the message of starlight.

CHAPTER **44** **THE STARS OF OUR NEIGHBORHOOD**

The general appearance of our sky may be compared to that of the hundreds of thousands of lights of Los Angeles as viewed from the Griffith Observatory, 800 feet above Hollywood. In the foreground between Los Feliz and Hollywood Boulevards, lights can be seen in the windows of homes. A little farther away these disappear, and the ordinary street lamps are the faintest lights visible. Fifteen to twenty miles away the only lights that can be seen are gigantic special lamps of high luminosity. Because many faint lamps cannot be seen, the average lamp visible over the whole expanse of the city is fully as bright as those used to illuminate the streets. An observer measuring the distances of the lights and using these distances to correct their apparent brightnesses, thus getting their absolute luminosities, might conclude wrongly that high candle power lamps are much more common than fainter ones. The same reasoning might easily lead a sky observer to make a similar error in concluding that most stars are of high luminosity.

Here and there in the city appear dark areas. In some there are no lamps, but in others a bit of smoke or haze has obscured the light. Where a large office building is completely illuminated, a cluster of "stars" appears. Most of the lamps give a yellowish light, but red ones are also quite common, and blue and green are found scattered through this earth-bound galaxy, reminding one of the similarly colored stars. Auto headlights provide meteors, and searchlight beams, when stationary, slightly resemble comets.

The eye gains immediately the impression of nearer and of farther lights. The observer senses that there must be millions of faint lights invisible at the greater distances. He thus has a greater advantage in looking at this ground-level "Milky Way" than has the astronomer who views the sky.

No immediate perception tells the scientist that some of the stars are farther than others. Indeed, only in recent generations has there been a general acceptance of the fact that the distances are not all the same. The year 1938 was the hundredth anniversary of the first measurement of the distance of a star. Until F. W. Bessel measured that quantity for the faint star 61 *Cygni*, many astronomers believed the stars to be much fainter than the sun. Present knowledge of stellar distances is the result of more than a hundred years of tedious, precise measurements. Dr. Frank Schlesinger made the greatest individual contribution to the field. Direct methods of measurement devised by him and his predecessors have been supplemented by spectroscopic and other procedures that make possible determinations of distances of more remote stars. Today we know accurately in some cases, approximately in others—the distances of more than 10,000 stars.

Our sun has sometimes been called a "dwarf" star because the average star visible to the naked eye gives more light than the sun. Recent studies, however, have turned to insignificant appearing stars and quite a number of them have been found to be near neighbors—stars that actually compare with the lights in homes.

Unlike the city lights, the stars slowly shift their positions. If an observer could watch for 100,000 years he would notice rather large changes in their configurations. As a general rule, the nearer stars appear to move faster than the distant ones. It is a relatively short and easy observation to find which stars have these larger proper motions. The comparative few thus sifted out are chosen for the more extended "parallax" observations that give their distances.

The results of these observations have been surprising. Despite the fact that many stars give thousands of times as much light as our sun, the majority—at least in our neighborhood—are less than 1/100 as luminous and are not visible to the naked eye.

Our "neighborhood" extends an almost unbelievable distance when measured by any other units than those used by astronomers. From the nearest star our sun would appear merely as one of the brighter stars. To provide a better grasp of the size of the earth's neighborhood, let us begin with more familiar distances. Light reflected by properly placed mirrors would travel $7\frac{1}{2}$ times around the earth in one second. It requires eight minutes to come from the sun. Therefore, the sun's distance is more than 3,600 times the earth's circumference. But this is just "next door" in the earth's neighborhood. The nearest known star is 270,000 times as far away as the sun, and it takes more than four years for its light to reach us. Light requires about 75,000 years to reach us from the most distant star in our own galaxy. The distance of 6,000,000,000,000 miles, which light travels in one year, is a light-year. We shall consider here

all the known stars whose light requires less than a dozen years to traverse space to us. The little volume occupied by these stars has a diameter of but 140,000,000,000,000 miles and compares with our galaxy as a drop does with an old-fashioned round laundry tub. We may have observed most of even the least luminous stars within this drop, and we can use their distribution to help guess about our galaxy as a whole. One of the main uses of the 200-inch telescope is to give us a drop with twice the diameter of the present one, eight times the volume, and probably about eight times as many stars.

One of the most essential things to learn about our galaxy is the range of luminosity from the smallest of the red dwarfs to the greatest of the hot blue stars. Another essential datum is the relative frequency of stars of various types. An examination of all the thousand million stars observable with our biggest telescopes and best photographic methods probably gives us a much inferior and more biased sample of our galaxy than is the small sample taken of our nearest stars! This may be surprising, for most people believe that the larger a sample, the more accurately it predicts. Other things being equal, that is true; but in a presidential election opinion poll, for example, a biased sample of a million gives a much less accurate prediction than a sample of a few thousand properly gathered.

If we had far larger telescopes, clearer skies, a steadier atmosphere, and more sensitive photographic plates, a larger star sample would be better. But consider how inaccurate such a sample would be at present. We find in our neighborhood, as one of the very nearest of all stars, a dwarf that gives only 1/500,000 the light our sun does. This little star, which is the companion of a brighter star, BD $+4°4048$, would not be observable at all if it were at the distance of the bright star Rigel. Most of the other stars found near us also give less than 1/100 as much light as our sun. We have no reason to believe that space near us is populated very differently from space in general within our galaxy. Therefore, we guess that for each distant star we observe, there are many too faint for our instruments.

One of the reasons for building the great telescope on Mount Palomar is to extend the observations of dwarf stars to greater distances. Such stars, instead of being extremely rare, as we believed four decades ago, probably are the most common type of all.

The stars nearest the sun.

The relative distances of these stars from the sun and their directions, so far as the plane of the equator is concerned, are shown. However, any flat diagram must have some distortion in representing positions of objects that do not lie in a plane. As a result, some stars that are much separated in a north and south direction appear close to each other.

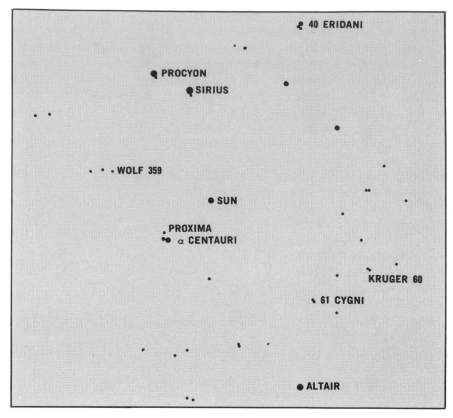

In the last few years the startling thought has come to us that perhaps the red dwarfs so far observed are not the faintest stars and that stars thousands of times less luminous even than they *may* be in our neighborhood in great numbers. Mount Palomar is beginning to give us the start of an answer to this conjecture. If such stars do exist, the new telescope will find a few close neighbors somewhat fainter than any yet known. When increase in observational power fails to extend materially the lower limit of luminosity, the evidence will favor a belief that at last we have observed almost all classes of stars.

None of the extremely luminous stars is found very close to us. Sirius, the brightest within a dozen light-years, gives but 30 times the light of our own sun. We are rather fortunate in this deficiency. For example, if Rigel were at the distance of Sirius, it would be 700 times as bright, or about midway in brightness between Venus and the full moon! Visible even in the daytime, it would interfere with night observations of faint objects. At the same distance the star *S Doradus*, in the Large Magellanic Cloud, would be quite comparable with the moon at its best.

Our little sample of stars within a dozen light-years contains only three stars that give more light than our sun. Procyon, which gives about six times as much, and one of the components of Alpha Centauri, which gives 1.1 times, are the only other very luminous stars, except Sirius, in the sphere. So we must consider our sun as being an ordinary star instead of a dwarf.

Our sun appears to have no companion star, although a bare possibility exists of some dwarf moving with it through space. Many other stars, however, are not single. Six of the closest systems are composed of two or more stars, instead of one. Of course, we have no idea whether any of them have planets. The Alpha Centauri system, the closest of all, has three stars. One is much like our sun. It gives 1.1 times as much light and has about the same size and temperature. The star nearest it is cooler and gives 0.3 the light of our sun. These two stars move around each other at a distance great enough so that each may well have its own system of planets, but close enough so that each would supply much sunlight to the other planets. Since the stars are differently colored, the color of daylight on such a planet would vary according to which star was above the horizon. The third star, Proxima Centauri, gives about 1/17,000 the light of our sun and though it is our nearest star, its existence was unknown until 1913. It is 100 times too faint to be seen without a telescope. Indeed, as viewed from the other stars of its own system, it would not be a conspicuous star of the sky.

Let us assume our sun to have a companion of the same luminosity as Proxima Centauri and at the same distance from us as Proxima is from its companions. There would be perhaps a thousand brighter stars in our sky than our own companion. It would be 12,000 times as far away as is our sun. It would take approximately 1,300,000 years to move around the sun. Since it would be traveling with our sun through space, it would change its position very slowly (one second of arc per year) among the other stars. If it were in the extreme southern part of our sky we might well not yet suspect its relationship to us. Thus it is quite within the bounds of possibility that some dwarf actually is a distant brother of the sun.

Another interesting system within the drop of our galaxy represented by a dozen light-years is that of Sirius. The smaller of the two companion stars is one of the most interesting stars ever observed, one too faint to be seen with the naked eye. It gives 1/400 the light of our sun and is almost as hot as its greater companion. Therefore, it must be very small. Its diameter, in fact, proves to be only about three times that of the earth. However, this tiny body is so massive that its gravitation moves its giant partner. Its density must be 2,500 times that of the densest substance known on earth. Yet it is a gas and made of the same elements we find here! For a while such white dwarfs were thought to be freaks, but when we consider that nearly 100 are known and that all of these are rather near us, we are convinced that they are quite common in our galaxy.

Faced with such a bewildering variety of sizes, brightnesses, and colors of stars, it is monumentally difficult to devise an adequate theory of their origins and lives. Until recently, these theories were little better than intelligent guesses. However, powerful computers have now made possible some reasonably reliable calculations of the interior structures and energy production characteristics of stars. Eventually these calculations will lead to a clearer understanding of the way stars change with time.

CHAPTER **45** FACTS ABOUT SOME NOTABLE STARS

There are stars with diameters of hundreds of millions of miles. Yet they appear as mere points of light in the most powerful telescope. Their distances are so great that it is impossible to magnify them so they will appear as disks.

In 1920, Professor Albert A. Michelson of the University of Chicago planned the first measurement of the angular diameter of a star by means of an interferometer. The name of this star was Betelgeuse. The news was important enough to appear on the first page of *The New York Times* of December 30, 1920. An article two columns long was devoted to this accomplishment, which was "regarded by scientists as a stupendous achievement."

It would be difficult and would take too long to give a detailed description of the interferometer and the mathematical calculations involved. However, the instrument can be briefly described. The first one used by Michelson involved covering the mirror of the 100-inch telescope of the Mount Wilson Observatory with an opaque cap containing two slits adjustable in width and distance apart. When such a telescope is turned to a star, there appears a series of light and dark interference fringes, which result from the alternate reinforcement and destruction of light waves coming from different parts of the star. When the slits are moved, a distance between them will be attained at which the fringes disappear. A formula gives the angular diameter of the star. If the distance of the star is known, its linear diameter can then be found.

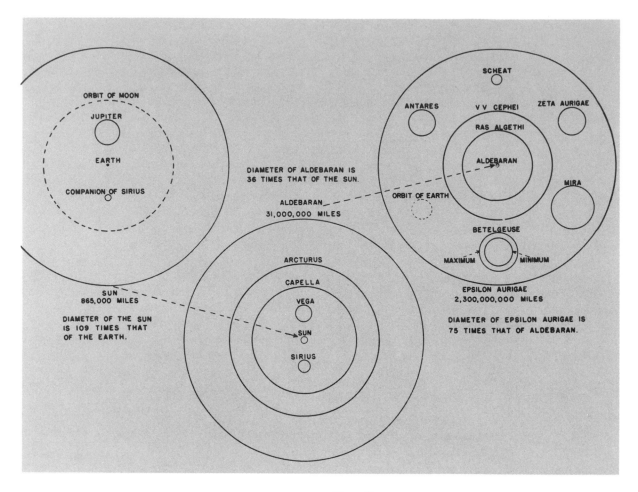

Size of largest star as compared with sun and earth.

Note that three different scales are used: (1) the sun as compared with the earth; (2) Aldebaran as compared with the sun; and (3) Epsilon Aurigae as compared with Aldebaran. Epsilon Aurigae, the largest star, is 2,700 times the size of the sun and nearly 300,000 times the size of the earth. These figures refer to diameters; for volumes, the numbers would have to be cubed.

Michelson improved on this device by using two small adjustable mirrors, which could be put as much as 20 feet apart. This interferometer was attached to the telescope. Staff members of the Mount Wilson Observatory, especially the late F. G. Pease, carried on this work and measured the diameters of a number of stars. But this method is good for only the closer and very largest stars. An equivalent electronic addition of photoelectric signals from two telescopes miles apart provides a much longer base line. In Australia this technique has been used to measure several star diameters.

The diameters of certain eclipsing double stars have been found from their periods, their distances apart, and the durations of the eclipse of one star of a pair by the other. The results in these cases and in those where the interferometer has been used are in harmony with the diameters computed from surface temperatures and luminosities. This third method, described in Chapter 40, can be briefly summarized in three sentences. The total radiation of a star is found from its apparent brightness and distance. The radiation per square mile depends on its temperature, which is found from its color and spectrum. By dividing the total radiation of the star by its radiation per square mile, we get its size.

Even if the angular diameter of a star is accurately known, its linear diameter will be uncertain if there is any uncertainty about its distance. Also, our knowledge of its real brightness is dependent on the accuracy of its measured distance. The determination of the distances of stars is one of the most important and most difficult problems in astronomy. The distance of the nearest star is about 25,000,000,000,000 miles.

Let us see how large that number is. Suppose we take a velocity of $\frac{3}{4}$ mile per second, the speed at which a bullet might be fired from a gun. If we multiply this by 500, we get 375 miles per second, which is about the speed of the fastest prominences erupting from the sun's surface. If this is multiplied by 500, the result is very close to the speed of light, 186,000 miles per second.

Light takes 500 seconds to travel from the sun to the earth, a distance of 93,000,000 miles. If this distance is multiplied by 500 twice, or by 250,000, the final result is nearly the distance to the nearest star. With the understanding that there is some uncertainty about the distances, which will affect the results, some facts will now be given about a few of the stars.

The sun is a typical star and gives us a standard with which we can compare other stars. Its diameter of 865,000 miles is 109 times that of the earth. Its volume is large enough to contain 1,300,000 earths. If the earth were placed at the center of the sun, the moon's orbit around the earth would take up only a little more than half the sun's diameter. This is shown in the first part of the diagram accompanying this chapter.

If the solar energy received by the earth in one second could be converted into power, its value at the usual rate of about 2 cents per kilowatt hour would be one billion dollars. The total energy sent out by the sun in all directions is about two billion times the amount received by the little speck called the earth. The sun radiates more energy in one second than the earth receives from it in 60 years.

Where does so much energy come from? Certainly not from combustion. A sun composed of coal and oxygen in the most favorable proportions could not have lasted as long as 2,000 years at its present rate of shining. A contraction of the sun could account for the solar radiation for some 25 million years, but this is inadequate. Geologists have found evidence that low forms of life have been on the earth for several hundred million years and that the earth is at least five billion years old.

Recent theoretical work indicates that the rate of solar radiation can be satisfactorily explained by a thermonuclear reaction, somewhat similar to that of a hydrogen bomb. In this reaction four hydrogen atoms are transformed into one helium atom, with the liberation of energy in the form of short-wave radiation. This is possible at the sun's central temperature of about 20,000,000° C. At this fantastically high temperature hydrogen atoms can collide so violently with each other that they may be transmuted into heavy hydrogen and eventually helium. The energy is passed on from atom to atom until it emerges at the surface of the sun as infrared, visible, and ultraviolet light. There appears to be enough hydrogen in the sun to continue the present rate of radiation for some tens of billions of years more.

By means of the spectroscope astronomers have determined the composition of the outer layers of the sun and of the stars. The solar atmosphere consists primarily of hydrogen and helium, but about two-thirds of the 92 known elements have been identified. Most of the stars have nearly the same composition as the sun.

Alpha Centauri, the nearest known stellar neighbor ($4\frac{1}{3}$ light-years away), is the third brightest star in the sky. A telescope reveals it as composed of two stars revolving around each other in a period of 80 years. The orbit is so highly eccentric and inclined to the plane of the sky that the apparent distance between the two stars varies from 2 to 22 seconds of arc. The actual distance varies from 11 to 35 astronomical units. An astronomical unit is 93,000,000 miles, the earth's distance from the sun. Thus when the two stars are closest together, their separation is about equal to Saturn's distance from the sun. When farthest apart, the stars are separated by a distance that is halfway between the distances of Neptune and Pluto from the sun.

The apparent magnitudes of the two components are 0.3 and 1.7, giving a combined magnitude of 0.1. When the distance is allowed for, the brighter component is found to have almost exactly the same luminosity as the sun. It has about the same temperature, size, and mass as the sun. The fainter component is slightly cooler, smaller, and fainter than the sun.

Two degrees away from Alpha Centauri is a faint star visible only in the telescope and known as Proxima Centauri. It was named Proxima, meaning nearest, when it was believed to be the nearest star, but its distance is probably about equal to that of Alpha Centauri, and it is considered a part of that star's system.

Alpha Centauri is not visible from most northern latitudes. A few degrees away from it is Beta Centauri, of magnitude 0.9. These two bright stars serve as pointers to the northern star of the Southern Cross. Alpha Centauri is known to the navigator as **Rigel Kentaurus**.

Sirius, the brightest appearing star in the sky (magnitude of −1.6), is the nearest star visible to the naked eye in northern latitudes. It is twice as distant as Alpha Centauri, being about 50 trillion miles, or nearly 9 light-years, away. It appears brighter, because it is a more luminous star. It gives out about 30 times as much light as the sun. Its diameter is nearly twice that of the sun and its temperature is higher, so that its surface is larger and each square mile radiates more energy than the same area on the sun.

Like Alpha Centauri, Sirius is a double star. The orbit of the two stars is a little more elliptical than that of Alpha Centauri, but the inclination of the orbital plane is considerably less and the apparent ellipse is a more open one. The angular separation of the components varies from 2 to 11 seconds of arc. Because of the brilliance of the brighter star, it is impossible to see the faint companion with any telescope when the stars are near their minimum distance.

Canopus is the second brightest star in the sky. Its magnitude of −0.9 makes it appear about half as bright as Sirius. Actually it is much more luminous, but it is much farther away. Its distance, which is uncertain, is at least 100 light-years. This would make Canopus about 2,000 times as luminous as the sun.

This star is so far south that it is not visible from most parts of the United States. It can be seen from the latitude of Los Angeles during the winter evenings when it and Sirius are near the meridian.

Vega appears of the same brightness as Alpha Centauri, having an apparent magnitude of 0.04. Its distance is about 27 light-years and its luminosity is about 60 times that of the sun. Its diameter is $2\frac{1}{2}$ times as great.

Capella consists of two stars revolving around each other in a period of 104 days. The distance between them is about the same as the distance of the earth from the sun. The two components are not visible as separate points of light in any telescope, because they are too close together. However, the interferometer reveals their presence and their angular separation. Also, the spectra of both stars can be photographed. As the stars revolve and alternately approach and recede from the earth, the lines of their spectra shift alternately to the violet and to the red ends of the spectrum. Allowing for a distance of about 42 light-years, we find that the two stars have a combined luminosity about 150 times that of the sun.

Arcturus is the star that was used to open the World's Fair in Chicago in 1933. It was selected because of its distance, which was thought to be 40 light-years at that time. The light reaching the earth in 1933 left the star about 1893, when there was another World's Fair in Chicago. A telescope brought the light to a photoelectric cell, which generated a tiny electric current. The current was amplified and used to turn on the lights of the Fair.

More recent measurements indicate that the distance of Arcturus is a little less than 40 light-years. Its diameter is about 20 million miles, or 23 times that of the sun. Its angular rate of change of position in the sky is one of the largest. In one year its position changes by 2.3 seconds of arc. Even at this rate it will take about 8,000 years to move five degrees, the distance separating the two pointers of the Big Dipper. Arcturus is moving in the direction of Spica, and the two stars will appear quite close together about 60,000 years from now. Spica is a more distant star and has a much smaller proper motion, so that its position in the sky will change by less than one degree in 60,000 years.

We can see why the expression "fixed stars" is used. The stars appear to be fixed in the sky and their positions do not change appreciably during many centuries. Yet we must realize that their real speeds are tremendous. They are moving as fast as the planets are, but they are millions of times farther away and their apparent motions are millions of times slower. Arcturus is moving across the sky with a speed of about 75 miles a second.

Rigel is so far away that its distance is quite uncertain. Using the available data, we get a figure of about 540 light-years. If that is correct, then the luminosity of Rigel is about 21,000 times that of the sun.

Procyon is a nearby star, being only 11 light-years away. With an apparent magnitude of 0.5, its luminosity is 6 times that of the sun. Its diameter is 1.7 times that of the sun.

Achernar is too far south to be visible from most

parts of the United States. It is about 70 light-years away and 200 times more luminous than the sun.

Beta Centauri is used with Alpha Centauri to find the Southern Cross. Its distance of 190 light-years yields a luminosity of 1,300 times that of the sun.

Altair is a nearby neighbor, being only 16 light-years away. The lines of its spectrum are very much widened, due to the rapid rotation of the star. The speed at the star's equator is something like 160 miles per second. To determine the period of rotation we need to know the star's size. Let us see how this can be found.

From the distance and the apparent magnitude of Altair we find that it gives out nine times as much light as the sun. Its spectrum shows that its surface temperature is about 8,500° Centigrade as compared with only 6,000° for the sun. The amount of energy varies as the fourth power of the temperature. Calculations show that each square mile of Altair gives off four times as much light as the same area on the sun. The surface area of Altair must be 9/4, or 2.25, times that of the sun. The diameter must be the square root of 2.25, or 1.5, times the solar diameter. This gives a circumference of around 4,000,000 miles. Dividing this by the speed of 160 miles per second, we get a period of 25,000 seconds, or seven hours. As far as the writers know, this is the shortest rotation period known for any star. The centrifugal force is so great that the equatorial diameter of Altair is estimated to be about half again as great as the polar diameter. If the star rotated a little faster it would probably break up into two masses, forming a double star.

Betelgeuse was the first star whose angular diameter was measured with the interferometer. The star is remarkable not only for its great size, but also because the size changes. The angular diameter was found to vary between 0″.034 and 0″.047. Using a value of 275 light-years for its distance, we find that the diameter varies from 260 to 360 million miles. This corresponds to 300 and 420 times the sun's diameter. The diameter of the earth's orbit around the sun is 186 million miles, and so the earth's orbit would easily fit inside Betelgeuse.

The brightness of Betelgeuse varies in semi-regular waves of 140 to 300 days, which appear to be superposed upon a still slower fluctuation with a period of six years. The range of variation is only one magnitude, which means that its maxi-mum light is $2\frac{1}{2}$ times its minimum brightness.

Acrux is the brightest star in the Southern Cross and is at the lower end of the main part of the Cross, which points toward the south celestial pole. Acrux is a coined word, the star being the Alpha star in Crux and also known as Alpha Crucis. It is a magnificent double star when seen in a telescope. The two stars are of magnitudes 1.6 and 2.1, and they are 5″ apart. Each component is revealed by the spectroscope as a binary, or double, star, and each binary has a period of one day. Using a distance of 220 light-years, we find a total luminosity of 1,400 suns for the stars that make up Acrux. The surface temperature of these stars is exceptionally high, being about 23,000° Centigrade.

Aldebaran, at a distance of 55 light-years, has a diameter of 31,000,000 miles, or 36 times that of the sun. Its temperature is 3,600°, and its luminosity is 100 times that of the sun. Aldebaran happens to be in line with a more distant cluster of stars, the Hyades. They are at a distance of 130 light-years and appear much fainter than Aldebaran.

Pollux is a little brighter than its twin, *Castor*. The two stars appear close to each other in the sky, but they are entirely unrelated. Pollux is about 33 light-years away, and Castor about 45. Castor itself is three pairs of twins. In 1719 it was first seen in the telescope as two stars, but it was not until 1803 that William Herschel recognized that they

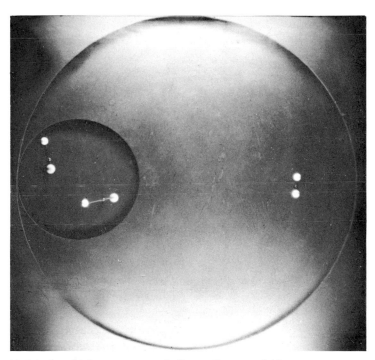

Model of multiple-star system similar to Castor and Mizar systems. Members of each pair revolve around each other. Two pairs revolve about one another, while their center of gravity and the third pair move more slowly in larger orbits. (Griffith Observatory.)

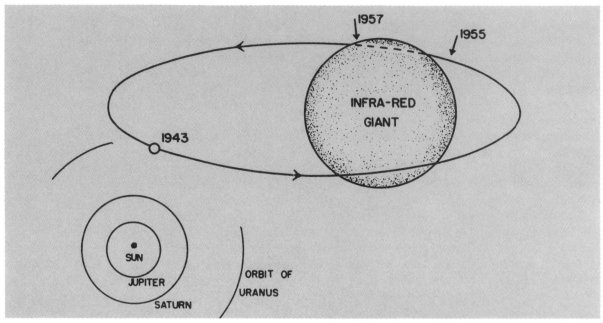

The orbit of Epsilon Aurigae.

The ellipse is the path of the small bright star which we call Epsilon around its huge dark companion, as seen from the direction of the earth. The positions indicated are for the beginning of the corresponding years. The orbits of Jupiter, Saturn, and Uranus are shown to scale. (Diagram by G. H. Herbig.)

were revolving around each other. In that year Herschel announced that after 25 years of observing Castor and five other pairs of stars, his measures gave conclusive evidence that they were real double stars. The period of the two components of Castor has been calculated as 380 years; but a complete revolution has not been observed. Around 1900 each component was found to be a spectroscopic binary, one with a period of 3 days and the other with a period of 9 days. Close to Castor in space and physically related to it is another spectroscopic binary, the third pair of twins, with a period of less than a day. Castor is indeed a remarkable multiple star system.

Spica is another spectroscopic binary, and its period is 4 days. The total amount of light from the pair is 1,000 times that of the sun, if the distance of 190 light-years is correct.

Antares is one of the few stars large enough to have its angular diameter measured with an interferometer. Using a distance from the earth of 220 light-years, we get a diameter of 245,000,000 miles. This is larger than the earth's orbit around the sun and equal to 285 times the sun's diameter. One astronomer has reported that when Antares emerged

from behind the moon after being occulted, the star took about 1/10 of a second to come to full brilliancy. In the case of an ordinary star, the end of an occultation is instantaneous, with the star flashing out very suddenly. The disk of Antares is not visible in the telescope, but it is large enough so that the moon takes an appreciable time to uncover it.

Antares has a green companion of the 7th magnitude and 3″ away. It can be seen only with a large telescope, because of the overpowering glare of Antares. At the time of an occultation, the companion is uncovered by the moon a few seconds before Antares.

Fomalhaut, at a distance of 23 light-years, is 11 times as luminous as the sun. It is the most southerly of the bright stars visible from the northern states.

Deneb is so far away that its distance is rather uncertain. Measurements made before 1940 give a distance of 400 light-years. This would give it a luminosity 4,800 times that of the sun.

Regulus, at a distance of 77 light-years, has a luminosity 150 times that of the sun. This is the 20th brightest star in the sky and completes our list

236

of the first-magnitude stars. Its apparent magnitude is 1.3, and the next faintest star has a magnitude of 1.5. However, there are a few more stars that will be mentioned here because of their size.

The largest star recognized at the present time is the huge dark member of the double star *Epsilon Aurigae*, which is 3° away from Capella. Epsilon Aurigae can be seen with the naked eye at the apex of a little triangle southeast of Capella. The smaller bright member of the double star is partially eclipsed by the giant every 27 years. The two stars revolve around each other in a period of that length. It takes the smaller bright star 190 days to decrease to one-half its normal brightness. Then it remains constant for 330 days. Finally, it takes 190 days to regain its usual brightness. Thus two years are required for this eclipse, which is only a grazing one. The smaller, brighter star just dips below one edge of the giant's atmosphere. The diameter of the giant is estimated to be 2,700 times that of the sun. If this star were placed at the position of the sun, its surface would lie midway between the orbits of Saturn and Uranus. But its distance of about 3,000 light-years and the feeble red glow it gives off make the giant itself invisible to us in the solar system.

VV Cephei is another system of eclipsing double stars. The period is about 20 years, and one star is estimated to be 1,220 times as big as the sun. *Zeta Aurigae* is a third eclipsing binary. Its period is 2 years and 9 months. One component has a diameter 300 times that of the sun.

Ras Algethi, or *Alpha Herculis*, at a distance of 800 light-years, is found to be 800 times larger than the sun. *Scheat*, or *Beta Pegasi*, is 160 light-years away and has a diameter that would reach from the earth to the sun.

Mira was the first variable star to be discovered. Most of the time it is invisible to the naked eye, but once about every 11 months it can be seen as a star of the 2nd or 3rd magnitude. It is about 1,000 times brighter at maximum than at minimum. Its distance of 165 light-years makes its diameter about 400,000,000 miles.

CHAPTER **46** **THE VISIT OF ARCTURUS TO THE SUN**

Five hundred thousand years ago the light of a very faint star, later to be named Arcturus, first fell on the uncomprehending retina of a troglodyte. The star was far to the north and, regularly, each 26,000 years took its turn as a very faint pole star.

If astronomers had been alive, this apparently unimportant star would have interested them very much, because their spectroscopes would have told them that it was rushing almost directly toward the earth with a speed of 74 miles per second. It was one of the fastest-moving stars of our galaxy.

They could have predicted that some day it would become very bright and would be far south of the position it had then. They could not have told how bright it would be at its maximum, nor how far away from us, for the most trifling error in measuring its sideways motion would have changed their predictions for the present time very much. But today's astronomers, who observe it almost at its closest point, can calculate easily and with good accuracy what our ancestors saw and what our descendants 15,000 generations removed will see. The procedure for such calculations is readily understandable.

At intervals of several decades astronomers have measured on their photographs the positions of important stars such as Arcturus. These measurements show the plate positions to an accuracy of about one fifty-thousandth of an inch. As a result astronomers know the angular speeds east or west (right ascension) and north or south (declination). Arcturus is moving west 1″.104 a year. In other words, it is changing westward by one degree in about 3,200 years. It is moving southward by 1″.999 per year. These components together carry it one-eighth of the distance between the Pointers of the Big Dipper in a thousand years.

One might think that to get its position 60,000 years ago it would be necessary only to multiply

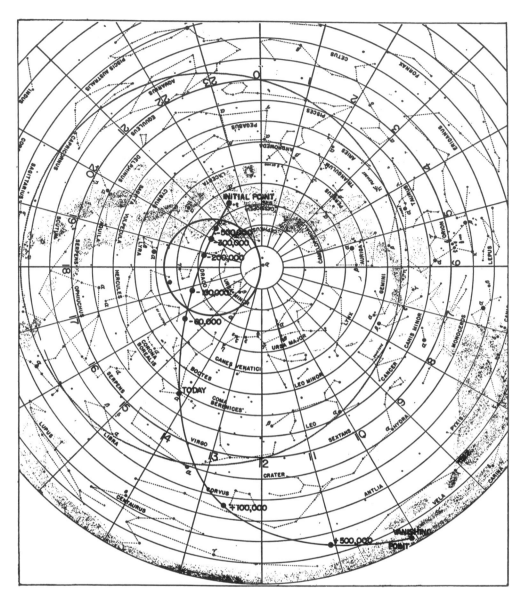

The path of Arcturus.

these quantities by 60,000 and add them to the present positions. That would do for, say, 50 years, but over centuries these angular velocities have changed greatly.

However, astronomers know two other things about Arcturus that allow them to trace its position back over the ages. Its distance is such that light requires 38.3 years to reach us. The spectroscope reveals that it is moving toward us with a velocity of 3 miles per second.

During a mere million years both Arcturus and the sun must have moved in almost perfectly straight lines through space and at constant speeds.

Even if we assume that they are moving around the center of our galaxy they move less than two degrees per million years in their gigantic orbits. Such a small arc of a circle is very nearly a straight line.

Using nothing more than trigonometry and elementary calculus, we can now solve the problem of its motion in this straight line through space.

We find that a million years ago it was so far away that it was not visible to the naked eye. It was then near the star Delta Cephei and was moving very slowly toward Alpha Cephei.

By 500,000 years ago it had become visible to the

238

naked eye and was at a distance of 213 light-years. Two hundred thousand years later its distance was about two-thirds of what it had been. It still was barely visible on a moonlit night. Two hundred thousand years ago it was fainter than the stars of the Big Dipper but after another 100,000 years it had brightened to surpass them.

The map on page 238 shows the path of Arcturus among the stars. On the map, also, is plotted a black circle around which the north celestial pole is carried by precession in 25,800 years.

Calculations similar to these show other less known stars to have been almost as interesting as Arcturus in the past or to be rushing toward us in paths that will bring them closer than any star is today. A glance through Schlesinger's star catalogue reveals many interesting tales.

For example, there is Schlesinger 1008, a 4th magnitude star, which today is almost twice as close as Arcturus. It is hurrying from us with a speed of 54 miles per second, and despite its low luminosity was at one time a rather bright star. It

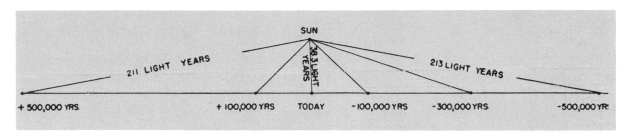

Path of Arcturus. The sun is considered as fixed.

Previous to 500,000 years ago the pole regularly passed the faint star that was to become so important later. However, for a long time after that date Arcturus was within the circle and crossed it again about 70,000 B.C.

At that time the pole was on the far side of its path, more than 40° from Arcturus. The closest approaches of the star to the pole occurred around 83,700 B.C., with Arcturus inside the circle, and 57,900 B.C., with it outside. At each of these times it was a brighter pole star than our present one, but was at a greater distance from the pole than we see Polaris in our sky.

Today Arcturus is almost at its minimum distance. Its speed *toward* us has decreased from 74 to 3 miles per second, although its speed has, of course, remained nearly constant in its approximately straight line through space. Its sideways speeds—that is, west and south—have increased as the speed toward us has decreased. The almost negligible side motion that an astronomer of 500,000 years ago would have found so difficult to measure has caused Arcturus to miss the sun. A few thousand years from now it will have ceased to approach us and will move on away. A few hundred thousand years from now it will vanish from the naked eye in a region too far south to be seen from the United States. Its visit to our neighborhood will have ended.

is moving about twice as fast as Arcturus among the stars, and years ago, when closer, had a proper motion as large as any of the few stars that seem to be runaways in our sky.

Schlesinger 2140 is five times as far away as Arcturus, and has a velocity away from us of 112 miles per second. Its proper motion is not large. At present it is of the 5th magnitude, but at one time in the past it was very bright.

S Carinae is moving away with a speed of 173 miles per second. Its proper motion is moderate, but its parallax is undetermined. It is a variable that changes four magnitudes in brightness. At one time in the past this wonderful star was brighter than Algol and, moreover, underwent a far greater variation of brightness in a complex cycle.

Schlesinger 7703 is today at about half the distance of Arcturus. It is approaching at 82 miles per second. However, it has a large side component of velocity. During the next few dozen thousand years it will become bright and will be one of the nearest of all stars.

Perhaps some day the complicated motions of Arcturus and the sun around the center of the galaxy will once more bring the two stars within hailing distance.

If so, it must be after the lapse of hundreds or even thousands of millions of years. Even stars may age in such an interval.

CHAPTER 47 CRUX, THE SOUTHERN CROSS

Few objects on the celestial sphere have carried as great an element of romance and appeal as does the Southern Cross. Unfortunately during the twentieth century the Cross does not rise in latitudes farther north than 26° and, therefore, is visible to few people living in the North Temperate Zone. This has not been true always. There was a time thousands of years ago when the Southern Cross was visible from England. It disappeared from view in the United States early in the Christian era, but will reappear to observers in southern United States before A.D. 10,000. This gradual north-south drift of the Southern Cross is a consequence of the precession of the earth, which produces a slow movement of the celestial poles among the stars on a circle with a radius of $23\frac{1}{2}°$. It takes the celestial poles 26,000 years to complete one cycle on this circle. For about two-thirds of the precessional cycle of 26,000 years, the Southern Cross rises each day as viewed from the latitude of, say, Los Angeles. At present, taking advantage of refraction of light, which makes objects near the horizon appear higher than they actually are, and using the extremely clear sky of Mount Wilson, near Los Angeles, and the slight dip of the horizon that the mile-high mountain affords, it is just possible, under the most favorable conditions, for astronomers there to see the northern star of the Cross. For about three thousand years more, the motion of the earth's north pole will be away from the region of the Cross and, therefore, it will seem to us to move farther and farther south before it begins again its 13,000-year journey toward the north. We of the twentieth century have, however, an advantage in the fact that stars halfway around the sky from the Cross are now visible to us, but will be hidden from our descendants.

The extreme beauty of the little group of stars that form the Southern Cross has caused tourists to watch for it as their ships have carried them toward the earth's equator, and quite often, in eagerness to be the first to see it, they have pointed out another cross to the west of Crux. This has

been done so often that the other group has received the name of False Cross. The False Cross is larger, but its stars are not so bright. In the Southern Cross are found, within a small area, three stars of the first to second magnitudes and a fourth one also quite bright. The very fact that the constellation is small makes these bright stars still more conspicuous, and the Cross has become a symbol to people of the Antipodes. It is found on flags and postage stamps, and visitors from the south attending a planetarium show in the United States are swept by a great nostalgia at its reproduction in the planetarium sky.

Though it has been the custom always to speak of the stars as the fixed stars, we know they are moving helter-skelter through space, somewhat like the motes of dust visible in a sunbeam. Some are rushing toward us with speeds of dozens of miles a second, some, away from us with speeds fully as great. Some are moving to the north, others to the south, some east, some west. Our constellations are changing continually and, given enough time, the sky will be wholly unrecognizable. One wonders, then, as he looks at the Southern Cross, whether it and other constellations were visible to our ancestors of a few thousand years ago. He wonders whether his descendants will be entranced by its beauty. It is not difficult at the present time for one to answer this. He can turn to a catalogue of the stars that specializes in the speeds at which they are moving toward or away from us —speeds determined by the spectroscope—and there he can find the rate of approach or recession. He can turn to other catalogues, such as Schlesinger's catalogue and Boss's catalogue, and from the various kinds of data get the distances of these stars and the directions of their motions across the sky. With these it takes only a few moments' calculation for him to tell approximately where they will be at the end of 1,000 years, 10,000, or even 100,000 years.

The diagram of the Southern Cross shows the positions of the six brightest stars. The sizes of the

Milky Way in region of Southern Cross. Cross is upright and near upper center. Below and to left of it is dark nebula called Coal Sack. Top star and right-hand star of Cross appear relatively faint because of their yellow color. (Harvard College Observatory.)

disks shown indicate their relative brightnesses. Through each of these six stars has been drawn an arrow pointed in the direction in which the proper motion is carrying it. The continuous parts of the arrows show approximately the paths that will be followed during the next 10,000 years. These paths, of course, are not related in any way to the precessional changes that were described earlier. The dotted parts reproduce the paths that have been followed during the last 10,000 years. For four of these stars the changes in position are quite negligible, even after 100 centuries. However, for the stars Gamma (γ) and Epsilon (ε) this does not hold. Gamma is moving almost directly toward the south and in 10,000 years will tend to make the Cross look

somewhat squattier than at present. It will not, however, deform it very badly during this time. Epsilon, which today looks to us like a misplaced central star, will move during the 10,000 years entirely outside the frame of the cross. Ten thousand years ago it was much closer to the center than it is now, although even then it was too far south to fulfill the function that the human eye tries to demand of it.

These paths show one extremely remarkable feature first noticed a generation ago by the great Dutch astronomer J. C. Kapteyn. During the time considered, four of them will move nearly the same distance in the sky and in almost the same direction. It would seem strange to find four stars this close

241

Motion of six brightest stars in Southern Cross. Disks show present positions, and ends of arrows show stars' positions 10,000 years ago and 10,000 years in future. Reason why four stars have same motion is explained in this chapter.

together, all with nearly identical paths, if there were not some common bond between them. The statistical evidence is strong enough to indicate that they have in common something that distinguishes them from other stars of our galaxy. The idea grows to a certainty when one examines the spectrum of each star and finds that all four of them are extremely hot stars, among the hottest that are catalogued by astronomers. Turning to Schlesinger's catalogue of parallaxes, we find they are rather distant stars, but nevertheless close enough to us for their distances to be measured by direct observation. The calculated distances differ from each other but the differences are but little more than the uncertainty of the results obtained.

If there were no more stars than just these four showing such similarities, we would feel sure that they were in reality a physical cluster of stars, and that we were close enough to them so that their distances from each other disguised their relationship. Kapteyn found that not only these four stars but some fainter ones in the constellation, and stars as far away as the Scorpion, which is easily visible from the United States, all have this same motion. He found also that most of the stars with this common motion are hot stars like the four described above. Ten of the stars of the Scorpion partake of the common spatial drift, leaving only three of its important ones with any widely divergent motion.

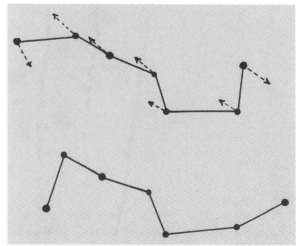

The Big Dipper now and 100,000 years hence. The arrows represent the relative motions of the stars during that time.

If our sun could be displaced from this moving cluster to a distance, say, 100 times that which it has, the cluster would appear much more normal, and a single glance at it then concentrated in a very small area of the sky would assure even the casual observer that it really is a cluster of related stars. Kapteyn gave the name Scorpius-Centaurus group to these stars. They are moving away from us with an average speed of about 15 miles per second and, therefore, in a few hundred thousand years will begin to take on the normal cluster appearance.

Other quite similar groups are known. Among these are the Taurus group and the Ursa Major (Big Dipper) group. If we were close enough to the Pleiades, to Coma Berenices, or to the Praesepe group, we would get from them very much the same appearance that we now get from the Scorpius-Centaurus group. However, the stars in these latter clusters are not so widely separated. Indeed, at least within the Pleiades, one still finds wisps of nebulosity occupying the space between the stars.

The observed facts tempt one to speculate and to form a hypothesis that originally all members of such a scattered cluster had a common origin within a relatively small volume of space. As time has passed, their small velocities with respect to each other gradually have caused them to separate until finally only the velocity of the group as a whole with respect to our galaxy remains to tell the story of their common relationship.

CHAPTER 48 THE PLEIADES

In the constellation of Taurus, the Bull, are located two of the most famous clusters of stars in the sky. The Pleiades (plee´ya-deez) look like a small reproduction of the Big Dipper and so they are sometimes incorrectly called the Little Dipper. However, the true Little Dipper is a much larger group than the Pleiades and is near the Big Dipper, the North Star marking the end of the handle of the Little Dipper. The other star cluster in Taurus is the Hyades (high´a-deez).

Contained within a circle having only twice the diameter of the moon's disk, the Pleiades since remote times have attracted the attention of people in all parts of the world. Countless myths and legends about them have been told. Most of these tales contain the idea that there were once seven stars where now only six are visible. The stars are very commonly called the Seven Sisters, but only six stars can be seen with the unaided eye under average conditions. A keen eye under favorable circumstances can see more.

Several years ago an advertisement of an optical company described a prehistoric cave-wall painting of the Pleiades, showing ten stars. The assumption was that cave men had better eyesight than modern man. However, there is evidence to the contrary. In *The System of the Stars*, by Agnes M. Clerke, it is stated that M. Maestlin, the tutor of Johannes Kepler, perceived fourteen, and mapped eleven Pleiades before the invention of the telescope. Also Miss Airy marked the positions of twelve with the naked eye. Two English astronomers, R. C. Car-

Hyades (lower left corner) and Pleiades (upper right). Five stars in Pleiades appear here as bright as or brighter than Aldebaran, brightest star in lower left corner. Visually, Aldebaran is more than five times brighter, but its reddish color does not affect photographic film as strongly as blue-white color of the principal Pleiades. (Paul Roques, at Griffith Observatory.)

Pleiades, a famous star cluster rich in myth and legend. Exposure time was long enough to reveal the filamentary nebulosity surrounding the stars. (Mount Wilson Observatory.)

rington and W. F. Denning, in the nineteenth century counted fourteen stars in the group without optical aid.

In 1937 it was found that the seventh brightest of the Pleiades had diminished by one-sixth of a magnitude in brightness during one year. It could not have been decreasing for very long at this rate, otherwise it would have been the brightest star in the sky less than half a century ago. The cluster is enveloped in a nebula, which might vary the brightness of a star by interposing streamers of different thickness. Whatever the explanation, the variability

lends some support to the legend that this star was once a conspicuous member of the Seven Sisters.

There is some inconsistency in the naming of the Pleiades. The seventh star, which has just been discussed, bears the name of Pleione, who in mythology was the mother of the Pleiades. The six stars easily visible to the naked eye are named after Atlas, the father, and five of the sisters. Two stars even fainter than Pleione are named after the other two sisters.

The magnitudes, names, and pronunciations of the nine brightest stars in the group are arranged

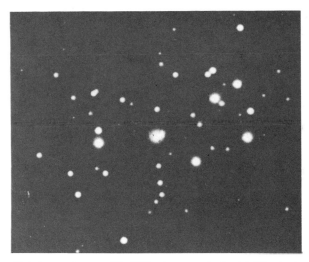

Pleiades on January 27, 1946, photographed with 25-minute exposure. (Paul Roques, at Griffith Observatory.)

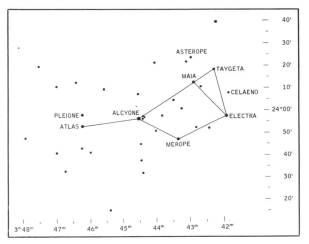

Map of brightest stars in Pleiades.

in order of brightness in the table that follows:

Magnitude	Star	Pronunciation
2.96	Alcyone	al-sigh′oh-nee
3.80	Atlas	at′las
3.81	Electra	ee-lek′tra
4.02	Maia	may′ya
4.25	Merope	mer′o-pee
4.37	Taygeta	tay-ij′ee-ta
5.18	Pleione	plee-oh′nee
5.43	Celaeno	see-lee′noh
5.85	Asterope	as-ter′oh-pee

There are two other stars in the Pleiades, which are slightly brighter than the 6th magnitude. Thus there are five stars between the 5th and 6th mag-

nitudes that are not visible to the average unaided eye. Each would be visible by itself, but the light of the brighter stars in the group hides the fainter ones. In a similar way, the four brightest satellites of Jupiter cannot be seen with the unaided eye because of the brilliance of the planet so close to them. Two of them are of the 5th magnitude and two of the 6th, so that they could be seen if they were not near Jupiter.

The second brightest star in the Pleiades was named after Atlas, who in mythology supported the heavens on his head and hands. Because a picture of Atlas supporting the world was often placed in the front of a collection of maps, such a volume is said to have been first named an atlas by the geographer Gerhardus Mercator. A mountain in northern Africa was named Mount Atlas, and from that was derived the name of the Atlantic Ocean, lying to the west of Africa.

No other part of the sky of equal area has been investigated as often and as carefully as that containing the Pleiades. In 1664 Robert Hooke, using a two-inch telescope, counted 78 stars in the group. In 1876 Max Wolf at the Paris Observatory recorded 625 stars. By means of four-hour exposures in 1887 Paul and Prosper Henry made a photographic chart showing 2,326 stars. The first careful measurements of the relative positions of 53 stars were made by the German astronomer F. W. Bessel during the twelve years from 1829 to 1841. Since that time many thousands of individual measures of the Pleiades have been made. It was discovered that while the configuration remains practically unchanged, the whole cluster is moving in a south-southeasterly direction at the rate of $5\frac{1}{2}''$ per century. At that rate it would take about 33,000 years for the stars to move half a degree, the angular diameter of the moon.

Thus the Pleiades may be compared to a flock of birds flying with the same speed in the same direction. Their change of position in the sky is called proper motion and is usually expressed in angular motion per year, which in this case would be $0''.055$. When allowance is made for the distance of the stars, as will be discussed later, we find they are moving across the sky at about 25 miles per second. They also are moving away from the solar system at the rate of about 4 miles per second. This is known as radial velocity and is measured with the spectroscope. Combining these motions, we get a space velocity of about 25 miles per second.

245

Not all of the stars in the region of the Pleiades partake of this common motion. In fact, most of the fainter stars are in the depths of space beyond the cluster and form the background upon which the Pleiades appear projected as seen from the earth. Of the several thousand stars in this small area, about 250 are now known to be members of the Pleiades system. When all the fainter stars are more carefully studied, it is possible that the total number of Pleiades will be found to be close to 500.

The brightest stars in the Pleiades have very nearly the same kind of spectra. For instance, the nine brightest have spectra that are either B5 or B8. This corresponds to a temperature of about 13,000° Centigrade. Fainter stars belong to classes A, F, G, and K. Stars of apparent magnitude 11 are of spectral class G, like that of our sun. These have a temperature of about 6,000°. We know how bright our sun would appear if it were at any given distance. We know how far away it would have to be to appear of the 11th magnitude, like the Pleiades stars of the same size and temperature as our sun. The steps in the reasoning are given in the next paragraph.

The absolute magnitude of the sun is + 5, which is the apparent magnitude it would have at the standard distance of about 33 light-years. The Pleiades stars that are like our sun are 6 magnitudes fainter than they would appear at a distance of 33 light-years. Stars of other classes show this same difference of 6 magnitudes between their apparent and absolute magnitudes. This corresponds to a difference in brightness of 250 times. Since the brightness decreases with the square of the distance, and since the square root of 250 is about 16, the Pleiades must be 16 times farther away than 33 light-years. This gives a result in round numbers of 500 light-years.

Now that we know the distance, we can find out how bright the Pleiades stars really are. Alcyone, the brightest, gives out 1,400 times more light than our sun. The faintest known members of the group are about 200 times less luminous than our sun. Thus there is a range in brightness of 280,000 times from the brightest to the faintest stars in the Pleiades. This is due to differences in temperature and in size, the extremes of temperature being 13,000° and 3,400°, and the diameters ranging from ten times to less than half the sun's diameter.

The stars are closer together in the Pleiades than in ordinary stellar space, but there is still plenty of room between them. On the average, it would take light more than a year to go from one star to its nearest neighbor. This space between the stars is not quite empty. Long-exposure photographs show that there are clouds around these stars. This nebulosity is shining by reflected or scattered light from these stars. The nebulous material must be made up largely of very small solid particles. The spectrum of the nebula is just the same as that of the stars embedded in it, showing that the clouds are shining by reflected starlight. This nebulosity cannot be seen with the unaided eye, but perhaps the fainter stars in the Pleiades may have given that appearance to Tennyson, who wrote:

"Many a night I saw the Pleiades rising through
   the mellow shade,
   Glitter like a swarm of fire-flies tangled in a
   silver braid."

CHAPTER **49** **THE HYADES**

About ten degrees southeast of the Pleiades is another cluster of stars, the Hyades. Mythologically the Hyades were daughters of Atlas and Aethra, and hence were half-sisters of the Pleiades. They were supposed to be seven in number, but their names were seldom used and have not been applied to the individual stars as in the Pleiades. This cluster has often been referred to as "the rainy Hyades," perhaps because they rise in the evening in the autumn, which marks the beginning of the rainy season in certain parts of the world. The daughters' names probably came from their brother Hyas, who was killed by a wild boar. Perhaps the many tears they shed over his death accounts for the watery reputation they have had.

The brighter members of the Hyades form a

letter "V," with the bright star Aldebaran at the upper left corner. The "V" marks the head of Taurus, the Bull, and Aldebaran marks one eye. This star does not belong to the Hyades cluster. It is about half as far away and happens to be in the same direction as the Hyades. It is moving southward nearly at right angles to the direction of motion of the cluster stars and with an apparent speed that is about twice as great.

reason why they should be moving along parallel paths. The convergence is only apparent and is due to perspective.

A few dashed lines have been drawn to indicate the convergence. These lines are slightly curved, because the diagram is a projection of part of the celestial sphere. It will be noted that the arrows farthest from the convergent point are longer than those nearest to it. The Pleiades would also show

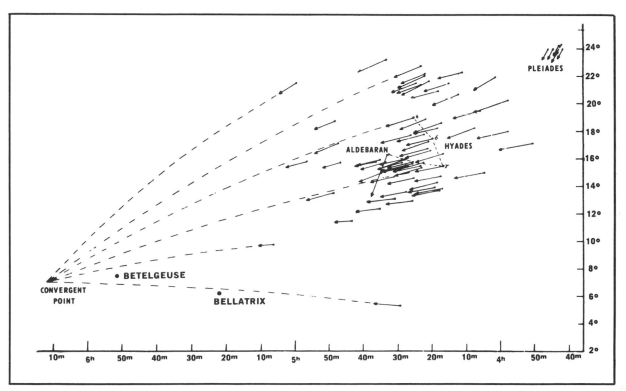

The convergence of the Hyades, or Taurus, cluster.

The V-shaped Hyades and neighboring stars in Taurus are converging toward a point in the sky just east of Betelgeuse. The lengths of the arrows represent the proper motions, or changes in apparent position, of the stars in an interval of 60,000 years. Aldebaran is not a member of this moving cluster. The positions and motions of the Pleiades are also shown.

About 150 stars have been found in this cluster. They are spread out over a much larger area in the sky than the Pleiades. The chart shows the six brightest stars of the Pleiades in the upper right corner and about 60 members of the Hyades. The arrows represent the proper motions of the stars in an interval of 60,000 years. The Hyades stars are moving in the same general direction, but their motions are not exactly parallel. In 1908 Lewis Boss found that these stars were converging on a point about -5° east of Betelgeuse. There is no reason why a group of stars should actually be moving in a converging stream, but there is a good

this convergence if they were spread out over a considerable area as the Hyades are. After the motions of the Pleiades have been observed for a long enough time, this convergent effect will finally show up.

The laws of perspective tell us that the direction from the observer to the convergent point is parallel to the true space motion of the group. In the diagram, a star in the cluster is at position A and is moving in the direction of the line AC. This line is parallel to the line EF from the earth to the convergent point. AC and EF are parallel, but they will meet the celestial sphere at the convergent point which is an infinite distance away. The angle AEF is the angular distance in the sky from the star to the convergent point. This is easily measured. In the case of Delta Tauri, it is 29°.1.

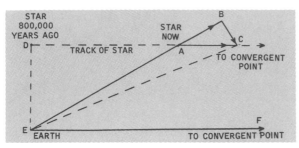

Finding the distance of a star in the Hyades.

Let AC represent the distance traveled by the star in one year. ABC is a right triangle constructed on AC as the hypotenuse. AB is a continuation of the line EA from the earth to the star and represents the distance that the star moves away from the earth in one year. By means of the spectroscope it is possible to measure the radial velocity of a star —that is, its velocity directly toward or away from the earth. Delta Tauri is found to be moving away from the solar system at a speed of 24 miles per second. Multiplying this by the number of seconds in a year, we find AB is about 750,000,000 miles.

In the triangle ABC the angle BAC is equal to angle AEF, since the angles are formed by the line BE cutting across the parallel lines, AC and EF. It is then possible to calculate the other sides of the triangle. AC is the actual path of the star in space in a year, but it may be thought of as consisting of two components. AB is away from the observer and BC is across the sky at right angles to the line of sight. Making use of angle BAC, we find BC is about 415,000,000 miles.

The angle AEC is the star's proper motion, or angular change of position, in one year. In the case of Delta Tauri it is 0″.115. From this angle we know that EA is about 1,800,000 times greater than BC. Hence the distance EA, from the earth to the star, is about 750,000,000,000,000 miles. Dividing this by 6,000,000,000,000 miles, which equal one light-year, we get a result of 125 light-years. Round numbers have been used in this example to make it simpler.

Thus we see the ingenious way in which the distance of this moving cluster has been deter-

Double cluster in Perseus. Like Hyades, this is another interesting "loose" cluster. (Yerkes Observatory.)

mined. Most of the stars in the cluster are contained in a roughly globular space about 35 light-years in diameter, the center of which is about 130 light-years distant. Their space velocity is about 28 miles per second. About 800,000 years ago the cluster was closest to the solar system, when its distance was half as great. After millions of years it will have shrunk to a telescopic star cluster smaller in apparent size than the Pleiades cluster is today. The common motion of the stars in a cluster suggests they have a common origin. The determination of the origin of star clusters is only one of many fascinating problems in astronomy.

Another extremely interesting "loose" cluster is the double cluster in Perseus, shown on page 248. On pages 292 and 293 is shown an entirely different type of cluster, the globular. It is illustrated by the examples found in Hercules and Sagittarius.

CHAPTER **50** **VARIABLE STARS**

The great nova of 1572 was observed by Tycho Brahe, whose description of it was quoted in the previous chapter. He had a pupil and assistant by the name of David Fabricius, who was a Dutch clergyman. In the early morning of August 13, 1596, Fabricius saw a star that he had never seen before and that he could not find in any catalogue. It was located in the constellation of Cetus, the Sea Monster, and appeared brighter than Hamal, a neighboring star of the 2nd magnitude. He observed it again in September and watched it grow fainter and finally disappear in October.

Fabricius probably thought this star was a nova. That may explain why the next recorded observation of the star did not occur until more than twelve years later. On February 15, 1609, Fabricius saw the star again. In the meantime, Bayer published his famous star chart "Uranometria" in 1603, in which he plotted this star as of the 4th magnitude. He was ignorant of its variability and assigned to it the Greek letter "omicron," and it is still called Omicron Ceti. It is also known as Mira, which is the Latin word for "wonderful." This was the name originally suggested by Fabricius because of its wonderful changes in light.

The next recorded observation of Mira after the one in 1609 was in 1631, when it was seen by Schickard. Then it was not seen again until 1638, when Holwarda noticed it while observing an eclipse of the moon. It disappeared quickly, but he found it again about eleven months later. Observations over many years finally showed that Mira varies in brightness with an average period of eleven months, but the interval between successive maxima ranges from about 320 to 370 days. At maximum it is usually of the 3rd or 4th magnitude, but it sometimes reaches the 2nd, equaling the North Star in brightness. At minimum it drops to the 9th magnitude, which is 3 magnitudes, or 16 times, fainter than the faintest stars visible to the naked eye.

When it was realized that Mira was not a nova, but had periodic variations, the fields of old novae were occasionally examined to see whether new outbursts might occur. In July, 1686, while G. Kirch of Berlin was examining the region of Nova Vulpeculae, discovered in 1670, he compared the surrounding part of the sky with the star charts. He noticed the absence of Chi Cygni, which had been marked on Bayer's chart as of the 4th magnitude. After watching the region for several months, he finally saw the star reappear. It was later found to have a longer period and a greater range in brightness than Mira. During $13\frac{1}{2}$ months its brightness varies by about 10,000 times, changing its magnitude from 14 to 4.

A third long-period variable star, R Hydrae, was found in 1704 by G. F. Maraldi, an Italian astronomer. It has a range of 6 magnitudes, but the most striking fact since found is the progressive decrease of its period. At the time of its discovery, the length of its cycle was 500 days. Now it is only 400 days.

No more variable stars were discovered during the next 78 years. Then in 1782 Koch found R Leonis, a long-period variable having many similarities to Mira. In each of the years 1784 and 1795 three new variables were discovered, but they were not of the long-period type. One of these discoveries, Alpha Herculis, was made by the famous astronomer William Herschel, but the other five were made by two remarkable young men in England, named Edward Pigott and John Goodricke. They were close friends and observed together.

In addition to determining the period of Algol, Goodricke discovered another eclipsing variable, Beta Lyrae, on September 10, 1784. On the very same night Pigott found Eta Aquilae to be variable. A few days later Goodricke discovered the variability of Delta Cephei, which has since become a very famous star.

The very promising career of John Goodricke was cut short in 1786, as a result of exposure while making astronomical observations. He died at the age of 21, only two weeks after he had been elected a fellow of the Royal Society. In three years he had given a great impetus to variable star astronomy by his pioneering work on three of the best known stars in the sky, Algol, Beta Lyrae, and Delta Cephei.

Goodricke's companion, Pigott, continued his work and in 1795 reported the discovery of two more variables, R Coronae Borealis, and R Scuti. The behavior of each was different from the others. Thus by 1800, in addition to a few novae, eleven stars had been found to be variable.

The rate of discovery gradually increased during the nineteenth century until about 1,000 variables were known by 1900. Celestial photography has been responsible for a tremendous increase during the twentieth century. The total number found to date is probably around 20,000.

The designation of these stars should be explained briefly. The first few variables discovered were naked-eye stars, each of which had already been designated by a Greek letter followed by the genitive case of the constellation in which it is located. A new system was invented for the fainter stars, which had no previous designation. The first variable discovered in a constellation receives the letter R, the next S, and so on to Z. This method was introduced by F. W. A. Argelander before 1850. Apparently he started with R because so many variable stars are red. Soon the nine letters from R to Z were exhausted in certain constellations. In 1881 Hartwig suggested using the double letters RR, RS, etc. This added 45 symbols and took care of the situation until about 1904, when double

letters from the first part of the alphabet were used. The letter J is omitted, probably to avoid confusion in the German system, where I and J are much alike. This addition provided for a total of 334 stars in each constellation, summarized as follows:

1. R	11. RS
2. S	. . .
. . .	18. RZ
9. Z	19. SS
10. RR	20. ST
. . .	. . .
26. SZ	79. AZ
. . .	80. BB
54. ZZ	81. BC
55. AA	. . .
56. AB	334. QZ

Even this extension was not enough. Finally it was agreed internationally to adopt a number system to supplement the letters. After 334 stars in a constellation have been designated by letters, the next variable found is called V335, followed by the genitive case of the constellation name. For example, the 334th and 335th variable stars found in Cygnus are designated QZ Cygni and V335 Cygni. There is no limit to the number of stars which can be designated by this numerical system.

Another system of nomenclature devised at the Harvard Observatory is also widely used. Six digits are employed to give the star's position for 1900. The first two give the hour of right ascension, the second two the minute of right ascension, and the last two the degree of declination. Thus the Harvard designation of R Leonis is 094211, because its right ascension was $9^h 42^m.2$ and its declination was $11° 54'$ N. If the declination is south, the two digits for it are put in italics.

There are many varieties of variable stars, and some are difficult to classify. Leaving out the novae and the eclipsing binaries, we will divide the periodic variables into those with short periods and those with long periods. Then we will consider stars that are semi-regular and those that are irregular in their light variations.

Variable stars with periods less than a day are often called cluster variables, because they were first found in great numbers in the globular clusters of stars just outside our galaxy. Recently many of them have been discovered in our galaxy. Stars having a period of about a week have been named

Cepheids, after the typical star, Delta Cephei, whose variability was found by Goodricke. The name "Cepheid" is now used to include the short-period variables of both types. When necessary to distinguish between the two, we can use the terms cluster variables and classical Cepheids.

Until recently the shortest period known among the cluster variables was that of CY Aquarii. It varies between the 10th and 11th magnitudes in only 1 hour and 28 minutes. When it is increasing in brightness, a change can be detected in about five minutes. A variable having a period of only 80 minutes was discovered in the southern constellation Phoenix by Dr. Olin J. Eggen, of the University of California, working in Australia. This star is at a distance of 120 light-years from the earth. Most of these cluster variables have a period of around half a day. Very few stars have periods between 21 hours and 2 days. After this gap the classical Cepheids begin, reaching a maximum number with a period of about one week. The longest period known among Cepheids in our galaxy is 45 days.

The typical light curve of a Cepheid shows a rise to maximum that is more rapid than the fall. The maximum is usually more sharply marked than the minimum. In some cases the rise and fall of brightness are about equally rapid, but no case has been found in which there is a slow rise and a rapid fall. The shape of each light curve changes very little. Minor fluctuations have been found in the period and form of the light curves of some cluster variables, but not of the Cepheids of longer period. The changes are of the order of a few seconds in a period of half a day or less.

Now that we have seen how these stars vary in light we would like to know why they do. That question is not so easily answered. In order to give even a brief explanation, it is necessary to give a few details about the spectra of these stars.

An ordinary star, which is fixed in brightness, has the same spectrum all the time. A variable star changes its spectrum by about one spectral class while it is going through its changes of light. Also, the spectrum and the period are related. A Cepheid with a period of half a day will show an A-type spectrum at maximum brightness and an F-type spectrum at minimum. For periods of four or five days the spectra vary between F and G, and for the longest periods they vary between G and K.

Since the spectral classes are directly connected with temperature, we find that the Cepheid stars are about 1,000° hotter when they are brightest than when they are faintest. This is confirmed by the changes of color. At minimum light a star appears redder than at maximum. This is detected by comparing the visual and photographic light curves. The range in variation for the visual curve is considerably less than that for the photographic curve. An ordinary photographic plate is not as sensitive to red light as the eye is, and so the drop in brightness from maximum to minimum light appears larger on photographs. Red stars are cooler than white stars just as a piece of red-hot iron is cooler than one that is white-hot.

The changes of light, then, are due to changes of temperature. The hotter a star is, the brighter it appears. In fact, the energy varies as the fourth power of the absolute temperature. For example, a star with a temperature of 6,000° radiates 16 times as much energy as a star of the same size having a temperature of 3,000°. Our next step is to try to explain why the temperature of a star can change.

The lines in the spectrum of a Cepheid show shifts of position in the same period as that of the light variations. According to Doppler's principle, a shift of the lines toward the violet end of the spectrum means that the source of light is approaching us. Similarly a shift toward the red occurs when the object is receding from us. The explanation of the rhythmical shifts of the spectral lines is that the surface of the star alternately expands and contracts, changing the temperature and hence the brightness.

Most Cepheids are excellent timekeepers. There are two kinds of oscillation used in timekeepers here on earth. One is the motion of a body under gravitational force, such as the pendulum of a clock. The other is the motion under an elastic force, such as the balance wheel of a watch. These two forces operate in a star. Gravitation pulls it together, and the elastic force of gas pressure keeps it from collapsing. Most stars are in equilibrium, one force balancing the other. Suppose, for some reason, that one force is greater than the other. If gravity is greater, the star contracts. This causes the gas pressure to increase and the temperature to rise. At a certain point the gas pressure becomes greater than gravity and the star begins to expand. The star expands beyond its normal condition of

251

equilibrium, like a pendulum drawn to one side and let go. During the expansion the pressure and the temperature decrease until the force of gravity again gets the upper hand and causes the star to contract.

Such a change in the size of a star is called pulsation. It is not known why some stars pulsate and others do not. Once started, the pulsation in a star should continue for a very long time. The pulsation theory was suggested by Harlow Shapley and developed mathematically by Sir Arthur Eddington. It is now generally accepted as the explanation of the Cepheid variation. In this model of a pulsating star the greatest brightness should be reached when the star is hottest and smallest. Since the amount of light radiated from a given area varies as the fourth power of the temperature, the increase in brightness per unit area more than compensates for the decreased surface area when the star is smallest.

However, the facts are that these stars are brightest, not when they are smallest, but at the time of most rapid expansion. Also, minimum light occurs at the time of the most rapid contraction. This discrepancy has been explained on the assumption that it takes some time for the effect of the internal heating to work its way out from the interior to the surface. The outer layers lag behind the interior, and this leads to waves running outward from the interior.

Now let us consider the variable stars of long period, of which Mira, the first variable star discovered, is an outstanding example. Most of them have periods between 200 and 400 days, with the greatest number around 275 days. Some have periods of about two years and one has a period of nearly four years. The light variation is only roughly periodic. The length of the cycle varies, and the brightness of the star at maximum may differ from one cycle to the next. All of the long-period variable stars are red, and most of them belong to spectral class M. About 10 percent belong to the rare classes, N, R, and S. In the typical stars, the bands of titanium oxide are very strong, indicating a low temperature. Near maximum light, most of these stars show bright hydrogen lines, instead of the dark lines found in the ordinary stellar spectra.

The light curves of these red variables show many different forms. In some the maxima and minima are of similar shape. Others have wide maxima and narrow minima, whereas still others have narrow maxima and wide minima. Like the Cepheids, the long-period variables are hotter at maximum than at minimum brightness. The range is about 600 degrees. This seems small, considering that these stars have an average light variation of about six magnitudes. The Cepheids have a range of 1,000 degrees, but a light variation of only one magnitude. The explanation is that at the low temperatures of these red stars a small change in total radiation results in a much larger change in visual radiation. When we consider the total radiation of a red variable, the range is only about one magnitude, the same as that for Cepheids.

The spectral lines shift in position, suggesting a pulsation. However, they show a maximum speed of expansion at minimum brightness, instead of at maximum brightness, as in the case of Cepheids. The explanation seems to be that the red variables are larger than the Cepheids. Only the denser part of the star pulsates as a whole. For the red stars the denser portion is a much smaller fraction of the whole star than is the case for Cepheids, so that there is an even greater lag of the outer layers behind the interior, as the waves run outward from the interior.

In addition to the Cepheids, whose periods are very regular, and the red variables, which are only roughly periodic, there are some stars whose variations are semi-regular and others that are very irregular. It is difficult to make a satisfactory classification, but a few examples will be given to show the great variety of variable stars.

There are about 25 stars belonging to the RV Tauri group, named after a typical member of the group. Its light curve consists of semi-regular fluctuations with a 79-day period, during which there are two maxima of almost equal height and two minima of unequal depth. These changes are superimposed on a slower wave of 1,300 days. Some stars in this group show frequent interchanges between primary and secondary minima.

In the RV Tauri group is R Scuti, discovered to be variable by Edward Pigott in 1795. Its period is the longest in the group, 146 days elapsing from one primary minimum to the next. Its spectrum varies from K at maximum to M at minimum. A few stars of this group show bright hydrogen lines at maximum. This is characteristic of the red variables. The RV Tauri stars seem to occupy an intermediate position between the Cepheids and the long-period variables.

Another group has been named after the star

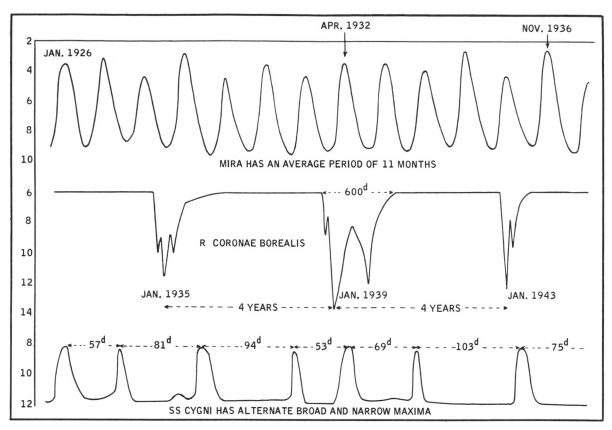

Light variations of three variable stars.

Magnitudes are indicated at the left. The time scale for the first two stars is about 8 times longer than that for SS Cygni, whose periods are indicated in days (*d*). The repetition of the 4-year interval between minima of R Coronae Borealis is accidental, its variations over a long period of time being very irregular.

U Geminorum. Normally it is very faint, of the 14th magnitude. It may continue this way for several months. Then suddenly it increases its brightness by 100 times, going up to the 9th magnitude in about one day. A few days later it rapidly drops back to its normal brightness. It is like a nova on a small scale, but it repeats its outbursts several times a year. A similar star is SS Cygni, which varies between the 12th and 8th magnitudes. The maxima are alternately wide and narrow, as a rule.

It is not possible to predict when these explosions will occur. The period of U Geminorum has varied between 62 and 257 days, with an average of 97 days. Other average intervals are 50 days for SS Cygni and 17 days for SU Ursae Majoris. There is an interesting relation between the period and the amount that the star changes in brightness. Stars with average periods between 10 and 20 days have a range of about three magnitudes. Those with periods between 50 and 70 days have a range of

four magnitudes. Still longer intervals occur with a range of five magnitudes. The longer the interval, the greater is the explosion. If we kept on increasing the interval and the range, we would finally come to the repeating novae, like RS Ophiuchi. If the relation should hold for ordinary novae, we might expect them to repeat their outbursts after intervals of thousands of years.

A star that has an upside-down light curve, as compared with U Geminorum, is R Coronae Borealis. It is normally of the sixth magnitude, and may remain there for many years, as it did from 1925 to 1934. Then it suddenly may decrease in light by as much as nine magnitudes. It is exceeded in range only by the novae, but its light curve is moving in the opposite direction. Its drop is not quite as rapid as the rise of a nova, usually taking weeks instead of days. The recovery is slower and may be accompanied by marked fluctuations. The times of minimum appear to be distributed absolutely at random, according to the laws of pure chance. There are about a dozen variable stars that behave in a similar way.

The spectrum of R Coronae Borealis is of class G, but it is peculiar in that the carbon lines are strong. It has been suggested that carbon in this stellar atmosphere behaves something like water in the earth's atmosphere. Both of them are transparent when they are in the form of gas. When the water vapor condenses to form droplets of water or ice crystals, the result is a cloud that is opaque. Similarly, it is possible that a change in temperature or some other cause would produce a condensation of the carbon vapor in the atmosphere of this star. The cloud formed in this way would obscure the star's light very effectively. The brightening of the star again would then be due to the gradual dissipation of the cloud.

minutes at a time. They are called "flare stars," since it has been suggested that the brightening may be a consequence of an eruption similar to solar flares. The accompanying curve, showing the variation of light of the flare star BD +19°5116 as observed by Paul Roques in 1954, is fairly representative of the group. The star doubled in brightness in about one minute, and then began immediately to fade, with some oscillations in brightness. By the end of fifteen minutes the star had returned to its former brightness. Other flare stars brighten even more rapidly, sometimes reaching four times the quiescent brightness.

Not all the different kinds of variable stars have been described, but enough have been presented to show what a variety there is and how many problems remain to be solved. With thousands of stars behaving in so many different ways,

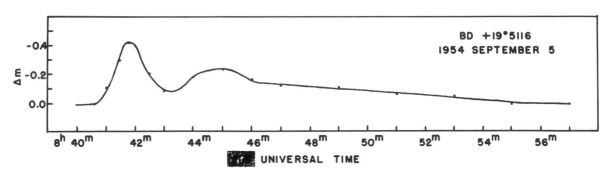

The brightening of the flare star BD +19°5116 observed photoelectrically.
(By Paul E. Roques at the Griffith Observatory in 1954.)

Many red stars vary in light in an unpredictable manner, except that the variations are within narrow limits. For instance, Betelgeuse shows semiregular waves of 140 to 300 days, which are superposed on a much slower fluctuation with a period of six years. The range of light variation never exceeds about one magnitude. Its spectral lines also oscillate in a six-year period, indicating an expansion and contraction.

Variations in light of the red stars are not restricted solely to the red giants or supergiants. Some of the intrinsically faint red stars have been found to increase suddenly in brightness for a few

the job is too big for the professional astronomers. Fortunately this is the kind of work that can be well done by trained amateurs. Numerous groups of observers throughout the world have been formed for the purpose of systematically watching variable stars. In 1911 a group of seven men formed the American Association of Variable Star Observers, abbreviated as A.A.V.S.O. Edward C. Pickering, former Director of the Harvard College Observatory, helped the group a great deal, but its guiding spirit for 25 years until his death in 1936 was William Tyler Olcott. Under his leadership this group has grown to a world-wide organization, which has accumulated more than one million observations of about 500 variable stars.

Olcott was trained for the legal profession, but never practiced law. He devoted his life to writing books on the stars, particularly for the layman. With Edmund W. Putnam he wrote "Field Book of the Skies," one of the most popular books of its kind. In describing variable star observing, Olcott wrote in one of the pamphlets of the A.A.V.S.O. as follows:

"The only requisites for the work are a telescope of three inches aperture or larger, a comprehensive star atlas, and a fair knowledge of the constellations. The work is entirely visual, involves no mathematics, and the details are quickly and easily mastered. It is merely a matter of perseverance, patience, and eye training to become an expert observer.

"The work is the most practical and valuable in which the amateur astronomer can engage. Each and every observation is of positive scientific value. One has the immense satisfaction of doing his bit to add to the sum total of scientific knowledge, something intellectually worth while."

Another reason for observing variable stars may lie in the reply of a scientist, who, when asked why he had given his life to research, said, "Because it's so much fun." Anyone interested in taking up this fascinating hobby can obtain information from the A.A.V.S.O., Harvard College Observatory, Cambridge, Mass.

CHAPTER **51** **EXPLODING STARS**

The brightest permanent object in the night sky, except the moon, is the planet Venus. It appears 12 times more brilliant than Sirius, the brightest star. However, in November, 1572, a star appeared that was brighter than Sirius and probably even exceeded Venus. It was extensively observed by the famous astronomer Tycho Brahe, who described it as follows:

"One evening, when I was contemplating, as usual, the celestial vault, whose aspect was so familiar to me, I saw, with inexpressible astonishment, near the zenith, in Cassiopeia, a radiant star of extraordinary magnitude. Struck with surprise, I could hardly believe my eyes. . . .

"The new star was destitute of a tail; no nebulosity surrounded it; it resembled in every way other stars of the first magnitude. Its brightness exceeded that of Sirius, and of Jupiter. It could only be compared with that of Venus. Persons gifted with good sight could distinguish this star

Bayer's star chart of Cassiopeia shows position of Tycho's nova of 1572. At its brightest, it was visible in daylight, but it is so faint today that telescopes have failed to identify it. It is probably about 1,600 million times fainter than it was at maximum brightness.

255

in daylight, even at noonday, when the sky was clear. At night, with a cloudy sky, when other stars were veiled, the new star often remained visible through tolerably thick clouds. The distances of this star from the other stars of Cassiopeia which I measured the following year with the greatest care, have convinced me of its complete immobility. From the month of December, 1572, its brightness began to diminish; it was then equal to Jupiter. In January, 1573, it became less brilliant than Jupiter; in February and March, equal to stars of the first order; in April and May, of the brightness of stars of the second order. The passage from the 5th to the 6th magnitude took place between December, 1573, and February, 1574. The following month the new star disappeared without leaving a trace visible to the naked eye, having shone for seventeen months."

This star was called a nova, coming from the Latin word meaning "new." The early observers without telescopes thought that such a star was newly created. It is now known that it is only temporarily a bright star, and so both terms, "nova" and "temporary star," are used. Later it will be shown why this can also be called an exploding star.

After the star of 1572, another nova appeared in 1604. It showed up in the constellation of Ophiuchus and was observed by Johannes Kepler, another famous astronomer. It became brighter than Jupiter but not as bright as Venus. It remained visible to the naked eye from October, 1604, until March, 1606, an interval of 17 months, the same length of time that Tycho's star was visible without a telescope. This was only three years before Galileo first pointed a telescope at the sky.

In 1941, Dr. Walter Baade, at Mount Wilson, found on red-sensitive photographs of the region a small fan-shaped patch of nebulosity, very near the position of Kepler's star. Practically invisible on blue photographs because of heavy obscuring clouds in the vicinity, it is quite conspicuous in red light. The nebulosity is apparently all that remains of the three-century-old supernova, although there is presumably a faint star somewhere within it to provide the illumination of the gaseous material.

Other fainter novae appeared, but nearly three centuries passed before another brilliant one was discovered. It was found in 1901 in the constellation of Perseus, and so it is referred to as Nova Persei 1901. The discoverer was Dr. T. D. Anderson, a Scottish clergyman and amateur astronomer. While he was walking home on the night of February 21-22, 1901, he noticed a new star of the third magnitude in the region between the famous variable star Algol and the brightest star of the constellation, Alpha Persei. He realized at once that it was a nova. In fact, he had been on the lookout for novae for many years and had already made the remarkable discovery of Nova Aurigae in 1891. He found that star when it was only slightly brighter than the 5th magnitude. Since there are more than a thousand stars brighter than the 5th magnitude, Dr. Anderson must have had an extraordinarily close acquaintance with the constellations to detect such a faint new star.

At discovery, Nova Persei was slightly brighter than the 3rd magnitude, about half as bright as the North Star. Dr. Anderson sent word of his discovery to the Greenwich Observatory, which spread the news to observatories all over the world. Two days later the star reached a magnitude of 0, making it as bright as Vega and Capella.

Fortunately the Perseus region of the sky had been photographed at the Harvard Observatory only two days before this nova was found. In the position of the nova there was found a faint star of the 13th magnitude. In two days the star had increased its brightness by 10 magnitudes, or 10,000 times. The total change from the 13th magnitude to 0 magnitude was made in four days and corresponds to an increase of 160,000 times. Earlier photographs showed that this star had been making small fluctuations in brightness between the 13th and 14th magnitudes.

As is always the case with such stars, Nova Persei immediately began to fade after its maximum light. It dropped about 4 magnitudes during the first two weeks. Then a series of oscillations began, with a period of about four days and an amplitude of a magnitude and a half. This stage lasted for several months while the star faded from naked-eye visibility. After 15 years it finally returned to its former state, where it is now fluctuating in brightness around the 13th magnitude.

The story of Nova Persei has two more startling chapters. Seven months after the outburst, long-exposure photographs showed a diffuse, extended cloud of faint light around the star. Further observations indicated that this nebulosity was moving outward in all directions from the star at a rate of about $1/6°$ per year. This may not seem very fast to the layman, but when the distance of the star was measured, it was found that the nebula was moving with approximately the speed of light. Such speed is out of the question for anything but light itself.

The explanation is quite simple. In the space surrounding the star was an invisible dark nebula, consisting of gas and dust. The star was intensely luminous for a few days and it sent out light waves that successively illuminated the surrounding nebula, like the widening ripples on a pond. As the expanding spherical shell of light passed over the clouds of diffuse matter, observers at a great distance saw an expanding ring of light. This unique phenomenon helps one to appreciate the immensity of interstellar distances. Although light travels with a speed of 186,000 miles a second, the expanding shell of light moved so slowly that a week or two had to elapse before the motion was noticeable.

The final chapter in this story came in 1916, when a tiny nebula was found around the star. It was also expanding, but so slowly that it had taken 15 years for it to be large enough to be detected with a telescope. This was a true shell of gas blown off from the star at a speed of about 750 miles a second. At the time of the outburst, in 1901, the spectroscope showed that gases were leaving the star with that high velocity. It was very satisfying to have this evidence of the spectroscope confirmed by the appearance in our larger telescopes of this

Nova Persei and its expanding nebula in 1917 (above) and 1949 (below). (Mount Wilson and Palomar Observatories.)

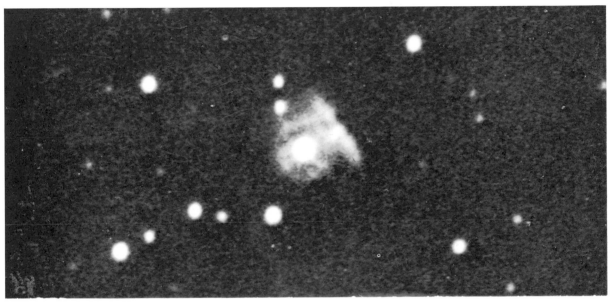

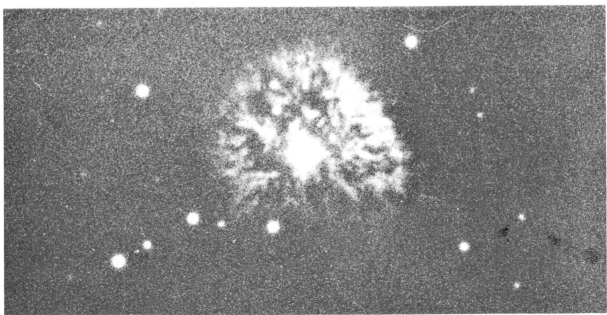

expanding shell of gas though it was 15 years later.

An expanding shell of gas was also observed around Nova Aquilae 1918, becoming visible within six months of the outburst. This nova was the brightest one of modern times, reaching a magnitude of −1.1, brighter than any star in the sky except Sirius. It was discovered independently by many observers on the night of June 8, 1918, when its magnitude was about +1. Two days later it was at maximum. It was afterward found to have been photographed on June 7 as of the 6th magnitude, and on June 5 as of about magnitude $10\frac{1}{2}$. For many years before 1918 it had appeared on occasional photographs, varying between the 10th and 11th magnitudes. Thus there was an increase in brightness of around 50,000 times in five days.

Nova Aquilae faded rapidly at first, and then more and more slowly. It was lost to the naked eye after eight months and reached its normal brightness after seven years.

Nova Herculis 1934 was discovered at 4:30 on the morning of December 13 by an amateur astronomer named Prentice. He was director of the Meteor Section of the British Astronomical Association and had been up all night observing meteors. He happened to look toward Hercules and saw a strange 3rd magnitude star about halfway between Vega and the head of Draco. Immediately he telephoned the Royal Observatory at Greenwich, with the result that a spectrogram of the nova was obtained before daybreak. Ten days passed before the star reached its maximum light at a magnitude of 1.4.

The changes in Nova Herculis were very leisurely at first. It dropped only three magnitudes in three months. Then during April 1935 it faded from the 5th to the 13th magnitude. During May and June it brightened slowly but steadily to the 7th magnitude. After that the final decline set in.

The most recent bright nova was Nova Puppis, which reached a maximum magnitude of 0.35 on November 11, 1942. Its discovery was reported by B. H. Dawson of the University Observatory, La Plata, Argentina, and it was found independently by other observers soon after. For example, in the early morning of November 10, when Dr. Edison Pettit of the Mount Wilson Observatory went out to get his newspaper in front of his house in Pasadena, he spotted the nova just above the southern horizon. Before daybreak he was able to measure its magnitude with his 6-inch telescope.

Most people in the northern latitudes did not see this nova, because it did not rise in November until well after midnight and did not get very high above the horizon. Also it faded very rapidly, disappearing from naked-eye view at the end of November. The pre-nova magnitude is uncertain, because it was too faint to show on photographs made of that region. It may have been as faint as the 18th magnitude. If so, Pettit estimates that during the first two months after its outburst, Nova Puppis radiated as much energy as it had in the previous 164,000 years. This is equal to the energy radiated by the sun in 18,300 years.

In the pre-nova stage, most of the stars are fainter than the 14th magnitude. About half of them show fluctuations in brightness of a magnitude or so before the outburst. The average range from minimum to maximum brightness is about 11 magnitudes, or 25,000 times. The initial rise amounts to 9 magnitudes, or 4,000 times, and is accomplished in two days. Then there is a pre-maximum halt, whose duration ranges from zero for the most rapid novae to many days for the slowest. The final rise through the last two magnitudes to maximum may take two days for a fast nova and two weeks for a slow one.

The maximum light lasts from a few hours to a few days. The next stage is called the early decline, extending from maximum to about 3.5 magnitudes below maximum. This change may be smooth or irregular, and takes from a few weeks to a few months. Following this is the transition, lasting from one to several years, during which the nova's light curve will do one of three things: (1) fluctuate sharply, with a rough period of several days per fluctuation, until the star is six magnitudes fainter than maximum; (2) drop to a deep minimum and then make a partial recovery, leveling off at a point 6 magnitudes fainter than maximum; (3) simply show a rather quick change toward a gentler slope. Whatever a nova does at 3.5 magnitudes below maximum, it comes out of this transitional stage at about 6 magnitudes below maximum.

The final decline is uneventful and takes several years for even the fastest novae. The average duration of decline of six well-observed fast novae is about eight years. One slow nova took about 30 years to reach a luminosity equal to that of the pre-nova stage. Small fluctuations of light may continue for years, but the important fact to be noted is

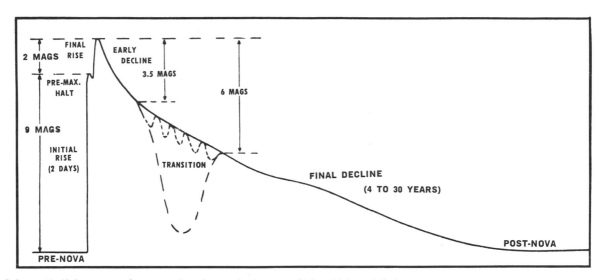

Schematic light curve of a nova, showing typical stages. (After McLaughlin.)
The time scale has been magnified in the early stages. During the transition a nova will follow either the continuous line or one of the dashed lines.

that the average magnitudes before and after the outburst are the same. After increasing 25,000 times, the nova finally gets back to the same luminosity it had always had.

We cannot be sure that a star has always been well-behaved before it becomes a nova. In fact, there are six cases of repeating novae: RS Ophiuchi, with maxima in 1898, 1933, and 1958; T Pyxidis, with maxima in 1890, 1902, 1920, and 1945; Nova Sagittarii 1919, with another maximum in 1901; T Coronae Borealis, 1866 and 1946; U Scorpii, 1863, 1906, and 1936; Nova Sagittae, 1913 and 1946. It is possible that all novae may repeat in intervals of thousands of years. It is not necessary to assume that every star in the course of time must become a nova in order to account for the observed frequency of novae. A special class of stars appears to be responsible for all nova activity, and these stars may become novae more than once. It is comforting to know that our sun does not belong to this class of stars, because all life on the earth would be destroyed if our sun should become a nova.

White lines have been drawn to connect the principal stars of Corona Borealis, the Northern Crown. Arrow points to T Coronae Borealis, as it appeared in 1946 after increasing about 1,500 times from eleventh magnitude to third. It immediately began to fade and disappear from naked-eye view. This star had a similar increase and decrease in brightness in 1866.

A peculiar star that has behaved somewhat like a nova is Eta Carinae. It was first noted by Edmund Halley in 1677 at the 4th magnitude. It varied in brightness between the 4th and 2nd magnitudes until the first part of the nineteenth century. First magnitude was reached in 1827, and a maximum of 0.2 was attained in 1838. There was a second

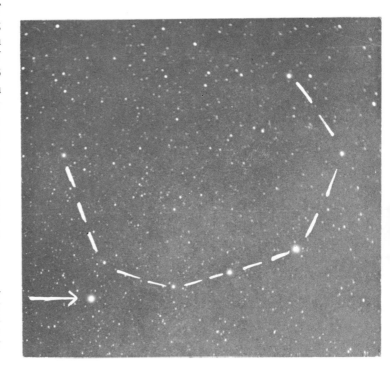

259

Spiral galaxy, NGC 4725, is a giant system of stars at a distance of 5½ million light-years. Arrows point to star of modest magnitude in 1931 (left) that on May 10, 1940 (right) appeared as a supernova, blazing up to a maximum luminosity of 30 million suns, or one-seventh the total luminosity of the galaxy itself. (Mount Wilson Observatory.)

maximum in 1843, when it was at its brightest recorded magnitude of —0.8. It remained brighter than the 1st magnitude for another 15 years. Then it began a steady decline, which stopped abruptly at the 7th magnitude in 1869. It finally dropped to the 8th magnitude, where it has remained nearly constant ever since.

Most of the observed novae are so far away that light takes more than a thousand years to come from them to us. Some are at distances of more than 5,000 light-years. Explosions that we see now may actually have occurred at the time when the pyramids were being built in Egypt, several thousand years before Christ. If those stars were as close to us as some of our bright naked-eye stars, their light at maximum would appear considerably greater than that of Venus.

The average absolute magnitude of a nova at maximum light is about —7. That is 11.8 magnitudes, or 50,000 times greater than the sun's luminosity. The fast novae are brighter than the average, and the slow novae are fainter. There is strong evidence for a duration-luminosity relation. High luminosity is related to a very rapid decline of light, and low luminosity to slow decline. This

relation seems to hold between the extreme values of —9 for Nova Aquilae 1918, a very fast nova, and —3.6 for a very slow nova called RT Serpentis.

In addition to the novae observed in our galaxy, similar objects have been found in other galaxies. The nearest of such extragalactic systems of stars are the Magellanic Clouds. These are two cloud-like objects near the south celestial pole. They are visible to the naked eye, but are too far south to be seen from the latitudes of the United States. One nova or nova-like star has been found in the Lesser Magellanic Cloud, and three novae have been found in the Greater Cloud. These systems are about 190,000 light-years away.

At a distance of more than two million light-years is the famous Andromeda galaxy. More than 100 novae have been recorded in this system in recent years. At least five novae have been found in the galaxy in Triangulum known as Messier 33. The average absolute magnitude of the well-observed novae in these exterior systems is about —7 at maximum light. This agrees with the value for novae in our galaxy.

The average apparent magnitude at maximum for novae in the Andromeda galaxy was found to be about 17 for slow novae and about 15 for fast novae. However, in 1885, a star in that galaxy reached an apparent magnitude of 7. This is roughly 10,000 times brighter than an ordinary nova. Such an object is called a supernova. Stars like this have been found in other galaxies, but they are much rarer than ordinary novae. It is estimated that although a galaxy may produce 30 or more common novae per year, a supernova appears in any particular galaxy about once every 500 years.

The average absolute magnitude of supernovae at maximum is about -14. If a supernova appeared at a distance of $32\frac{1}{2}$ light-years, it would have this apparent magnitude and would appear more than three times brighter than the full moon. Whether a supernova has been seen in our galaxy is not absolutely certain. The famous novae of 1572 and 1604 may have been supernovae in our galaxy, but at great distances from us. Since we do not know their distances, we cannot tell whether they belong to this special class of novae. Their light curves indicate that possibility. In general, supernovae show broader maxima, and slower and steadier declines after maximum. The bright novae of 1572 and 1604 seemed to show these characteristics.

Within the last few years, evidence has been accumulating that there was a very bright supernova in 1054, which was responsible for the formation of the Crab Nebula. Recent translations of ancient Chinese chronicles show that a "guest-star" appeared in the constellation of Taurus on July 4, 1054, and remained visible to the naked eye until April 17, 1056. The account reads in part as follows: "It was visible by day, like Venus; pointed rays shot out from it on all sides; the color was reddish-white. Altogether it was visible for 23 days." Presumably this means that it could be seen in the daytime for 23 days.

The exact position of the star was not given, but it was stated as being near the star we call Zeta Tauri. At a distance of $1\frac{1}{4}°$ from Zeta Tauri is the Crab Nebula. It does not look much like a crab, but that name has persisted for a century. Photographs made over an interval of many years show that the nebula is expanding. If the rate of expansion were precisely constant, the nebula should have started about 800 years ago. The nova appeared 891 years ago, but in view of the uncertainties of the measures and the possibility of a small

change in the speed of expansion, the results agree closely enough to probably settle the question.

The spectrum of the Crab Nebula consists of double bright lines. This is because the part of the nebula on the near side is approaching us, while the part on the far side is receding. The separation of the members of a pair of spectral lines is a measure of the linear speed of expansion. The results give an average speed of over 800 miles a second. Combining this linear velocity in the line of sight with the angular velocity across the line of sight, we can calculate the distance. This is based on the assumption that the nebula is expanding at the same rate in all directions. This method gives a distance of about 4,000 light-years.

The recent development of radio astronomy, together with the construction of highly directional radio antennae that enable radio astronomers to pinpoint sources of radio emission in the sky, have shown that the Crab Nebula is one of the strongest sources of radio noise in the galaxy. Detailed study has shown that the emission from the nebula is probably "synchrotron" radiation, so called because it was first observed in synchrotron beams, and explained by the rapid whirling of electrons in the strong magnetic fields of such machines. The conclusion is that there must be great numbers of free electrons in the Crab Nebula, acted on by a strong magnetic field that is also effective in producing the very complex filamentary structure of the gases of the nebula.

The Crab Nebula is one of the most interesting physical systems in astronomy. The central star, which is probably the object that exploded, has been found to be an optical and radio pulsar emitting 30 pulses every second. This is the shortest period found for any known pulsar at the present time. According to current theory, the central star is a highly collapsed neutron star with a radius of 5 or 6 miles and a density of at least a hundred million million times that of water. Pulsars are discussed in more detail in Chapter 59. It is probable that each is the result of a supernova outburst, and that the pulsation is associated with the rotation of the neutron star. The Crab Nebula is also an intense source of X-rays and cosmic rays; indeed, practically all of the physical processes of interest to astronomers are to be found in the Crab.

There are about 300 nebulae that are spherical in shape and are called planetary nebulae because

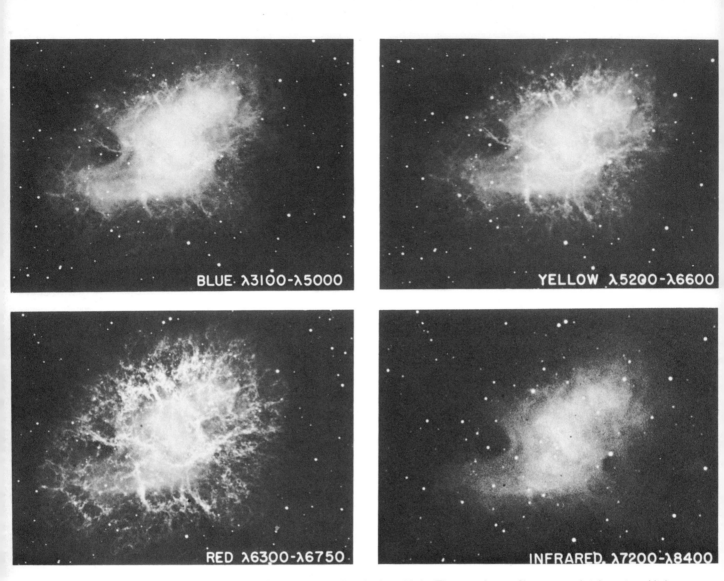

BLUE λ3100-λ5000

YELLOW λ5200-λ6600

RED λ6300-λ6750

INFRARED λ7200-λ8400

Crab Nebula in Taurus, photographed in blue, yellow, red, and infrared light. The ragged outer filaments are brightest in red light, and are produced by glowing gases that are expanding outward from the center of the nebula. The light of the inner part, which is about equally bright in the four photographs, is produced by electrons moving with speeds close to the velocity of light in a strong magnetic field. Similar glows are observed from electrons in high energy accelerators; hence its name, "synchrotron radiation." (Mount Wilson and Palomar Observatories.)

they present disks like planets. Each nebula is lighted by a star at its center. It is very tempting to believe that these planetary nebulae are the wrecks of old novae. However, the resemblance is only superficial. There is a doubling of the lines in the spectra of some planetary nebulae, indicating an expansion, but the velocity is very much smaller than that of the nebula around a nova. The nebular shells around novae are usually very short-lived, but the planetary nebulae have not shown visible changes in many years. The Crab Nebula seems to be entirely different from the planetary nebulae.

It shows a velocity of expansion like that of a nova, amounting to over 800 miles a second. The velocity of a typical planetary nebula is only about 15 miles a second.

Space does not permit a description and explanation of the remarkable spectral changes that accompany the outburst of a nova. Briefly, during the interval of a nova's rise to maximum luminosity, it usually shows a continuous spectrum crossed by narrow absorption lines. These lines are displaced toward the violet end of the spectrum by amounts which, according to the Doppler prin-

Dumbbell Nebula in Vulpecula. This planetary nebula looks like shell of gas around old nova, but resemblance is superficial. (Photographed through 120-inch telescope, Lick Observatory.)

ciple, correspond to velocities of approach ranging from a few hundred to over a thousand miles per second. This indicates that the stellar surface swells up like a balloon. The star grows larger and becomes brighter.

After maximum light, broad emission bands appear. These are really lines that have been symmetrically widened about their normal places. They are produced by light coming from an expanding shell of gas, the front part approaching us and the rear part receding. On the violet edge of each emission band is an absorption line. It comes from that part of the shell between us and the star. It is shifted toward the violet by an amount cor-

responding to the full velocity of expansion. When the nova has faded considerably, the bright lines characteristic of gaseous nebulae appear, widened into bands. These fade out after a few years.

The behavior of a nova was described very concisely in a famous cablegram sent by the astronomer Dr. J. Hartmann: "Nova problem solved. Star swells up, bursts." The nova problem is still far from being completely solved, because we do not know what makes a star explode. Nevertheless, much progress has been made during the twentieth century in the study of these stellar explosions, which take place on a scale and with a speed almost beyond comprehension.

263

CHAPTER 52 THE BIRTHS AND DEATHS OF STARS

What is undoubtedly the most important fact emerging out of astrophysical research of the first half of the twentieth century is the realization that the seemingly permanent stars in the sky are continually evolving. They shine as stars for a while, then die, to be succeeded by the next generation. Just as the population of a city is a mixture of generations, so is the galaxy composed of a mixture of young, middle-aged, and old stars.

While it is true that astronomers are still not entirely sure of all the stages that a star passes through during its existence, particularly the later stages, yet the broad outline that has emerged is undoubtedly correct, merely requiring the filling-in of details.

Stars are being continually born in those regions of the galaxy replete with the swirling clouds of interstellar gas and dust that we see as the bright or dark nebulae on photographs of star fields. As these clouds drift around, driven by forces such as the pressure of radiation from surrounding stars, denser regions make their appearance, some of which become dense enough to start contracting toward their own center under gravitational attraction. Astronomers point to relatively small dark "globules" in these clouds as objects that may be undergoing this process of compaction at the present time. As the globule contracts, it will get hotter, the dust within it will break up into atoms, and it will begin to give off energy from its periphery in the form of light. A star is born. As it continues to contract and get hotter, its color goes from red to yellow to white. Sometimes these young stars are so deeply embedded in the surrounding dusty and gassy nebula that they are completely hidden from view so that the only evidence we have of their existence is the infrared glow they produce in the nebulae. In yet other cases the star can be seen but with an erratic variability, indicating that it has not yet settled down.

The contraction and heating of the young star will continue until the temperature at the hot center reaches about 15 million degrees Centigrade. The length of time required to reach this state depends entirely on the amount of material inside the star. If it is equal in mass to the sun, it will take about 50 million years. On the other hand, if it is ten times the solar mass, its much greater gravitational attraction toward its center means that it will contract much faster, taking only 200,000 years. Correspondingly, a star lighter than the sun will take hundreds of millions of years.

The central temperature of 15 million degrees is important, because at such temperatures the fusion of nuclei becomes a dominant source of energy. About 90 percent of the atoms in a typical star are hydrogen atoms, most of the rest are helium, with only small traces of all the other elements we are familiar with, such as carbon, nitrogen, oxygen, iron, etc. In a cold gas, atoms in the gas are made up of a nucleus with one or more electrons moving in orbits around the nucleus, just as the planets move in orbits around the sun. In the case of the solar system, it is the gravitational attraction between the planet and the sun that keeps the planet in its orbit. In the atom, the electron is held in its orbit by an electrical attraction between the positively charged nucleus and the negatively charged electron. At the elevated temperatures in a star's interior, collisions between the atoms are everywhere violent enough to strip away these orbiting electrons, leaving bare nuclei embedded in a sea of free electrons. In most regions in a star's interior, the positively charged nuclei repel each other, and they keep apart. However, when the central temperature becomes high enough, collisions between hydrogen nuclei become violent enough to overcome this repulsion, and they fuse together to form nuclei of greater mass. A series of such hydrogen fusions will produce, in turn, deuterium, then eventually helium. In the process of producing the helium nucleus, four hydrogen nuclei will have disappeared. It turns out that the combined mass of four hydrogen nuclei is slightly greater than the mass of the helium nucleus that has been created. The mass that has been lost goes into energy in accordance with Einstein's famous equation, $E=mc^2$, which expresses the relationship between mass (m) that is converted into energy (E). Everybody, in this age of nuclear energy and nuclear bombs, knows of the tremendous amounts of energy that are released in such a process. As an example, if 1/28 of an ounce of matter turns into energy, that energy is capable of lifting a mass weighing 7 million tons one mile into the air against the earth's gravitational attraction. Calculation shows that, to keep a star like the sun radiating energy at its current rate, 564 million tons of hydrogen must be converted into 560 million tons of helium every second, the 4 million tons of matter that is lost being converted into energy. This is such a great mass loss that one might imagine that the sun would soon waste away. However, there is so much material in the sun that only 0.1 percent of the sun's mass would be lost in 1.5 billion years.

While it appears certain that nuclear fusion processes must be going on in the central regions of stars, including the sun, during the past few years a note of concern has appeared that the current picture we have of solar energy production may be overly simplified. The problem is that when hydrogen nuclei fuse on their way toward producing helium, not only is energy released but the process also produces tiny, almost massless particles called neutrinos. Being almost massless and having no charge, these neutrinos can easily pass right through the sun out into interplanetary space and presumably are passing through the earth. Therefore, a test of the validity of the hydrogen fusion process theory as the source of solar energy would be the detection of these neutrinos. Now, these neutrinos are extremely difficult to detect. It turns out that the best method is to look for argon created when a neutrino strikes an atom of chlorine. Accordingly, Raymond Davis of the Brookhaven National Laboratory has been trying to do just that. He built an enormous tank 5,000 feet below ground in a gold mine in South Dakota and filled it with a hundred thousand gallons of perchloroethylene (C_2Cl_4), a cleaning fluid. The reason why it was put so far underground was to get away from the effect of cosmic rays. He then used very sensitive techniques to count the number of argon atoms created in his tank after a certain period of time. The problem is that Davis has not been able to detect nearly the number of argon atoms that he should have if astrophysicists are correct. The reason for his failure is still being hotly debated. It is possible, as some maintain, that we do not know enough about the behavior of neutrinos. Others suggest that stars like the sun have internal conditions that shut off hydrogen fusion processes from time to time. Since it takes millions of years for the radiation produced in the sun's core to struggle its way to the surface, any periodic fluctuations in energy production would be smoothed out, and not noticed from the earth.

The surface temperature, diameter, and total energy output of a star when it is "burning" hydrogen in its core all depend entirely on its overall mass. A massive star will be very bright, larger than the sun, and blue, while a low mass star will be relatively faint, smaller than the sun, and red. At this stage in a star's evolution it will be located at a particular point on the main sequence that is determined by its mass (see illustration on page 207). Massive stars are located at the top of the main sequence, while stars low in mass are at the bottom.

As long as a star has any hydrogen left in its hot central core, it will stay essentially fixed on the main sequence. This is the longest stage in the life history of a star, so that it is not surprising that we find most observable stars on the main sequence. The actual length of time that a star will continue to have hydrogen in its core will, obviously, depend on how fast it is using the hydrogen up. Some very luminous blue main sequence stars are consuming their hydrogen 1,000 times as fast as does the sun, and cannot exist as main sequence stars for longer than 100 million years. In contrast, the sun will take over ten billion years. Since current estimates put the age of the solar system at 4½ billion years, this means that the sun is nearly half-way through its life as a reliable source of energy.

When a main sequence star has eventually consumed all of its core hydrogen, a very dramatic change takes place. The core, which is now largely helium, is no longer producing energy via nuclear reactions, so it starts to contract under the pressure of the overlying gas. In contracting, however, it obeys the law of physics pertaining to compressing gases and becomes hotter, emitting even more energy than before. The outer part of the star, in response to this extra energy flowing through it, will expand tremendously. Calculation shows that this expansion is so great that the surface temperature actually drops, despite the fact that the total energy output of the star is greater than when it was on the main sequence. The star has turned into a red giant. Some of the more massive red giants become so large that if they were substituted for the sun, the earth would find itself in orbit inside the star!

The details of the events that take place after a star has become a red giant are still not known with as much certainty. It *is* known that, when the increasingly hot helium core reaches a temperature of 100 million degrees, the helium nuclei themselves begin to fuse together, producing heavier nuclei, particularly carbon. This is called the "helium flash" because it takes place quite rapidly. The extra energy has been shown, by computer modeling, to actually reverse the behavior that the star has been exhibiting up to that point, making it smaller and less luminous.

The star is now beginning to reach the end of its life. An object like the sun will probably continue to contract, becoming a white dwarf about the size of the earth, then cooling very slowly over billions of years until it becomes a black cinder. Calculations show that every star less than 1½ times the mass of the sun will suffer a similar fate. The numerous stars more massive than this become unstable if they try to become white dwarfs. As consequence of this instability, part of the mass of the star is blasted away. One possible result is the production of a planetary nebula, described in Chapter 54. However, if the star is more than eight times the mass of the sun, a much more dramatic supernova explosion takes place. Chapter 51 describes what we observe during such an event. These observations tell us what is going on outside the star, but not in the interior. There is as yet no unanimity among astrophysicists regarding the processes that are going on there. One scenario that looks promising goes as follows: At the very high temperature prevailing at the center of a massive star, the fusion of nuclei can proceed to create heavier and heavier nuclei, culminating in the development of a central core of iron nuclei. Now it turns out that when iron nuclei fuse with, say, helium nuclei, they *take up* energy rather than produce energy. The result is that the core contracts. But, just as before, the contraction heats

up the core, in this case sufficiently to break up the iron nuclei, resulting in a further loss of energy, further contraction, and further heating. Within seconds the central temperature becomes fantastically high, so much so that protons and electrons in the collapsing material around the core can combine to form neutrons. The core actually collapses so violently that it rebounds, colliding with the outer layers that are in the process of collapsing. The resulting shock bangs nuclei together so violently that nuclei heavier than iron are produced, while at the same time the violence of the shock blows the outer part of the star completely off, leaving only a core of neutrons. The heavy elements produced by the shock are spewed out into space along with the rest of the star's envelope, to eventually enrich the surrounding interstellar matter with heavy elements. Thus, when the succeeding generation of stars is created, the new stars will start their lives possessing a slightly enriched composition. This will have a profound influence on the mix of elements found in various parts of the galaxy as it ages.

The central core of neutrons is all that remains after the supernova explosion. While astronomers of a generation ago felt that conditions inside a white dwarf were extreme, the conditions in the interior of this core (a "neutron star") are even more fantastic. A mass that may be as great as two suns may be compressed into a volume only 10 to 12 miles in diameter. A teaspoonful of this matter could weigh a billion tons!

Obviously, an object only 12 miles in diameter at stellar distances could never be detected. How, then, could one ever prove that neutron stars actually exist? This is where the fascinating story of the pulsars comes in.

The discovery of pulsars was undoubtedly the most important discovery of the 1960s. They were discovered in 1967 at Cambridge, England, when radio astronomers were using a radio telescope specially designed to investigate rapid variations of signal strength. They were actually looking for the effect of the earth's atmosphere on radio signals from space (analogous to the "twinkling" of stars at visual wavelengths), but instead discovered pulsars. The first in the constellation Vulpecula proved to be radiating extremely sharp pulses of radio waves with a startlingly regular interval between pulses that is a fantastically short 1⅓ seconds. About 300 pulsars are now known, each with its characteristic, regular beat, ranging in period from 4 seconds for the slowest down to hundredths of a second for the most rapid.

A unique method has been used to estimate the distances to pulsars. The radio observations show that shortwave radio pulses from a pulsar arrive a little later than longwave pulses. Such a retardation of a radio pulse could be the consequence of the presence of interstellar electrons in the path between the pulsar and the earth. From the measured retardation of waves from the pulsar in Vulpecula, for example, its distance turns out to be around 400 light-years. Their distribution in space is also concentrated toward the Milky Way plane, so that pulsars are definitely all members of the galaxy. Optical astronomers have found faint, bluish stars near the reported positions of some of the radio sources, but initial attempts to prove that they were the pulsars proved inconclusive.

It was apparent from the beginning that the object responsible for the pulses was a new kind of object not previously recognized. Any regular event like the pulses must be a consequence of either stellar pulsation, rotation, or orbital motion of a binary star. The last possibility could be eliminated since the period is too short. A pulsation (increase and decrease in size of the star) of a white dwarf would have a period of less than one minute, nowhere near the short period observed. Similarly, if a white dwarf rotated with a period of less than one second it would fly apart. So it was evident that the source must be much smaller than a white dwarf.

Another reason why it was believed that the source of the pulses must be very small was the very short duration of individual pulses, which is typically about a thousandth of a second. Thus the radius of the object cannot be greater than the distance light travels in a thousandth of a second, which is 186 miles, otherwise the delay-time of the pulse from various parts of the object's disk would give a pulse duration that would be too long. Therefore, essentially by a process of elimination of alternative possibilities, it is believed that a pulsar is an extremely small rotating object.

The suggestion that was being made, that there

Brilliant object in photograph at left is a pulsar in the Crab Nebula. Flashing at the rate of 33 times per second, it was captured on film in both its "on" state and its "off" state (at right) through a rapid-scanning technique developed by the Lick Observatory. Object at top left in both photos is a regular star. (Lick Observatory.)

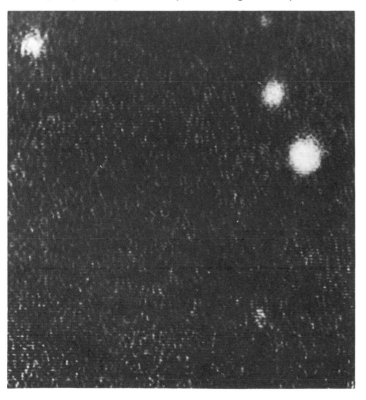

267

was a connection between supernovae and pulsars, was given a big boost when a pulsar was discovered by radio astronomers in the center of the Crab Nebula in the constellation Taurus. It was found to have a fantastically short period between pulses of only 1/30 of a second. The Crab is the result of a supernova explosion that occurred approximately 6,500 years ago. The tremendous outburst of light that accompanied this explosion reached the earth in 1054, producing an object in our sky bright enough to be seen in full daylight, according to ancient Chinese records. The nebula that we see surrounding the star today is made up of gas ejected at the time of the outburst. A faint blue star at the center of the Crab is believed to be the remnant of the star that blew up. This was confirmed in 1969 when it was found that this remnant star was pulsating in optical light at precisely the same rate as its radio pulses. Interestingly, all of the visible light from this star is concentrated in the pulses; between pulses the star is completely invisible. Therefore, since current theory suggests that this remnant is a neutron star, it is natural to conclude that all pulsars are rotating neutron stars.

It turns out that the rapidly spinning pulsar at the center of the Crab is converting at least ten percent of its rotational energy into the energy of fast electrons and protons that emerges as X-rays and cosmic rays and produces the bright glow that we see in the nebula. Observations tending to strengthen this possibility were announced late in 1968, namely that the Crab pulsar and at least five others are very slowly decreasing the rate of their pulsations. For the Crab pulsar the loss of rotational energy is exactly equal to the energy required to keep the nebula glowing. According to this reasoning, therefore, the Crab Nebula pulsar is the youngest we have yet discovered, while longer period pulsars are the result of supernova explosions that took place so long ago that the ejected material has long ago dissipated into space.

There is still considerable uncertainty regarding exactly why a rotating neutron star produces pulses. They are believed to be somehow connected with an extremely large magnetic field that theory shows is associated with the star. One theory assumes that the north and south poles of this magnetic field are tipped relative to the axis of rotation, and that beams of radio waves ejected at these poles sweep by the earth as the neutron star rotates.

Theoretical calculations of the expected structures of neutron stars have shown that their upper limit of mass is about three times the mass of the sun, and that if the remnant core is more massive than this, its collapse will proceed without limit until it becomes almost infinite in density and infinitesimal in size. When this happens, Einstein's general theory of relativity predicts that it will become invisible, that is, a black hole. Astronomers have known for years that the general theory of relativity is correct. For instance, it predicts that light emitted from massive objects is shifted toward longer wavelengths (the "gravitational red-shift"). The shift becomes the greater the denser the emitting body. For ordinary stars like the sun the shift is extremely small, while for white dwarfs it is still small but measurable. As an example, the companion star of the double star 40 Eridani is a white dwarf. From the orbit we know what the speed of the white dwarf in the line of sight should be at any instant. However, when this speed is calculated from the locations of lines in its spectrum, an excess speed of recession of 13 miles per second is found, attributable to a shift of the locations of the lines to longer wavelengths by the very strong gravitational field of the white dwarf. This shift to longer wavelengths would become greater and greater the stronger the gravitational field, until in the limit the wavelength would become infinite—that is, there would be no emitted radiation.

Another prediction of the theory of relativity is that the paths of light waves passing massive objects are bent slightly toward that object. In the case of the sun, the predicted shift in direction has been confirmed during solar eclipses, when the glare of the sun is blotted out by the moon and the positions of stars appearing nearly in the same direction as the sun can be measured. The light is actually traveling in a straight line through space; it is the space itself that is curved in the vicinity of the sun. This curvature becomes greater with increased gravitational attraction, until, in the limit, the resultant curvature of the paths of the light attempting to leave the surface of the star becomes so great that the light is turned back onto the surface.

The size of a star of given mass with a gravitational attraction strong enough to prevent radiation from escaping can be easily calculated. It was actually done in 1916 by Karl Schwarzschild, shortly after Einstein had produced his general theory. It turns out that a star with three times the solar mass should have a radius of 8 miles. The radius varies directly as the mass; thus, a 30-solar-mass star would have to have a radius of 80 miles.

Since black holes are invisible, one has a problem proving that they actually exist. Fortunately, we can take advantage of the fact that the black hole still exerts a gravitational force on its surroundings, and look for double stars involving black holes. A number of possibilities have been found. The most interesting is the intense X-ray source Cygnus X-1, so-called because it was the first X-ray source found in the constellation Cygnus by orbiting X-ray telescopes. There is a faint, ninth magnitude star at the same position. Its spectral lines shift back and forth in a regular period of 5.6 days, showing that it is a component of a double star. Its spectrum also shows that it must be a massive blue supergiant. In order to have an orbital period of only 5.6 days, its companion star must be almost in contact and have a mass eight times the mass of the sun. The strong X-ray remission must be coming from the companion. According to current ideas, the companion is so close to the supergiant that it is drawing material from it. The captured material circulates around the companion, going faster and faster as it is drawn downward toward its surface and emitting X-rays as it becomes denser. In order for all this to happen, the companion must be very small. Since its mass is eight solar masses, it cannot be a neutron star, and may be a black hole. Further studies during the next few years will tell us whether this is correct.

Filamentary nebula in Cygnus, also known as "The Loop," probably the remnant of a supernova explosion. The symmetry in this complex veil of nebulosity suggests an expanding shell of gas whose origin was in an ancient explosion of truly cosmic proportions. Its measured rate of expansion points to a beginning about 25,000 years ago, from a center located about 2,500 light-years away. Recent observations of this object with X-ray detectors mounted in a rocket have located a point-like source at the center of the loop. (Palomar Observatory, California Institute of Technology.)

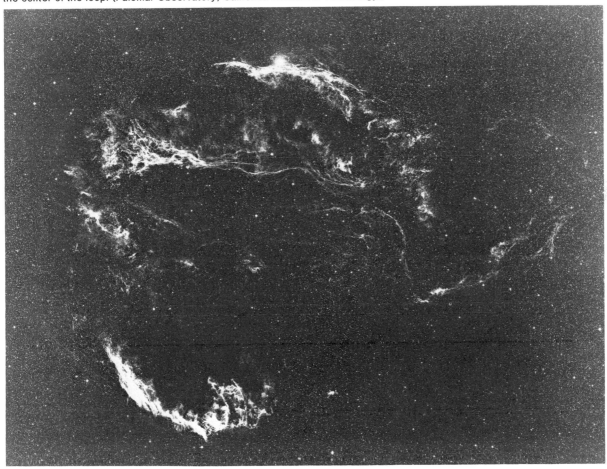

CHAPTER **53** THE GALAXY AROUND US

When one looks at the sky on a clear, moonless night, particularly in late summer, one sees, in addition to stars and, possibly, one or more planets, a hazy band of light, stretching from the constellation Scorpius through Cygnus and Cassiopeia, that we call the Milky Way. Continued observations over the course of the year show that the center line of the Milky Way divides the celestial sphere neatly into two equal halves. Galileo with his crude telescope was the first to show that the Milky Way is actually composed of myriads of stars. We now know that it is the visual evidence of the galaxy around us. Since the Milky Way is a band across the sky, the galaxy must be a flattened disk, with the sun not too far from its central plane.

If you have ever been lost in a forest of trees you will appreciate the problem facing astronomers wishing to determine the size, structure, and motion of the galaxy. Part of the problem is that the sun is embedded within the mass of stars that make up the galaxy. Another, even more frustrating problem, is that the flat, disk-like galaxy we are in is strewn with disorganized interstellar matter, described in more detail below. Some of this matter is in the form of dust, which blocks our view of distant stars in the same way that fog prevents our seeing more than a block or two down a foggy street. Thus, at the turn of the century when an attempt was made to derive the size of the galaxy by photographing and counting stars on successively longer exposure photographs, it turned out that the numbers thinned out in all directions, leading to the conclusion that the sun was very near the center of the galaxy, which was found to be a flat disk with a diameter of 5,000 parsecs (one parsec is 206,000 times the distance of the earth from the sun). This was considered to be the full extent of the entire universe, because in those days there was no conception that the spiral galaxies (they were called nebulae) were as far away as they have proved to be.

270

When we consider the billions of stars that make up the galaxy, it is difficult for us, with our earth-bound conception of distances, to grasp the tremendous distances that actually exist between stars and the extreme emptiness of space. If one were to construct a model, locating the moon one inch away from the earth, the *nearest* star to the sun would be 1,500 miles away. The accompanying diagram will help to give a broad overview of the probable number of stars in the galaxy. The first circle (A) represents a cross section of a sphere with a radius of 10 light-years and with the sun at the center. One light-year, the distance

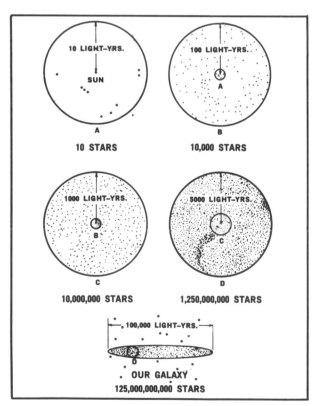

Our sun among the stars.

The circles represent cross sections of spheres centered at the sun. Our galaxy is shown edgewise, surrounded by globular clusters.

traveled by light in one year (about six trillion miles) is about 63,000 times the earth's distance from the sun. It would be represented almost exactly by one mile on a map on which the earth was one inch from the sun. It is obvious that the position of the earth cannot be shown on the diagram and that the dots representing the sun and stars are not made to scale.

The number of stars within a given distance of the sun is uncertain, because there may be some very faint stars in our neighborhood whose distances have not yet been determined. However, we shall not be far off if we use the round numbers of 10 stars in a sphere with a radius of 10 light-years. Assuming the same density throughout our galaxy, we can show graphically how the number of stars would increase as we include larger volumes of space.

In the second part of the diagram the scale has been changed so that circle A is 1/10 its former size. Therefore, the second large circle (B) represents a cross section of a sphere with a radius of 100 light-years. Since this radius is 10 times the first radius, and since the volume of a sphere varies as the cube of the radius, the second sphere has a volume 1,000 times that of the first and should contain about 10,000 stars. In a similar manner, circle C represents a cross section of a sphere with a radius of 1,000 light-years containing 10,000,000 stars.

The radius of circle D is only 5 times that of circle C, so that the volume increases in this case by only 125 times, making the number of stars in this sphere 1,250,000,000. Finally, we reach the limits of our galaxy, the main body of which can be pictured as shaped like a cart wheel, with a diameter of about 100,000 light-years and a thickness of about 10,000 light-years. The thickness probably is greatest at the center and decreases toward the rim of the wheel. The density of stars rapidly decreases toward the edges, and individual stars have been found beyond the main body, in addition to at least 118 globular clusters that surround the galaxy. Individual globular clusters differ enormously in the total number of stars included in the cluster. Some comprise as few as 10,000 stars, others well over a million. Thus it is difficult to estimate the total number of stars in the galaxy. If we assume its volume is 100 times that of the sphere having a radius of 5,000 light-years (circle D), we arrive at the figure of 125 billion (125,000,000,000) stars. This figure is very uncertain, but is consistent with estimates made from other data.

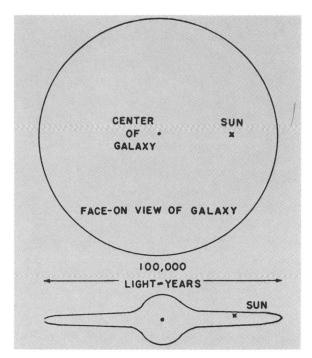

Edge-on view of galaxy.

A billion is a familiar figure today, with so many billions of dollars being spent every year. However, few of us realize how large that number is. One billion minutes equal about 2,000 years. Counting at the rate of 125 stars a minute, it would take about 2,000 years to count all the stars estimated to be in our galaxy.

The Milky Way is our edge-on, inside view of this system of stars. The light of the Milky Way is produced by the combined light of many very faint, closely crowded stars, which cannot be resolved by the naked eye. The apparent crowding of the stars is due to their arrangement in the galaxy's great flattened disk, like a cart wheel. The sun is about two-thirds of the way out from the hub to the rim of this great wheel of stars. Because of this eccentric position, we are looking toward the rim whenever we face any part of the Milky Way except in the general region of Sagittarius, which contains the nucleus. The stars toward the rim appear faint because they are so distant, and they appear so numerous because they extend all along the spokes of the wheel.

The space between the stars is practically a vacuum. But it is not entirely empty. In addition to the diffuse nebulae, which can be photographed as bright and dark clouds, the spectrograph reveals the presence of interstellar gas. Also a general haze

271

A diffuse Nebula (NGC 7635) in Cassiopeia as photographed in the 200-inch Hale Telescope at the Palomar Observatory. (Wide World Photos.)

Stars in one region of Milky Way photographed on blue-sensitive plate (top) and on red-sensitive plate (bottom). Note striking difference in amount of detail recorded. (Mount Wilson Observatory.)

POLARIS   VEGA

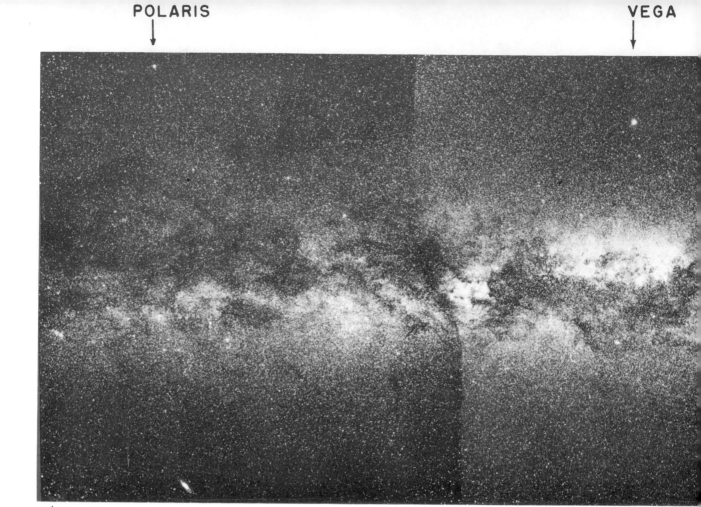

ANDROMEDA GALAXY
DOUBLE CLUSTER IN PERSEUS

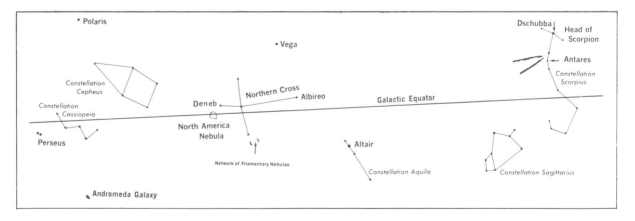

has been detected by its dimming and reddening effects.

When the sun is near the horizon, it appears dim because of the greater thickness of atmosphere through which its light passes than when it is up higher. The light goes through many miles of the

274

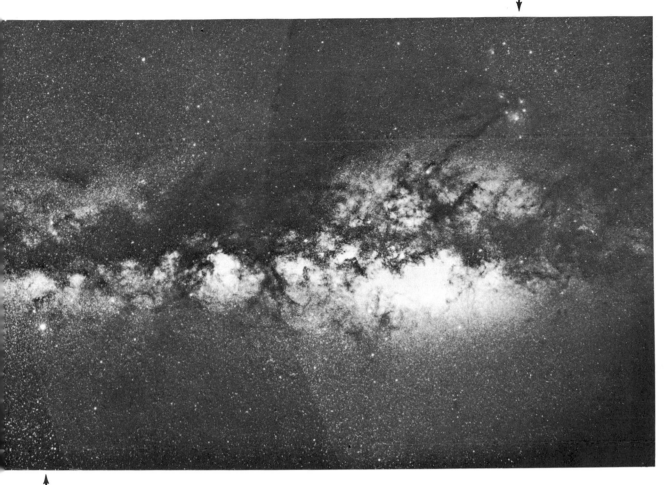

ALTAIR

Mosaic chart of Milky Way from Perseus to Scorpius, consisting of four photographs made on yellow D plates. (Mount Wilson and Palomar Observatories.)

lower air levels, which are the densest and contain the most dust. There is a reddening of the sunlight, because the gas and dust particles scatter the blue light more than the red light. Most of the shorter waves of light are scattered to form the blue of the sky. Only the longer waves penetrate the atmosphere to reach our eyes, and they make the sun itself appear orange or red.

The ordinary photographic film is sensitive to blue light and will not record the details of a distant landscape as clearly as film sensitive to the yellow and red region of the spectrum. This principle has been applied with striking effects in photographing the Milky Way. The two accompanying photographs were taken by Walter Baade with the 100-inch telescope of the Mount Wilson Observatory. They are both of exactly the same region in the Milky Way within five degrees of the direction of the center of our galaxy.

The first photograph was taken on a blue-sensitive plate, and the other on a plate sensitive to the yellow and red region of the spectrum. The exposure times were comparable. The great excess of stars on the red plate as compared to the blue proves the existence of a general haze in space. It must be realized that this haze is very much thinner than any in our atmosphere. Less than 100 miles of bluish atmospheric haze will hide anything beyond. Many trillions of miles of the dust and gas in space only dim and redden slightly the stars beyond.

The composite chart above of the summer

275

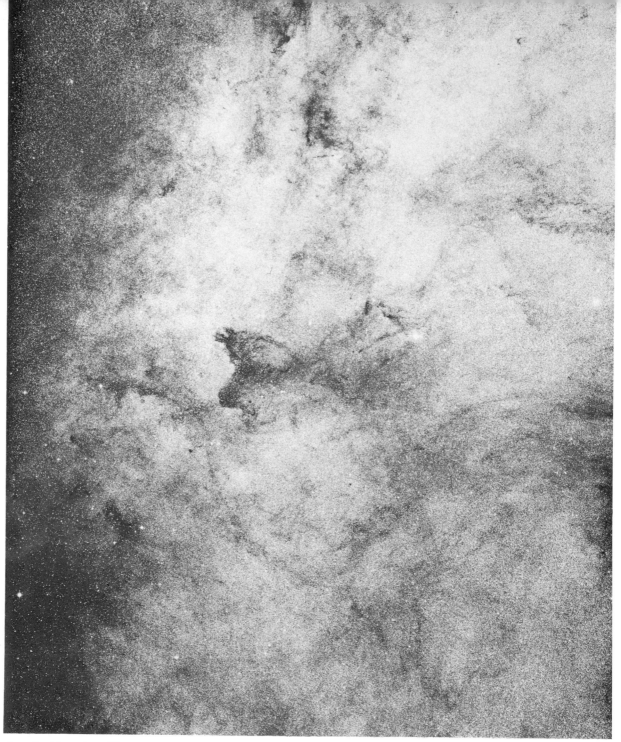

Star cloud in region of Sagittarius, photographed in red light with 48-inch Schmidt telescope. Center of our galaxy is in this direction. (Mount Wilson and Palomar Observatories.)

Milky Way from Perseus to Scorpius was made by matching, cutting, and pasting together a series of photographs made with a 5-inch Ross lens. Many interesting details can be seen. The position of Deneb is indicated in the line drawing. Just below

Deneb is the North America Nebula. Below and to the right are the Network and Filamentary Nebulae. Above them in a nearly vertical line are the three stars that form the short arm of the Northern Cross. Vega and Altair are marked on the

276

Open star cluster in a diffuse nebula in Serpens. Open star clusters and diffuse nebulosities are parts of our own galaxy. (120-inch telescope, Lick Observatory.)

drawing. About halfway between them and a little to the left of a line connecting them is Albireo, at the bottom of the Northern Cross.

Scorpius is at the extreme right of the photographic chart. The three stars in the upper right corner mark the head of the Scorpion. The middle one is Dschubba, a navigational star. Antares is not so easy to identify, because of the nebulosity near it. Also, it is a red star that does not show very well on the blue-sensitive plates with which

Trifid Nebula in Sagittarius. (120-inch telescope, Lick Observatory.)

these pictures were made. Dark lanes of gas and dust form a letter "V," lying on its side. The region where these lanes meet is just above and slightly to the left of Antares.

The black horizontal line drawn on the chart is the central line of the Milky Way and is called the galactic equator. This is inclined 62° to the celestial equator and crosses in the constellation of Aquila.

About one-third of the Milky Way is included in the composite photograph. The part beyond Cepheus at the left end passes through Cassiopeia and Perseus. Auriga, Gemini, and Orion are not in the picture. This part is visible on a winter evening. It appears much less conspicuous, being toward the rim, than the region in Sagittarius halfway around the galactic equator, which is toward the center of the galaxy.

In general, the stars that appear brightest are nearest to us. They are almost equally distributed in all directions from us. The stars that appear fainter are generally more distant, and most of them lie in the directions of the rim of the wheel-shaped galaxy. Thus the apparent crowding of the stars along the Milky Way does not mean that they are closer to each other in space. They simply extend out to greater distances in the directions of the Milky Way.

A system of stars cannot be highly flattened without turning around an axis at a rapid rate. Such rotation has been found in every spiral galaxy where the necessary measurements have been possible. The detection of the rotation of our own galaxy has not been easy, and the methods are too involved to be described here. But it is now well established that the galaxy is rotating about an axis at right angles to the plane of the Milky Way (see diagram of edge-on view).

If all the material in the galaxy were fairly uniformly distributed, all parts would go around in the same period. It would be like the disk of a phonograph record, with a point on the rim traveling with a greater speed than a point near the center, but both points having the same period. Observations show that the galaxy is not rotating in this way. The stars near the center of the galaxy are moving faster than those farther out. The arrangement is like that of the solar system, where the period of revolution of a planet increases with distance from the sun. Thus Mercury, the closest planet to the sun, has a period of three months and moves with a speed of 30 miles a second. Pluto,

the most distant planet, is about 100 times farther from the sun than Mercury. It has a period of nearly 250 years and a speed of only 3 miles a second.

The sun is about 30,000 light-years from the center of the galaxy. It is ⅔ of the way out from the center to the rim. It is moving with the stars in its neighborhood at a speed of more than 150 miles a second around the nucleus of the galaxy. It takes the sun and its neighbors about 250 million years to get around once. This means that, if the galaxy is 4½ billion years old, it has rotated on its axis 18 times during its history.

As we will see in Chapter 58, photographs of distant galaxies show that most of the highly flat-

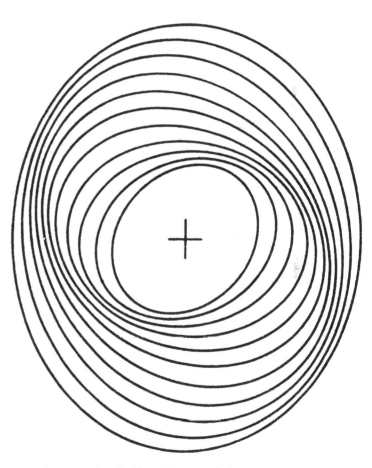

According to the Density Wave Theory, spiral arms are produced in a rotating galaxy if the paths of stars and associated interstellar matter around the galaxy center depart from perfect circles. Places where these paths squeeze the interstellar mater together are places where the bright blue stars in the arms are being born.

tened galaxies show spiral arms extending outward from a bright central bulge. Accordingly, it is natural to expect that our galaxy should have a similar structure. However, it has been very difficult to determine what its structure actually is. This is because interstellar dimming by dust has prevented our seeing more than a relatively small portion of the galaxy in the immediate vicinity of the sun. Fortunately, the interstellar matter becomes progressively more transparent as one goes to longer and longer wavelengths so that radio waves can be received from all parts of the galaxy. Advantage is also taken of the fact that the interstellar hydrogen emits a very important "line" at the radio wavelength of 21 centimeters. Except for its long wavelength, it is exactly the same as the spectral lines observed in absorption or emission in optical spectra. It had been predicted in 1944 by H. C. van de Hulst in Holland, and was observed for the first time by H. I. Ewen and E. M. Purcell at Harvard University. This radiation is from "neutral" hydrogen in the interstellar gas and reveals vast clouds of such hydrogen. The hydrogen line has proved to be of inestimable value in the study of the spiral structure of the galaxy, since the hydrogen producing the line is primarily concentrated along the arms of the spiral.

If the observation of hydrogen emission of 21-centimeter wavelength from a spiral arm could be coupled with an estimate of the distance to the emitting gas, then a series of such observations in various directions in the Milky Way plane could be combined to trace the orientation of an arm relative to the sun and to the center of the galaxy. The distance of the emitting gas can be estimated, since fortunately the galaxy does not rotate on its axis like a solid wheel, but rather in a manner reminiscent of the revolutions of the planets around the sun. Thus, as we look out from the vicinity of the sun in certain directions, the stars must appear to be approaching us, while in other directions they will be receding from us. The relationship between the distance of a particular object and its velocity of approach or recession can be calculated once the velocity distribution of the galaxy as a whole is known. The velocity in the line of sight of clouds of emitting hydrogen in an arm can, in turn, be measured from the shift of the wavelength of the 21-centimeter line from its normal position. This wavelength shift is produced by the same Doppler effect used by optical astrono-

mers to calculate velocities from lines in the stellar spectra. Lengthy and detailed observations in various directions in the Milky Way plane, notably by observers in Holland and Australia, have resulted in maps of the spiral arms of the galaxy that are remarkably complete. An example of these maps is shown on page 297.

In addition to emissions by hydrogen atoms, radio astronomers are finding increasing numbers of molecules in clouds strewn between the stars. To date, emissions by over four dozen molecules have been detected, and many more will undoubtedly be discovered through continued search. The list includes molecules of water, ammonia, formaldehyde, and more complex substances. They all involve those elements we know to be the most abundant in these clouds, such as hydrogen, carbon, nitrogen, and oxygen, with occasionally an additional atom of other elements such as silicon and sulphur. It is still a matter of debate as to where these molecules come from. Some theories suggest that they are born in the denser circumstellar envelopes being slowly ejected from giant red stars; other theories point to the interstellar grains and the possibility that molecules are the debris left over when dust grains are destroyed by collisions, etc.

There have been many theories advanced in attempts to explain the spiral structure of galaxies. Some propose that arms are being spewed out from the central bulges like Fourth of July pinwheels; others maintain that material in the arms is moving inwards toward the central bulge. Neither approach has been particularly successful. A more promising approach makes use of the way stars and the interstellar material move as the galaxy rotates. In general, the motion is in circles around the galactic center, but individual motions deviate significantly from this circular motion, so that at various places in the disk the material crowds together to form a region of increased density. The calculations are very difficult, but do suggest that two or more waves of compression ("density waves") may form, spiraling out from the galactic center. As the interstellar matter becomes compressed in the wave, denser nebulae will result, favoring the birth of stars. It is undoubtedly significant that the main reason why spiral arms are so conspicuous in the photographs of distant galaxies is that they are replete with bright, blue, young stars.

CHAPTER 54 THE NEBULAE OF OUR GALAXY

The word "nebula" comes from the Latin and means "cloud." Until a generation ago all hazy, cloudlike objects that seemed fixed among the stars were called nebulae. However, with the improved methods and larger telescopes of the twentieth century, it was found that many of the so-called nebulae were systems of stars outside of our galaxy. These became known as extragalactic nebulae. Since most of them are spiral in shape, they are also called spiral nebulae. But it seems preferable to call them galaxies, because that simple term describes their true nature and differentiates them from the ordinary nebulae composed of gas and fine dust.

We are concerned here not with these external systems, but with the nebulae of our own galaxy, which may be divided into two main groups, planetary and diffuse nebulae. Planetary nebulae were so named because many have circular or elliptical disks that look a little like planets when

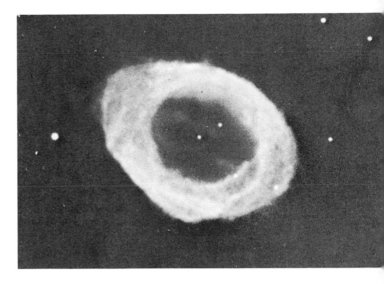

Ring Nebula in Lyra. (Lick Observatory.)

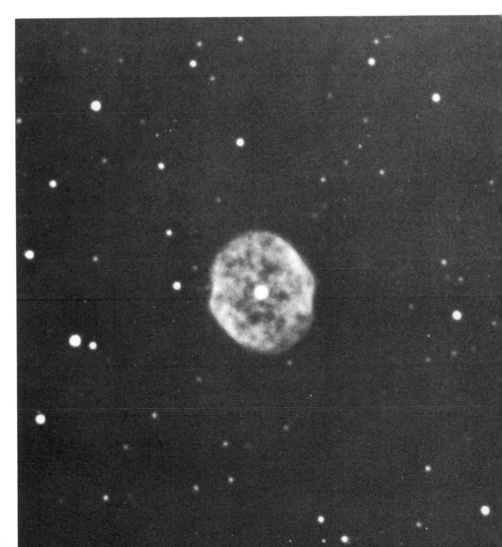

Planetary nebula in Camelopardus.
(Mount Wilson Observatory.)

seen in a telescope. About 500 of them have been found. Most of them lie between 3,000 and 30,000 light-years away. Their sizes differ, but a typical one has a diameter of a trillion miles, or more than 10,000 times the distance from the earth to the sun. Yet the total mass of such a nebula is less than that of the sun. Thus the density is exceedingly low.

None of the planetary nebulae are visible to the unaided eye. The nearest one has an angular diam-eter of about 15′, half that of the moon. The most distant ones appear little larger than stars, from which they can be distinguished by their bright-line spectra. In most cases—and perhaps in all, if we could see it—there is a hot star at the center of the nebula. This star is so hot that most of its energy is given out in the form of invisible ultra-violet light. This is absorbed by the atoms of the gases making up the nebula and is re-emitted in

Great Nebula in Orion. (Lick Observatory.)

visible form. The result is that the total amount of visible light radiated by the nebula is nearly fifty times greater than that emitted by the star itself. This process by which the gaseous nebulae shine is called fluorescence. It is what makes certain minerals, which are dull in white light, shine with bright color in ultraviolet light.

The spectra of planetary nebulae consist of bright lines, which are produced by the glowing gases. Several lines do not match in position with any appearing in spectra made from sources on the earth. They were once attributed to a hypothetical element called "nebulium." It is now known that these nebular lines result from ionized oxygen, shining under conditions not readily attainable on the earth.

Many of these nebulae are brighter at the circumference, due to the fact that the dominating fluorescence occurs at some distance from the central star, forming a glowing shell. This shell appears to us as a glowing ring because a greater thickness of glowing gases is in the line of sight at the apparent edge of the shell. The distance between the central star and the glowing shell is such that fluorescence of the appropriate ion is favored. The spectral lines are often inclined in such a way as to suggest a rotation around the shortest axis of the nebula. A doubling of the lines indicates that the nebula is expanding. It is so nearly transparent that the light comes to the earth from both the front and rear. An expansion shifts the lines toward the violet for the gas on the nearer side and toward the red for that on the farther side.

The fact that novae, or exploding stars, send out expanding shells of gas suggests that planetary nebulae were formed in the same way. However, the resemblance is only superficial. The velocity of expansion of a typical planetary nebula is only about 15 miles a second, whereas that of a nova may exceed 1,000 miles a second.

The second group of galactic nebulae, the diffuse type, show a great variety of size, shape, and brightness. The brightest one is the Great Nebula in Orion, located in the middle of Orion's sword. In a telescope it appears as a luminous greenish cloud around a group of four stars. A long-exposure photograph with a wide-angle camera shows that it spreads over the whole region of Orion.

Like the planetary nebulae, the diffuse nebulae are illuminated by nearby stars. In the absence of such stars, the nebulae are very nearly dark. The extent of the illumination of the nebulae is directly proportional to the apparent brightness of the star associated with it. The spectroscope shows that there are two ways in which these masses of gas and dust are illuminated. When the exciting star is hotter than about 18,000° Centigrade, the spectrum consists chiefly of bright lines like those of a planetary nebula. The process is fluorescence. However, when the exciting star is cooler than about 18,000°, the spectrum consists chiefly of dark lines and is similar to that of the star. In this case, the illumination is simply reflected starlight.

Dark nebulae, which have no stars nearby to illuminate them, dim the light of whatever lies behind them. They are especially conspicuous in photographs in which they obscure parts of the Milky Way or of bright nebulae. The S-shaped nebula in Ophiuchus is a famous example of a dark nebula.

Gas and dust are not only massed together in conspicuous bright and dark nebulae, but also are scattered thinly between the stars in all parts of our galaxy. The gas is detected by the spectroscope. In the spectra of many spectroscopic binaries the lines of ionized calcium do not partake of the periodic shift of the other lines caused by the star's velocity. These stationary or detached lines, as they are sometimes called, are produced by gas atoms in the space between the stars. They have also been found in the spectra of nonbinary stars having a sufficient velocity to separate the stellar lines from the interstellar ones. The interstellar lines are stronger in the spectra of distant stars than in those of nearer stars. Thus these lines give an indication of the distance of a star. In addition to calcium, several other interstellar gases have been found, including sodium, titanium, and potassium. A distance of about 10,000 light-years is necessary to produce an interstellar sodium line equal in intensity to a sodium line made by a gas flame in the laboratory. Thus the path through a fraction of an inch of gas flame has as many effective sodium atoms as in an interstellar light path of 10,000 light-years.

In addition to lines produced by interstellar atoms, a growing number of molecules are being found in interstellar space, particularly by radio astronomers. They are described in Chapter 59.

The dust of interstellar space is detected by its effect on making stars appear dimmer and redder than their normal colors. The particles scatter the blue light of the stars more than their red light, in the same way that our atmosphere does when it reddens the light of the setting sun. These particles must be very small. If they were as much as a

Part of "Veil" Nebula, a vast filamentary nebula in Cygnus, photographed through 120-inch telescope. (Lick Observatory.)

Filamentary Nebula in Cygnus. The trail of a meteor appears along the right edge. (Mount Wilson Observatory.)

Milky Way in northern part of Cygnus. Brightest star near top is Deneb. To left of it is North America Nebula. In lower left corner are Network Nebula and Filamentary Nebula. Cygnus is also known as Northern Cross. Albireo, bottom star of Cross, is not included in picture. Cross is tilted in picture, with small arm of Cross running from lower left to upper right. (Photographed by F. E. Ross at Flagstaff, Ariz., with a Ross wide-angle 5-inch lens.)

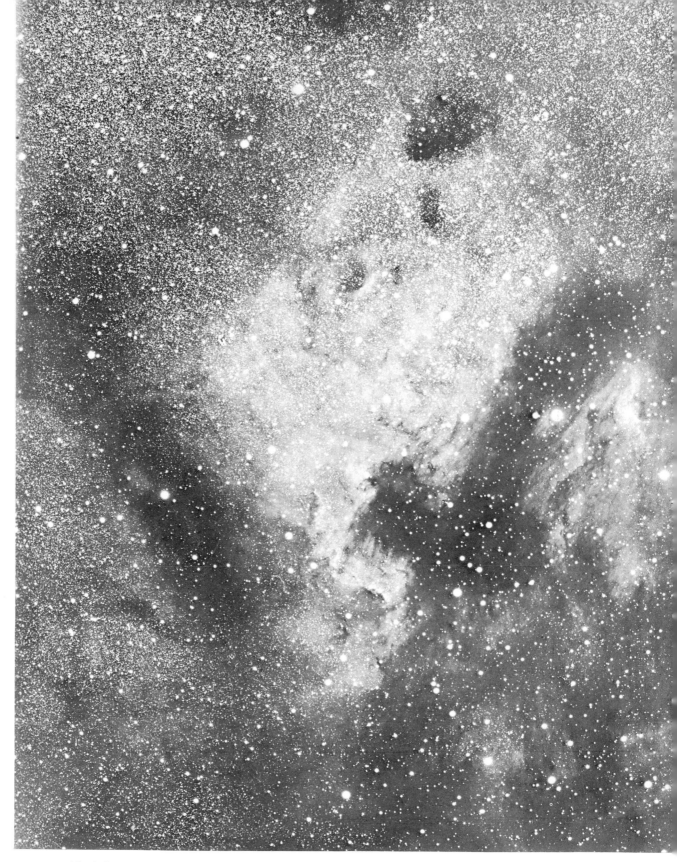

North America Nebula in Cygnus. (Photographed through 120-inch reflector, Lick Observatory.)

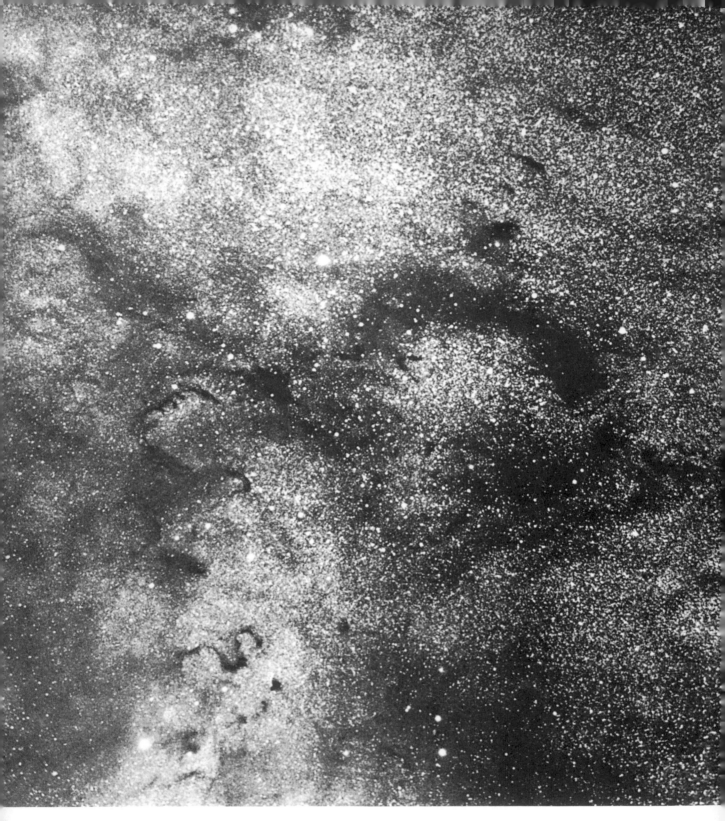

Milky Way in constellation of Ophiuchus. Sharpness of apparent holes proves absolutely that they cannot be due to a lack of stars. Each vacancy would have to be a tunnel of vacancy out through space and pointed directly toward earth to give us this appearance. It is inconceivable that these regions can be due to anything else than obscuring matter that hides the stars beyond. Appearance of comparatively close stars within the dark regions is in full accord with such an interpretation. Note sharp S at lower left, a larger view of which is shown in next photograph. (Mount Wilson Observatory.)

thousandth of an inch in diameter, they would obstruct the light without changing its color. But if they were atoms and molecules, they would cause greater reddening than is observed. Their sizes must lie somewhere between a thousandth and a hundred-thousandth of an inch, roughly comparable to the size of smoke particles. The heating effect of starlight on these particles is quite insignificant. They are probably only two or three degrees above absolute zero, or at a temperature of about −270° Centigrade. They are not metallic, but are probably composed primarily of ice. The recent discovery of their ability to polarize the light of distant stars shows that they are not all spherical. A large number must be elongated and aligned preferentially in a direction parallel to the Milky Way plane, presumably under the influence of a weak magnetic field existing along the spiral arms of the galaxy.

Although the interstellar particles are very small and are so widely scattered that the space between the stars is a better vacuum than can be produced in a laboratory, the total amount of such material is estimated to be about as much as is in all the stars of our galaxy.

While some interstellar material probably dates back to the earliest epochs in the history of the galaxy, much of it has resulted from ejections of material from evolving stars. According to current theories, successive generations of stars are continuously condensing out of these clouds.

S-shaped nebula in Ophiuchus. (Mount Wilson Observatory.)

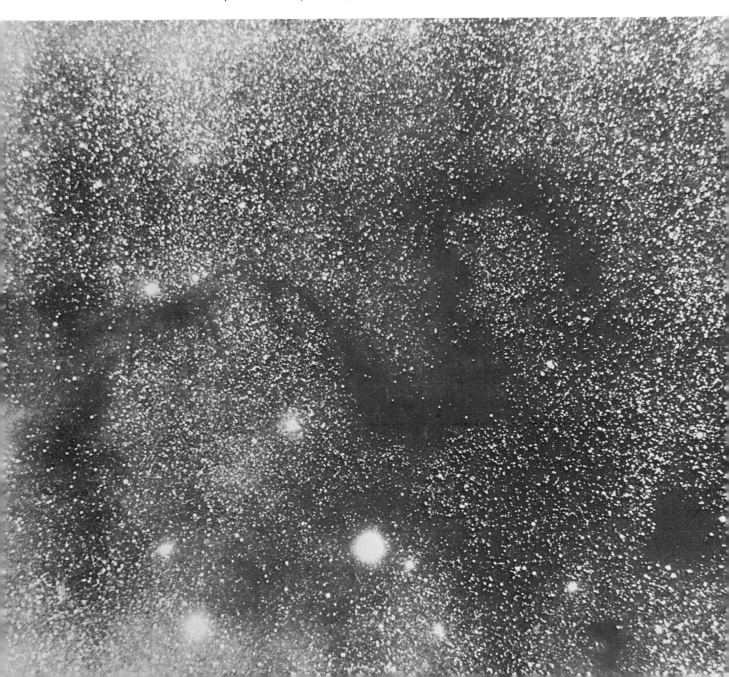

Region of Rho Ophiuchi. A very large nebula has hidden more distant stars through all central part of photograph. Wisps of the nebula can be seen extending out into regions that are crowded with stars. Where this nebula has been either entangled in or close to very bright stars, gases have been made luminous by radiation, resulting in bright nebular material observed. Between these bright parts the general dark mass is observable only by its obscuring effect. (Mount Wilson Observatory.)

CHAPTER 55 A WORLD OF THE HERCULES CLUSTER

The conditions favorable to the development of life on the earth have resulted largely from the earth's size and its distance from the sun. But a planet moving around some other star might be subject to entirely different conditions, even though it was of the earth's size and received the same average amount of radiation. Consider a planet moving around one of the components of the "double double" star in Lyra. This planet would have a secondary sun, much too bright to be called a star, though giving far less light than the primary sun. Conspicuous in the night sky would be a pair of extremely bright stars, the other components of the "double double." Disregarding the probability that its orbit would be unstable, another interestingly placed planet would be one that moves around one star of a close pair, the stars of which have contrasting colors, such as red and blue. The inhabitants of this planet would have a red day and a blue day. They also would have a white day, with blue shadows and red shadows, when their two suns were simultaneously in the sky. Still another interesting position for a planet would be around some star located very near to the Nebula of Orion. A large part of the sky would be covered by that irregular cloud of milky light. A planet moving around a star lying entirely outside the plane of our galaxy would see our system somewhat as we see the Magellanic Clouds. There would be no bright stars in the sky, and astronomy would develop very far before it became possible to measure the distances of any of these celestial objects.

Of all such peculiarly placed planets, one moving around a central star of a globular cluster, such as the Great Cluster of Hercules, might provide the most interesting existence for its inhabitants. Let us consider that cluster and from observed facts deduce the spectacle to be observed from such a world. The cluster is about 200,000,000,000,000,000 miles away. In other words, light traveling 186,000 miles per second would reach us nearly 35,000 years after it left this object. At such a distance even the largest telescopes cannot reveal to us any stars that give as little light as does the sun. The only ones we can photograph are the giants, some of which give more than a thousand times the sun's radiation.

To the naked eye and through binoculars the cluster appears as a faint star. With a small telescope it begins to lose its starlike appearance, and a bright nucleus is seen to be surrounded by a faint haze. The diameter of the disk thus revealed is about one-sixth of a degree. The 6-inch reflecting telescopes, so commonly built by amateurs, show a few separate stars, and increasing sizes of telescopes show more stars. Photographs secured with great reflectors reveal an immense ball of stars.

Astronomers have learned that stars with certain peculiarities of spectra all emit somewhat the same amount of light. An examination of the spectra of the brightest stars of the cluster has thus given their luminosities, and from the luminosities their distance. Also, there are a number of stars whose light varies in a certain definite way. There is a known relationship between the length of time that it takes the light to complete its cycle of variation and the average amount of light emitted by the star. From these stars also, astronomers can determine

Great Cluster of Hercules. (Mount Wilson Observatory.)

the distance of the cluster in which they are located. The distance of the Hercules Cluster stated earlier is that derived by these two methods.

An object having an angular diameter of one-sixth of a degree has a distance from the observer about 345 times the actual diameter. Therefore, this cluster is 600,000,000,000,000 miles, or 100 light-years, across. Probably some stars lie even outside this sphere. We do not know how many stars exist in the ball. More than 50,000 are bright enough for observation, but this does not touch at all on the number of "dwarf" stars such as those

so common in the neighborhood of our sun. Of course, we do not *know* whether such dwarfs exist in the cluster, but it seems likely they do, for the bright stars have a range quite similar to that found in ordinary space.

The bright stars near the center of the cluster are crowded together much more closely than in the outer parts. A calculation tells us that their *average* distance from one another is about one-sixth the distance of the nearest of all stars to the sun. Any dwarfs that may exist would cut down this distance between stars. But even with this "crowding," there

292

is sufficient space so that stars should be able to develop planetary systems without hindrance from their neighbors.

With these facts we can describe the sky as viewed from a planet whose parent star lies near the center of the cluster. The sun of this planet would be moving a dozen or so miles per second in a sort of orbit around this center. All other stars also would be moving in a somewhat similar manner. Instead of constellations remaining almost unchanged through thousands of years, as they do in our sky, marked shifts in positions would be observed in a single generation. The idea of a fixed sphere of stars could never prevail as it did for

Globular star cluster in Sagittarius. This object, known as Messier 22, is one of about 100 globular clusters, each of which contains at least 50,000 stars. Nearest cluster is about 20,000 light-years away, and the most distant one is ten times farther away. They lie outside of Milky Way and form in conjunction with some isolated stars, a roughly spherical halo around the center of our galaxy. (Mount Wilson Observatory.)

thousands of years among our ancestors. Most of the time the nearest of the larger stars would be about one-sixth as far away as is our nearest neighbor and would be far more luminous. Often a star approaching through centuries would grow brighter and brighter until, when nearest, it would be about forty times as bright as is the planet Venus at its best. Once in a while some star would become much brighter even than this, perhaps rivaling the full moon. A telescope would show readily the actual disk of the star, growing through the years of approach and then diminishing until the largest glass showed only a point of light. Under such conditions the most elementary of cultures could not believe all the stars to be at one distance.

Usually there might be two dozen stars brighter than Venus scattered over the sky of this planet. Thousands of stars as prominent as our Vega and Arcturus would be visible. There would be a splendor to the sky so far surpassing our own as to be almost unimaginable.

Even the dullest of primitive "men" would raise their eyes to the heavens in awe. It would be impossible to forget them at any time. The development of our own civilization has been affected greatly by the stars. On the Herculean planet they would dominate everything. The efforts to know and to understand them would lead to a more rapid development of science than occurred here on earth. With other conditions equally favorable, the "men" of that planet, after an interval of many fewer centuries, would progress to a point beyond where we are today. Furthermore, scientific knowledge would be far more widespread. It is impossible under such conditions for any normal man to ignore his universe.

So far everything that has been considered is far more favorable to the race of Hercules. But now a change appears in the picture. Let us consider our sky on a moonlit night. Only the brightest stars are visible to the naked eye. Astronomers cease certain types of work because they cannot observe faint objects. On the world of Hercules the difficulty would be far greater. The stars would produce a brighter glare in the sky than does our full moon and, as a result, photographic plates would fog before stars much fainter than those visible to our naked eye could be recorded on them.

Their astronomers would chart the inner parts of their cluster. They would learn distances, velocities, masses, and other data for thousands of stars, but the most distant objects of their "universe" would be considered by us as close stars. They would measure temperatures, pressures, and radial velocities, and would chart the chemical constitutions, but large gaseous nebulae would be unknown. They would observe the actual disks of more than one star, but would know nothing of the Magellanic Clouds. Their universe would be bound up in a little ball, which to us in our superior position seems insignificant. Speculations about "curved space" and about an "expanding universe" would not arise. Rapidly they would develop their science to a point somewhat beyond the average of ours today and then would face a danger of stagnation.

There might be one help for them. Millennia might cause their sun to drift with its little family away from the center of the sphere. Gradually one half of their sky would appear less and less important till finally it dawned upon their astronomers that the one side of their "universe" presented a marked thinning of the stars. They would decide that it was spherical and finite. Then if the outward drift continued, other such balls of stars would appear—island universes to them.

Finally they might grasp the boundless point of view that our telescopes have made us comprehend.

However, when this fantasy is finished, there lurks a fear that perhaps we, ourselves, actually have observed little more of the observable than have the Herculean astronomers whose universe has seemed so small.

CHAPTER 56 RADIO ASTRONOMY

The hero of a science fiction story a quarter of a century ago invented a telescope that did not use ordinary light. Instead, it used short radio waves to produce the images. With it he accomplished all sorts of unbelievable observations. In part, his feat already is a fact. We *do* have powerful telescopes that use radio waves. They do not make images of the objects at which they are pointed, but they do record the data in other ways. If it were necessary, a scanning device could be used with them to produce visual images. Optical telescopes have many advantages over radio telescopes, but the latter have enough peculiar advantages of their own to rate as one of the greatest astronomical advances.

Like so many other important developments, radio astronomy began accidentally. However, it did involve a scientist with the imagination and training necessary to "take advantage of the breaks." In December, 1931, Karl G. Jansky of the Bell Telephone Laboratories was experimenting with short-wave reception. He was using a wave length of 15 meters. Almost always there is some "noise" in radio reception. Jansky observed a diurnal variation in its intensity and also in its direction, although the aerial he used then was incapable of determining the direction accurately.

If the period of the noise had been the solar day, it could have originated either in the sun or in the earth's atmosphere. However, he found the period to be 23 hours and 56 minutes. This is the period of the earth's rotation with respect to the stars. The waves certainly were coming from interstellar

The 600-inch aluminum mirror radio telescope on Naval Research Laboratory, Washington, D.C. The mounting of this telescope is a U.S. Navy gun mount weighing 28 tons. (Official U.S. Navy Photograph.)

space. His aerial had enough directional effect to indicate that they were coming, at least approximately, from the center of the galaxy. This was the most likely source after the sun had been eliminated. The failure of his attempts to find a solar-day period, as we realize on looking back, was due to the fact that there was a minimum of sunspots at the time. Our sun radiates such waves far more intensely near years of sunspot maximum.

Nine years later Grote Reber, another American, began a series of experiments using waves as short as 9 centimeters. He obtained nothing from the shortest wave lengths, but with a wave length of a little less than 2 meters he began to locate areas of strong radiation in the sky. His receiver could determine directions within perhaps half a dozen degrees. This made it possible for him to map, roughly, contours of equal intensity in the Milky Way. With waves of a third this length he secured less definite but confirmatory evidence. J. S. Hey, S. J. Parsons, and J. W. Phillips in England secured somewhat the same results. The greatest intensity was observed in the direction of the center of our galaxy, beyond the stars we see in Sagittarius. The contours followed the great "rift" of the Milky Way to Cygnus, where a secondary maximum was observed.

World War II brought radar, with much more sensitive directional receivers. In radar, short waves are sent out to be reflected back from a target. Early in 1946 the United States Army bounced radar waves back from the moon. In Hungary, Z. Bay did the same thing. Soon others repeated the experiment.

J. S. Hey and G. S. Stewart in 1945 observed some radar echoes that occurred during the daytime. The next year J. P. M. Prentice, A. C. B. Lovell, and C. J. Banwell at Jodrell Bank in England investigated these echoes and found by a combination of visual and radar observations that they were due to daytime meteor showers. The intensity of showers from radiants fairly near to the direction of the sun is remarkable. The continuance of this work is adding much to our knowledge of meteors.

The valuable early results led quickly to great improvements in short-wave receivers and to the construction of radio telescopes. England, Australia, and the United States have been the leaders.

The 85-foot radio telescope at National Radio Astronomy Observatory, Green Bank, W. Va. (Blaw-Knox Corporation.)

The purpose of a radio telescope is to receive extremely weak signals from a tiny area in the sky and to amplify them sufficiently so that their intensities may be recorded.

or the photographic plate. Radio reflectors concentrate radio waves on what radio astronomers call the "feed." Two things affect the gain in concentration—the size of the reflector and the

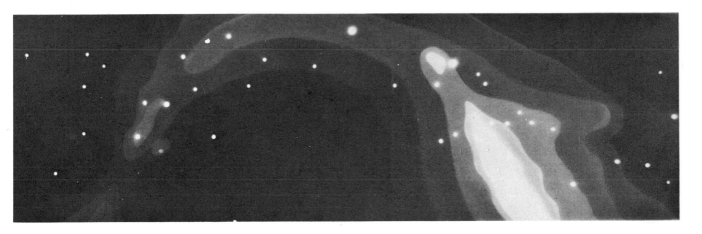

How Milky Way would look (above) if our eyes were sensitive to radio waves of 1.2 meters. This corresponds to radio map (below) made by observers at Ohio State University. Note extreme brightness of central region of galaxy in direction of galactic nucleus, and second maximum of brightness in direction of Cygnus, as one looks along spiral arm in which sun is embedded. At this wave length a number of the discrete radio sources can be seen. (Dr. John Kraus, Ohio State University.)

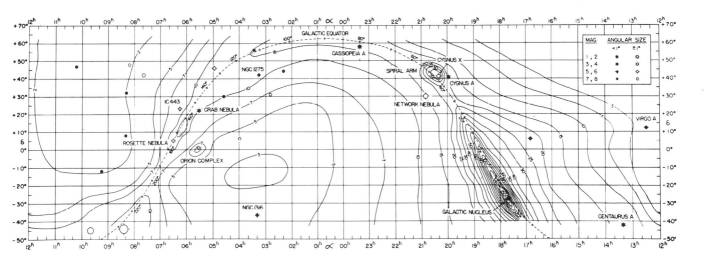

The waves are extremely weak as received on the earth. Therefore, the reflector must be of such a shape as to reflect them almost to a point. At this point (the focus) an antenna receives the waves and feeds them to a powerful short-wave radio receiver. Usually the reflectors are concave and follow the same curvature (a paraboloid) as do the mirrors of our reflecting telescopes. Indeed, their purpose is identical with that of such mirrors. Mirrors concentrate enough light to affect the eye

accuracy with which it conforms to the proper shape. Usually a parabolic wire basket is used, instead of a solid surface, to reduce weight.

One of the most important characteristics of any kind of astronomical telescope is its resolving power—the ability to separate two objects close together in the sky. Theoretically a one-inch telescope, using ordinary light, will separate objects that are $4\frac{1}{2}$ seconds of arc apart. A 10-inch telescope should separate objects ten times closer, and

297

the 200-inch Hale telescope at Palomar should separate stars that are only 0″.022 apart. Actually this limit never is reached because of trouble from the earth's atmosphere and unavoidable imperfections in every telescope.

As we go to wave lengths longer than those of visible light, the necessary arc between barely separable stars becomes greater. With radio telescopes it is many times as great as for ordinary light. However, we can gain back part of the loss in resolution by building reflectors with much greater diameters. The 200-inch diameter of the mirror of the Hale telescope is by far the greatest in existence for optical telescopes. However, 85-foot paraboloidal reflectors are increasingly common for radio telescopes, one of 250-foot aperture is in operation at Jodrell Bank, England, while a 300-foot partially steerable antenna is located at the National Radio Astronomy Observatory at Greenbank, West Virginia, and a fixed dish of 1000-foot aperture pointing to the zenith is at Arecibo, Puerto Rico.

Resolution at radio wave lengths is further improved by the use of an interferometer. This instrument has been used with optical telescopes since 1920 to measure the angular diameter of nearby stars far too small to be measured by ordinary astronomical means. Interferometers improve the resolution of radio telescopes by combining the signals received from long strings of reflectors, known as arrays. Especially in Australia, great arrays of these reflectors have been set up. The arrays have lengths up to 1,500 feet. With them, directions from which waves are coming can be determined much more accurately than with any single radio telescope. These interferometers have been used principally to locate on the disk of the sun the positions of flares of very intense radiation. The sun as observed by this radiation appears entirely different from the sun we photograph. Ordinary light from the sun varies but little in intensity from time to time. Radio waves from local areas of the sun vary a thousandfold. This finding came as a great surprise.

The development of very precise atomic clocks during the past few years has made it possible for radio astronomers at two widely separated observatories to record on magnetic tape the signals received at their radio telescopes while simultaneously recording on the tapes precise time marks. When the two tapes are sub-

sequently played back and compared by computer, they yield the same results obtained by an optical or radio interferometer. The two observatories can be many miles apart, with very high resultant resolution. By 1972 baselines as long as 6600 miles had been used, combining tapes made in Australia and in California. Such combined efforts by radio astronomers in different countries result in what have been called *Very Long Baseline Interferometers*, or VLB. The resultant resolutions are much better than can be achieved by the largest optical telescopes, approaching a thousandth of a second of arc!

Intensive work during the past 15 years or so has revealed that the radiations received at radio wave lengths can be divided into two groups. The first group is called thermal, because the intensity of radiation depends on the temperature of the source. This radiation is produced by the hydrogen in the interstellar gas of the spiral arms of the galaxy when the gas is in the vicinity of very hot stars. The ultraviolet radiation from the hot stars knocks electrons loose from hydrogen atoms, ionizing the atoms; the resulting free electrons can radiate or absorb radio waves when the electrons are in the vicinity of ions. It is interesting that, at a wave length of 15 meters, this band of ionized hydrogen in the Milky Way plane absorbs radio waves from more distant regions, whereas at much shorter wave lengths the hydrogen appears as a series of bright clouds surrounding the hottest stars of the Milky Way plane.

The other kind of radio emission is called nonthermal, because it does not depend primarily on the temperature of the source. There may be several ways by which the stars or the interstellar material can emit nonthermal radiation, but at the present time only one mechanism has been definitely established. It produces what is called synchrotron radiation, so named because the emitting particles behave similarly to the electrons in a synchrotron accelerator in the physics laboratory. In the accelerator, the electrons emit the radiation when they spiral around the lines of force of the magnetic field in the machine. Similarly, it is proposed that the synchrotron radiation from the galaxy is produced by the interaction of free electrons with various magnetic fields of the galaxy.

We receive nonthermal radiation from sources both inside and outside our galaxy. That from within our galaxy has three sources. The first is a halo (corona) surrounding the entire galaxy. Emission from a similar corona surrounding the Andromeda Galaxy has been detected. Because the radiation is spread over a range of wave lengths, the study of its intensity distribution has to be carried out in a series of steps, tuning the radio receiver to a series of wave lengths in succession. For example, if the receiver is tuned to a wave length of 3.5 meters, considerable differences are found in the brightness of the coronal emission from various directions. However, if the details of the emission pattern are smoothed out, it is found that the corona can be approximated by a series of concentric zones of diminishing intensity, all with the same flattened spheroidal shape. The center of the corona is located at the center of the galaxy, and the shortest axis of the spheroidal system is perpendicular to the Milky Way plane. As we go outward from the center, the intensity of the radiation from the zones diminishes. It is still somewhat uncertain why the corona emits a nonthermal radio emission; the best guess is that it is a synchrotron emission produced by an extremely tenuous halo of electrons, which are being influenced by a weak general magnetic field possessed by the galaxy as a whole.

The second nonthermal source is called the "spiral arm" component. This has been recognized as being distinctly different from the thermal emissions from the spiral arms mentioned above. Its observation is quite strictly limited to the Milky Way plane, and its intensity is greatest both toward the center of the galaxy and in directions that would carry the line of sight tangential to a spiral arm lying between the center and the sun. Theoretical calculations have shown that the observations can be explained if it is assumed that an individual spiral arm is circular in cross section, with a diameter of 1,500 light-years. Again, the spiral arm component is probably synchrotron emission.

Probably the most interesting of the nonthermal sources is the third, which comprises all the so-called discrete sources. These radio sources are so small that most of their measured diameters are equal to the resolving power of the particular receiver being used. In other words, the observed diameters are spurious, just as are the observed diameters of stars through the ordinary telescope. Unlike the spiral arm radiation this is not strictly limited to directions related to our Milky Way. In 1952, B. Y. Mills in Australia found that the dis-

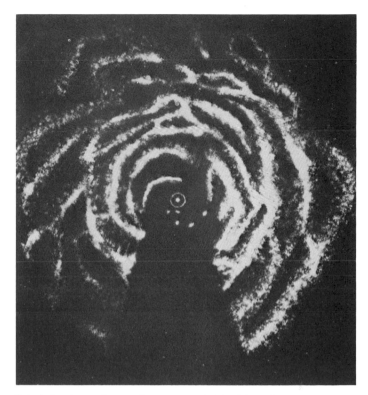

Spiral structure of our galaxy as revealed by 21-centimeter surveys. Center of galaxy is marked with large circle, and location of sun with small circle (above). (Division of Radio Physics. Commonwealth Scientific and Industrial Research Organization, Sydney, Australia.)

crete sources could be divided into two classes. Class I is probably produced by objects in our own galaxy, since they are concentrated close to the Milky Way plane. Class II objects are oriented in random directions and are probably outside our galaxy. See Chapter 59.

The Class I discrete sources are still somewhat of a mystery, although several have been identified with unusual types of nebulosity that can be observed optically. The most definite identification is with the Crab Nebula, which is an intense nonthermal source. The Crab Nebula is believed to be the result of a supernova outburst, and it has been suggested that most of the nonthermal discrete sources of the galaxy may be associated with gases ejected during such outbursts; several of the other sources can be identified with wisps of nebulosity or ringlike nebulosities that may be the remains of such outbursts.

299

CHAPTER 57 THE MAGELLANIC CLOUDS

Of all objects in the southern sky, the Southern Cross holds first place in the hearts of those who gaze casually at the firmament, but the Magellanic Clouds certainly stand not lower than second. Large and faintly nebulous in appearance, they seem like fragments broken from the brightest parts of the Milky Way.

In a sense, these clouds are clusters of stars, but when we examine them in any detail at all, we find they are very different from other clusters. Until a comparatively few years ago, all that astronomers knew about them was that they were the largest clusters to be found, that their distances were so great that the vast majority of their stars were unresolvable even in the largest telescope, and that they contained nebulae as well as stars.

Among the remarkable developments of the twentieth century have been those which have combined observation and theory to give us the distances of very remote objects. These estimates have been based on several different principles. One method requires that we determine the temperatures of the stars by the general type of spectrum, and that the pressures make certain small modifications in the spectral lines, from which it has been possible to determine fairly accurately the total amount of light given by a star. Theory has been correlated with the direct observations of stars close enough for their distances to be measured by direct means, and then calculations have been extended out to the faintest stars whose spectra can be photographed with the largest telescopes.

A second method of determining the distances of stars is derived from the fact that the type of variable stars called Cepheids have periods of variation of luminosity that are related to the amount of light they give. If such a star is close enough so that we can observe it as an individual, even though we do not get enough light to determine the details of its spectrum, the period of the variation, through giving us its total luminosity, must furnish us with a quite good estimate of the distance (see page 209). Other methods have been developed based on absorption of light in space, mean proper motions, and other factors, and they probably are fairly good for determination of the average distance of a large group of stars. However, unlike the methods to which we gave first consideration, and which were based on the peculiarities of individual stars, these depend on characteristics of interstellar space and of the organization of our galaxy. In comparison with the first two, they tell little or nothing about any one star.

If an astronomer observes the period of light

Magellanic Clouds. These two clouds form triangle with star Achernar (lower right), whose image is much overexposed. (Harvard College Observatory.)

Large Magellanic Cloud. (Harvard College Observatory station at Arequipa, Peru.)

Detail of part of Large Magellanic Cloud. (Harvard College Observatory station of Bloemfontein, South Africa.)

The great 30 Doradus Nebula in the Large Magellanic Cloud (Kitt Peak National Observatory.)

variation of a Cepheid that is a member of a star cluster, this period, giving the distance of the individual star, must give also the distance of the cluster. In large groups of stars he may find many Cepheids and, through averaging their distances, be able to eliminate to a considerable extent the error in distance of any one such star. This is the method that has been applied to the globular clusters, to the Magellanic Clouds, and to the nearer of the other galaxies of space.

The smaller of the Magellanic Clouds is so distant that we see it by light that started on its journey toward us 206,000 years ago. In other words, this cloud is 206,000 light-years away. The larger cloud is 179,000 light-years distant. These distances are more than twice the diameter of our own galaxy, placing the clouds definitely outside the galaxy. They are actually the nearest of the major extragalactic objects. Only one other galaxy is as close as twice the distance to the clouds, and this is a faint dwarf elliptical system. The giant spiral galaxy in Andromeda is over ten times as far away. Thus the Magellanic Clouds provide an unexcelled opportunity to study intrinsically fainter component stars, such as are beyond the reach of even the largest telescopes if they are sought in the Andromeda galaxy.

Since the clouds are located only some 20° from the South Celestial Pole, an observer must be in the Southern Hemisphere to observe them comfortably. Two great radio telescopes in Australia—the 210-foot-diameter dish at Parkes and an enormous interferometer array, with arms a mile long, at Hoskinstown—have provided invaluable data at radio wavelengths. The general recognition of the importance of studying the clouds in as much detail as possible has been one of the strong stimuli to the construction of larger optical telescopes in the Southern Hemisphere. The superb observing conditions found in the arid regions of Chile have prompted the location of most of these telescopes in that country. At present, a 144-inch telescope is under construction at the European Southern Observatory on La Silla in Chile; a 150-inch reflector is being built on Cerro Tololo, also in Chile; and plans are being laid to construct a 150-inch telescope in Australia as a joint British-Australian venture.

In spite of the present shortage of suitable telescopes, much work has already been done on the structures and motions of the clouds. A conspicuous feature of the large cloud is the central system of stars—about 6° in diameter, which means that its diameter is 18,000 light-years. The central system is actually a rotating flat disk, reminiscent of the flattened structure of our galaxy. A star at a point 8,000 light-years from the center of the disk travels around the center with a velocity of 36 miles per second, so that it describes one revolution in a quarter of a billion years. Most of the highly luminous supergiant stars in the large cloud are located in the central system, which also contains the conspicuously bright bar, 3° in length, located slightly south of its center. A less extremely flattened disk of open clusters surrounds the central system, extending out to a diameter of 14°, and this disk of clusters in its turn is embedded in an apparently spherical halo of globular clusters, with a 24° diameter.

The small cloud is approximately half the size of the large cloud, showing a central body that may also contain a bar. The small cloud too is in rotation. We do not know the precise masses of the two clouds, principally because uncertainty about the tips of their flattened disks relative to the line of sight affects estimates of their observed velocities of rotation. The best estimates place the masses of the large and small clouds at, respectively, six billion and one-half billion times the mass of the sun. These are relatively small masses when compared with the mass of our galaxy, which is 100 billion times the mass of the sun.

Just as radio astronomers have been able to map the distribution of interstellar hydrogen within spiral arms of our own galaxy, so too have the large antennae in Australia revealed the presence of great amounts of hydrogen between the stars in the Magellanic Clouds. In the large cloud the hydrogen shows a marked clumpy structure, and there is evidence for the existence of spiral arms. There is a surprisingly large amount of interstellar hydrogen in the small cloud, possibly as much as 25 percent of its mass. In the large cloud, the amount of interstellar hydrogen is more nearly comparable to that found in the vicinity of the sun in our galaxy. The hydrogen in both clouds extends well beyond the visible images, and indeed a bridge of hydrogen even extends from the one to the other.

The clouds are close enough to us to permit detailed study of the brighter stars, either singly or in groups. There is every expectation that these studies will shed increasingly valuable light on the difficult problem of the birth and evolution of bright stars, particularly when the great telescopes now being planned for the Southern Hemisphere are put into operation.

CHAPTER 58   THE GALAXIES OF SPACE

When one contemplates the progress that has been made since the days of Galileo in our concept of the universe and of humanity's place in it, a single theme keeps reappearing. As each new development expands our view of the universe, it reminds us again that our environment is far from being unique. Thus, at one time we were convinced that the earth was at the center of the universe; it turned out that it was merely one of the smaller planets in orbit around the sun. The sun, in its turn, was found to be a quite average star in a galaxy composed of billions of companions around its center. Finally, over the past sixty years we have come to realize that our galaxy, enormous as it is, is just one among at least a billion galaxies.

We are very fortunate that the sun is located where it is in the galaxy, out in the suburbs where the obscuring interstellar dust and gas is constrained to a relatively thin layer. So, while we have difficulty discerning the true structure of our galaxy hidden behind this obscuration, when we look out into space above or below this layer, our line of sight quickly breaks out "into the clear," and the only limitation to our seeing progressively farther is our ability to gather for inspection the progressively fainter light from progressively more distant objects.

When we leave our galaxy behind on a "thought" trip into outer space, the first objects we meet are the two Magellanic Clouds, described in the previous chapter. They might be called satellites of our galaxy, because of their proximity (less than 200,000 light-years away from the sun) and their smaller diameters (less than 30,000 light-years). A similar triple system consists of the great spiral in Andromeda and its two small companions. The distance of this system is estimated to be about 2,300,000 light-years. It is the most distant object that can be seen with the unaided eye. On a clear, moonless night, the nucleus can be seen as a faint patch of light about half the size of the full moon. Photographs with a telescope show the spiral arms extending to a diameter of about 120,000 light-years. By refined methods it is possible to detect fainter extensions to nearly double this diameter. More than 100 globular clusters of stars are distributed over this larger area. Thus the Andromeda Galaxy is seen to be comparable to our own galaxy.

The diagram on page 274 shows our galaxy and the relative distances of some of our neighbors. Their positions are indicated, but no attempt is made to show their sizes or shapes. They are generally designated by their numbers in certain catalogues made by C. Messier and J. L. E. Dreyer. M 31, for instance, is No. 31 in Messier's catalogue, and refers to the Great Spiral Galaxy in Andromeda, which is shown in the photograph from the Lick Observatory. The other two systems take their designations from Dreyer's New General Catalogue (abbreviated as NGC) and two supplements called Index Catalogues (IC). These systems of stars form a local group distributed through a volume of space whose longest diameter is more than two million light-years.

With the 200-inch telescope at Palomar, nearly a billion galaxies could be photographed, if long

Great Galaxy in Andromeda. This is nearest large system of stars comparable with our own. Its distance is about 2,300,000 light-years. A few smaller galaxies are nearer. (Lick Observatory.)

exposures could be made of all parts of the sky. Most of those that have been recorded are spiral in shape, but some are elliptical or spindle-shaped, with no apparent internal structure. A few are irregular clouds. Their spectra show that they are composed of stars that are similar to stars we are familiar with in our own galaxy. The elliptical galaxies and the central bright regions of the spirals seem to be composed of old stars, similar to those found in globular clusters, whereas the young blue stars and diffuse nebulosities are found along the arms of the spirals.

The distribution of galaxies in space is far from uniform. Local concentrations into clusters of galaxies are quite common. Almost 3,000 clusters have been catalogued as a result of the Palomar Sky Survey. Some of these clusters are very large, comprising over 500 member galaxies. One peculiarity of the spiral galaxies found in rich clusters is that many of them lack the usual young stars and

diffuse nebulosities in their arms, suggesting that the arms may have been swept clean of diffuse matter as the result of collisions between the more closely spaced galaxies in the cluster. When such a collision between two galaxies takes place, it has little or no effect on the member stars; the two populations merely pass through each other. The fate of the diffuse matter between the stars, however, is quite different. It apparently is swept out of the galaxies and left behind in the space between them as they recede. It used to be thought that certain intense radio sources might represent colliding galaxies, but this hypothesis is no longer considered likely.

Aside from the tendency to form local clusters, galaxies seem to be distributed rather uniformly in space out to the most distant objects reached by the 200-inch telescope. The accuracy of this statement depends upon the precision of the methods of determining the distances of remote galaxies, and

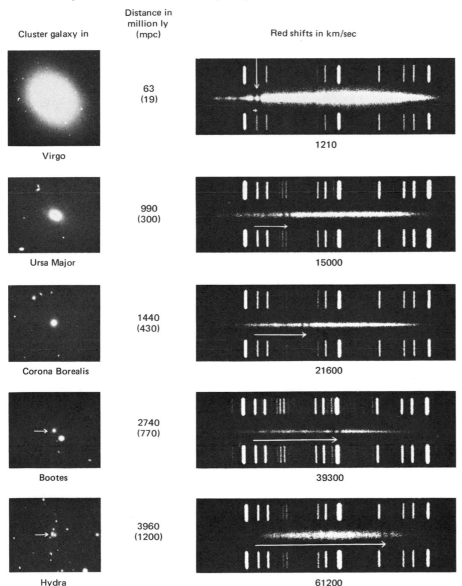

Cluster galaxy in	Distance in million ly (mpc)	Red shifts in km/sec
Virgo	63 (19)	1210
Ursa Major	990 (300)	15000
Corona Borealis	1440 (430)	21600
Bootes	2740 (770)	39300
Hydra	3960 (1200)	61200

The redshift-distance relation. (Palomar Observatory, California Institute of Technology.)

Resolution of Andromeda Galaxy into stars. This is upper portion of galaxy as it appears in preceding photograph. (Mount Wilson and Palomar Observatories.)

these methods are admittedly still only approximate. For the nearer galaxies with resolved stars, one can make use of the Cepheid variable stars in the galaxies, also the brightest stars, and the supernovae observed in them. The distances of more remote systems can be estimated from the size of the galaxies or their total brightnesses. Finally, statistical studies have to be made of the brightnesses of the member galaxies of the most remote of the clusters.

When E. P. Hubble was conducting his pioneer-ing study of the galaxies around 1929, he made a discovery of fundamental importance, namely, that the spectra of galaxies are shifted progressively farther to longer wave lengths as one goes to more remote objects. This "red shift" is usually interpreted as a velocity effect, indicating that the universe of galaxies is expanding in all directions.

Several theories have been advanced to explain the apparent expansion of the universe. One group of theories proposes that the universe was at one time in a highly compacted state and underwent a

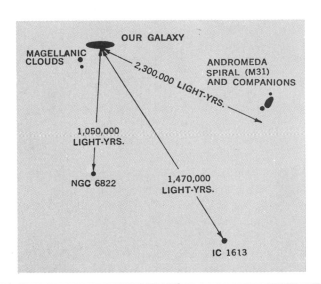

Our galaxy and its neighbors.
All are in about the same plane except NGC 6822, which has been swung into that plane to show its relative distance.

Spiral galaxy, Messier 81, in Ursa Major. This galaxy is observed obliquely. It has a very large stellar nucleus. (120-inch telescope, Lick Observatory.)

Spiral galaxy in Coma Berenices, seen edge-on. Notice dark line of obscuring matter. (Mount Wilson and Palomar Observatories.)

308

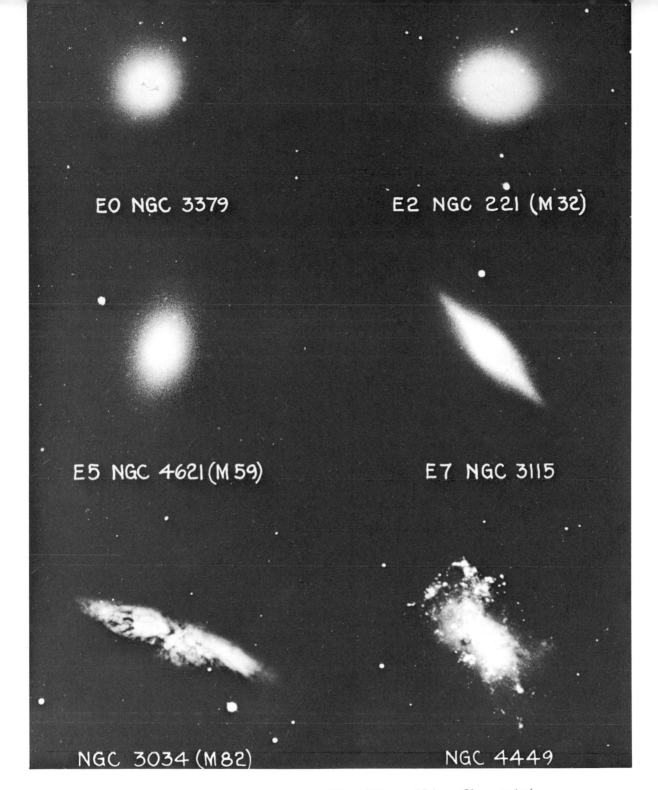

EO NGC 3379

E2 NGC 221 (M 32)

E5 NGC 4621 (M 59)

E7 NGC 3115

NGC 3034 (M 82)

NGC 4449

Six types of elliptical and irregular galaxies. (Mount Wilson and Palomar Observatories.)

"big bang." When in the compacted state, all of the molecules and atoms of matter were broken down into their fundamental particles, and the lighter elements were formed out of this primeval mixture of fundamental particles and radiation within the first half hour following the onset of the expansion. After about a quarter of a billion years the mass had expanded to such a state that it broke up into

Sa NGC 4594

SBa NGC 2859

Sb NGC 2841

SBb NGC 5850

Sc NGC 5457 (M101)

SBc NGC 7479

Six types of normal and barred spiral galaxies. Normal spiral nebulae (S) have a pronounced bright central region from which the spiral arms seem to originate. Classes Sa, Sb, and Sc are distinguished here. Barred spirals are divided into three classes SBa, SBb, and SBc, depending on the size of the central region in relation to the bar. (Mount Wilson and Palomar Observatories.)

millions of pieces that evolved into the galaxies. In the late 1940s George Gamow predicted that, if this is what actually happened, the universe should still be filled with the radiation that accompanied the "big bang," but now cooled to a low three degress absolute (Kelvin) as consequence of the

Cluster of faint galaxies in Corona Borealis. (Mount Wilson and Palomar Observatories.)

universe's expansion. The discovery of this residual energy by Penzias and Wilson in 1965 provides the strongest evidence extant that the universe was once denser and hotter than it is today.

Some theories maintain that the expansion of the universe will continue forever; others suggest that the rate of expansion will gradually decrease and will be followed by a contraction that will lead back eventually to the highly compacted mass with which the process started, and a subsequent reexpansion. A number of tests have been proposed to help decide which of these behaviors is correct. Most of these tests tax the capabilities of the largest telescopes to the limit, and are still indecisive. Probably the space telescope and the larger ground-based telescopes currently being designed

Galaxies of space. The four nebulous objects appear in constellation Leo and are about 28,000,000 light-years from our galaxy. (100-inch telescope, Mount Wilson Observatory.)

will eventually succeed in settling the issue. In the meantime, the most suggestive evidence we have today is provided by estimates of the universe's mean density. This is because the gravitational attractions between adjacent galaxies must be inevitably slowing the rate of the universe's expansion. If these gravitational attractions are strong enough, the expansion will eventually stop, to be followed by a contraction. Otherwise, the expansion will continue forever. It is the latter fate that is strongly suggested by recent calculations of the amount of matter in the universe, which fails by the tremendous factor of thirty of being sufficient to stop the expansion. Admittedly, we may not be seeing everything that the universe is made of. There may be dark invisible clouds, little clusters of stars, or intergalactic dust that would change our ideas profoundly. The last word has not yet been said on the subject.

CHAPTER **59** **THE MYSTERY OF THE QUASARS**

Just as the visits of the astronauts to the moon and the planetary flybys by unmanned satellites have revolutionized our understanding of the structure and evolution of the solar system, so has the discovery of quasars raised hopes that their study will shed light on the structure of the universe. In the meantime, however, the quasars have raised as much controversy as solutions. The former will have to be resolved before quasars can take their correct place in the cosmological scheme of things.

The discovery of quasars involved the efforts of both radio and optical observers. The Class II discrete radio sources mentioned in Chapter 56 are probably all located outside the galaxy. While over 2,000 of these sources have been mapped, it is interesting that only a few have yet been positively identified with optical objects. Some are produced by spiral galaxies, particularly those with well-developed arms. An example is the Andromeda Galaxy. Sources associated with spiral galaxies, however, are quite weak. The extreme distances of most of the intense-radio galaxies became recognized in 1954 when a faint optical counterpart of an intense radio source was identified in Cygnus. These distant galaxies are up to a million times brighter than our own. Usually, though not always, the radio emission comes from two centers approximately similar in size and intensity, separated by a distance often several times their diameters and lying on opposite sides of the galaxy. The radiation shows a polarization indicating the presence of an all-embracing magnetic field. The visible galaxy associated with the radio sources is usually an intrinsically bright elliptical type, seen at a great distance. It often has an extended envelope, with unusual lanes or patches of absorbing material, and bright emission lines in its spectrum showing evidence of an unusual disturbance. It is fre-

In this striking "photograph" of cold atomic hydrogen in the sky, the degree of blackness is proportioned to the amount of gas. Interstellar gas had been thought to exist in clouds, randomly distributed in space and velocity, until Carl Heiles at Berkeley showed with this photograph that the gas resides primarily in long or delicately-connected filaments. Heiles fed 140,000 measurements by an 85-foot diameter radio telescope at the University of California's Hat Creek Observatory into a computer to produce the magnetic tape that generated the photograph. The photograph covers a maximum of 65 degrees on either side of the Galactic plane, or Milky Way, represented by the central horizontal band. The large blank at left is the southern sky, invisible from the observatory in California and the small rectangular blank areas, above the Milky Way, and near the middle, is the North celestial pole.

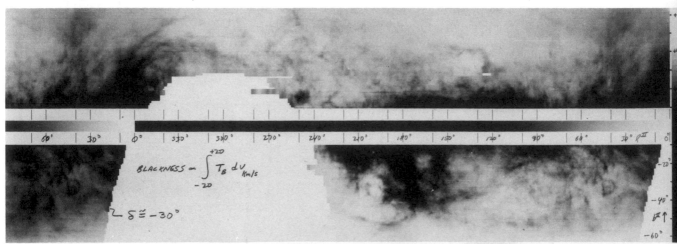

quently among the brightest members of a very remote cluster of galaxies.

It is probable that all the strongest sources of Class II are produced by unusual and violent events. Should this really be the case, a calculation shows that the optical astronomer will never be able to "see" most of the Class II sources with his present equipment. Were a source, similar to the source in Cygnus, located at such a distance from us that the radio emission was close to the limit of detectability of the 250-foot radio telescope, then its optical magnitude would be such that it would be beyond the limit of the 200-inch telescope at Palomar. Thus one has the interesting situation that the radio astronomer can penetrate deeper into space than can the optical astronomer, at least as far as these unusual radio sources are concerned.

It is therefore natural that research activity in this area should be intense. Radio telescopes of increasing size have been constructed to reach fainter objects and to determine their positions with increasing precision. By 1960, powerful radio interferometers had ascertained positions for a number of these sources, with an uncertainty of only 5 seconds of arc. Searches for optical images by astronomers using the great Palomar telescope revealed faint, starlike objects at the same positions. These quasistellar radio sources have acquired the name *quasars*. By 1973 over two hundred had been identified. On the basis of the incompleteness of the search, it is estimated that there may be as many as a million brighter than the 20th magnitude. Spectra of the brighter among these optical images revealed broad emission lines at wave lengths not usually seen in the spectra of emission-line stars. This was very puzzling for a while, but the problem was resolved when it was discovered in 1963 that the emission lines were

Photograph of the now famous Stephan's Quintet showing the possible interaction of its members. (Palomar Observatory, California Institute of Technology.)

The Seyfert Galaxy NGC4151. The nucleus of this galaxy is so bright that if it were very far away, it would probably be mistaken for a quasar. There is strong evidence for the violent ejection of gas from the nucleus of this galaxy. (Palomar Observatory, California Institute of Technology.)

A photograph of the irregular galaxy M82 taken in the red light emitted by hot hydrogen atoms shows large fans of gas above and below the spindle-shaped galaxy. This gas has apparently been ejected from the galaxy's center. (Palomar Observatory, California Institute of Technology.)

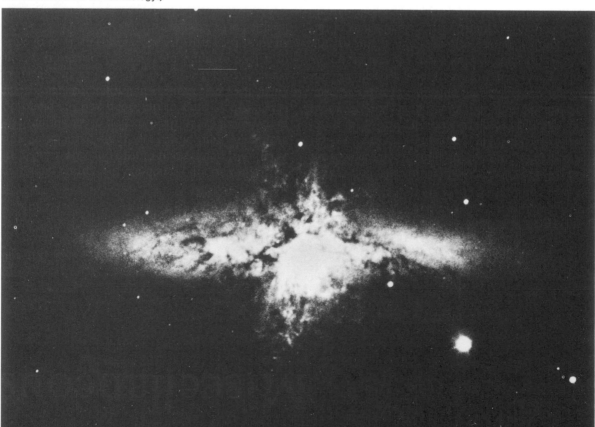

The Exploding Galaxy M82. Huge filaments of gas erupting from this galaxy are evidence of a colossal explosion that occurred roughly two million years ago at its center. (Palomar Observatory, California Institute of Technology.)

315

The fixed radio telescope of the Arecibo Observatory in Arecibo, Puerto Rico, has an aperture of 1000 feet. The Observatory is part of the National Astronomy and Ionosphere Center, operated by Cornell University under contract with the National Science Foundation.

typical after all, but were shifted drastically to much longer wave lengths. These red shifts have proved to be a universal characteristic of quasars. Some of the shifts are so great that they indicate the quasar is receding from us at 9/10 the velocity of light—if the usual interpretation is made that the red shift is a consequence of recession.

As observations of quasars have multiplied, several features have emerged that have not yet been explained. Quasars are quite blue in color, similar to white dwarfs, old novae, and other stars of great age. Their color has been a great aid in the discovery of further members of the group. Indeed, through the use of the color criterion, several quasars have been found that lack the strong radio emission noted above but that do have marked spectral shifts.

From studies of the relative strengths of the bright emission lines characteristic of quasars, it has been concluded that they are roughly similar in chemical composition to gaseous nebulae in our own galaxy, with effective temperatures ranging from 15,000° C to 30,000° C, depending on the method of calculation.

Shortly after the first identification of quasars, a study of photographic records—some as much as seventy years old—revealed variations in light intensity. This remarkable discovery has since been amply confirmed. Sometimes changes take place in just a few weeks. The variation is frequently quite large, amounting to as much as a factor of 2. The changes seem to be erratic and unpredictable. Variations in intensity of the radio spectra of quasars have also been reported, though these are not as abrupt as the optical changes.

When very great red shifts were first found in conjunction with quasars, it was assumed by most astronomers that they were indicative of an expanding universe. The extreme degree of these shifts meant that quasars must be the most remote objects yet identified. This immediately generated intense interest among cosmologists and physicists, because quasars would thus offer the very real possibility of shedding invaluable light on several aspects of the past history and present structure of the universe. The light from the most remote of the quasars may have started out on its tremendous journey to us when the universe was only a tenth of its present age. It was also hoped that they might help settle the controversy over the nature of our universe, that is, whether it is a steady-state universe or one evolving in time.

Almost transcending these important questions, however, are the intriguing and baffling questions of the nature of the quasars themselves, and their source of energy. It was immediately recognized that if they are as remote as their red shifts indicate, their optical energy output would be 100 times the output of a normal galaxy! This conclusion would not be too difficult to accept, were it not for the fact that they show the variations in brightness mentioned above. This sets an upper limit to the diameter of the emitting object as being the distance a beam of light travels in the same interval, i.e., in some instances as small as a few light-weeks!

Several astronomers, notably Halton Arp, have argued that the red shifts exhibited by quasars are not consequences of great distances, but rather that some other process, possibly gravitational, shifts the lines in quasars that are actually at distances quite similar to those of ordinary galaxies. They point to certain quasars that seem to be connected to nearby galaxies by luminous bridges, and suggest that the quasars may have been ejected violently from the associated galaxies. However, all such quasars are red-shifted; none show blue shifts, as might be expected if their ejections had been in random directions.

Despite such arguments, the consensus at the present time is that quasars are extremely remote. To have the consequent energy output, they would have to convert a hundred million solar masses in a million years if we invoke the relation $E = mc^2$. Only about one percent of the mass is lost by nuclear processes even if one starts with hydrogen and converts it all to iron, so that at least ten billion solar masses would have to be packed into an object only a few light weeks in diameter. Many theories have been advanced to explain the tremendous energy output. These range from an extension of supernova theories, through theories involving gravitational collapse, to suggestions that they are the result of a collision of an object composed of ordinary matter with another composed of antimatter. None of these theories have been entirely satisfactory. Recently theorists have been exploring the possibility that processes exist by which quasars could be turning all of their mass into energy. If such processes exist, then the whole mass would be quite comparable to masses found in the nuclei of galaxies, and with densities that average only around 10^{-8} grams per cubic centimeter.

The proponents of the more popular interpretation argue that, if quasars were more numerous in the early universe, we should see more of them with large than with small redshifts, since larger redshifts imply greater distance and an earlier epoch. This is exactly what is found. As one goes to greater redshifts, the number of known quasars increases, just as expected if one is looking back in time. Furthermore, no quasars are found with redshifts beyond a certain maximum, which is to be expected if they did not make their appearance in the universe until it had reached a certain age after the occurrence of the big bang.

A few quasars showing smaller redshifts have been found to be the nuclei of galaxies. Examples are the quasars 4C37,43 observed by Stockton in Hawaii, and 3C48, studied by Gunn of the Mount Wilson and Palomar Observatories and Wampler at the Lick Observatory. Since a smaller redshift implies a smaller distance and thus a later time in the history of the universe, the possibility thus exists that quasars develop into galaxies as they age.

An interesting development was the discovery in 1979 of a double quasar, consisting of two small blue objects equal in brightness separated by about 6 seconds of arc. Their spectra and redshifts

The Exploding Galaxy NGC1275. This remarkable photograph shows filaments of gas erupting from the center of a very distorted galaxy. (Kitt Peak National Observatory.)

are almost identical. The interesting suggestion has been made that there is actually only *one* quasar, and that the two images that we see are the consequence of the bending of its light around the two sides of an intervening dark, massive object that acts as a gravitational lens. The bending is similar to the bending that light of a distant star experiences when it grazes the surface of the sun, as has been observed during total solar eclipses. They are both examples of the gravitational bending of light predicted by the general theory of relativity.

It has been suggested that quasars themselves might be further examples of the effect of general relativity, which predicts shifts to longer wavelengths of the light emitted by small, dense objects. However, when calculations are made of the size and mass of an object showing the wavelength shift of a quasar, it turns out that the object would collapse into a black hole in a few seconds. It is possible that further study will resolve this dilemma—only time will tell.

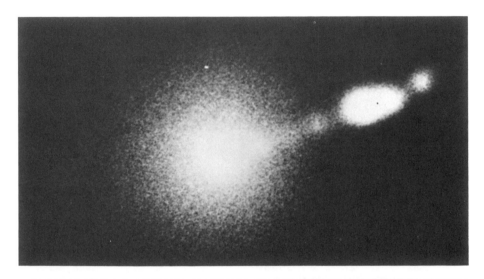

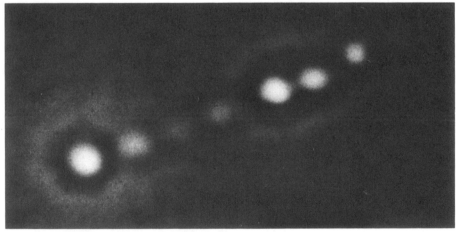

The "Jet" in M87. A short time-exposure reveals a luminous jet surging out of the nucleus of M87. The upper view is a single, ordinary photograph of the jet. The lower view is a computer-enhanced exposure made from several photographic plates. (H.C. Arp; Hale Observatories.)

318

SECTION EIGHT

Miscellaneous

CHAPTER **60** **THE NATURE OF RADIATION**

Many of the facts explained in this book were learned from a study of the radiation received from celestial objects. Light is only one very small section of the radiation received from the stars.

We usually think of light as being a kind of wave, although we cannot be entirely sure this is true. Ordinary waves are of two kinds. Consider a coil spring whose ends are fastened. We can grab it at some point near one end and pull it toward the point where it is fastened. This releases part of the spring from tension but stretches the other part. If now we let go, the spring goes back to its former position and on beyond it, then returns again. If there were no loss of energy through friction or other cause, the spring would keep up this vibration forever. Of course, that is an impossible condition. If the spring is a long one, we will have a wave of compressed spring, followed by a bit of stretched material, traveling along its whole length. Such a wave is called a *longitudinal wave*. Sound waves in the air or in a metal rod are the best known examples of longitudinal waves.

If, instead of stretching the spring along its length, we pull a point of it to one side, it takes up a different kind of wave, with its particles moving perpendicularly to the direction of the spring. This is what the strings of the violin and the piano do. Also, it is essentially what water particles do in water waves. The water particles of the ocean or of a lake or pond move mostly up and down instead of forward with the wave. This kind of wave is called a *transverse wave*.

Electromagnetic Waves. Light waves are transverse waves, and included with them are radio waves, heat waves, ultraviolet light, X-rays, gamma rays, and cosmic rays. All of these seven kinds of waves are really one kind, differing only in their lengths or in the number given off each second. The name *electromagnetic waves* is given equally to all these seven. This is because all of these waves actually are superposed electric and magnetic fields traveling together through space at the incredible speed of 186,000 miles per second. To explain just what is meant by the preceding sentence would be

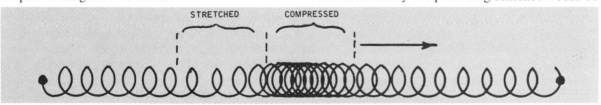

Longitudinal waves.

The compression of a part of the spring and its release cause a longitudinal wave to travel along it. If the spring is fastened rigidly at both ends, the wave is reflected from the far end. These waves resemble sound waves.

319

rather difficult and would involve a great deal of mathematics.

However, these properties can be shown in the laboratory because these waves can be changed by passing them through powerful electric or magnetic fields and also because they affect certain delicate electrical instruments. To such observations must be added the fact that radio waves are produced by electrical means in our broadcasting instruments.

reach those only about 0.00003 inch long, which give us our deepest red. These also affect photographic plates. Waves a little more than half this long give us the sensation of violet and have lost nearly all their heating effect. Waves shorter than these give no sensation of light but do affect photographic plates, produce sunburn, and have other chemical effects.

Waves of the order of one one-thousandth the length of light waves are called *X-rays*. Waves very

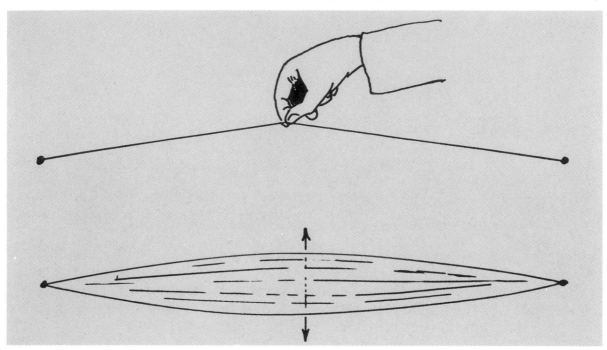

Transverse waves.

The strings of a harp, violin, piano, and other similar instruments; waves on the surface of water and light waves—all are transverse waves.

The waves that produce our vision are a very narrow band of the whole group, with the longest ones less than twice as long as the shortest. In different words, we state the same thing when we say that less than one octave of electromagnetic waves gives us the sensation of light.

The longest of the observed radio waves are more than twenty-five miles long. The ordinary broadcasting band uses waves in the general neighborhood of 1,000 feet long. The shortest radio waves in special electrical instruments are of the order of a tenth of an inch. By this time they also are observable from their heating effect and are called *heat waves*. As we go through many octaves of heat waves of shorter and shorter length, we finally

much shorter than X-rays are called *gamma rays*. These are produced in large quantities when the nuclei of atoms are broken up. *Cosmic rays* are much shorter even than gamma rays. It must be remembered that all of these seem to differ only in their lengths and that they travel at the same speed of 186,000 miles per second. The very longest radio waves are given out at the rate of about 7,000 waves a second. Waves we see as the color red are given out at the rate of 400,000,000,000,000 each second. The rate increases for X-rays, still more for gamma rays, and reaches the greatest number for cosmic rays.

Bending, or Refraction, of Light. The bending of a ray when it goes from the vacuum of space into air or from air to water or glass depends upon two facts. The first of these two facts already has been stated for transverse waves: that they always travel in a direction perpendicular to the front of the wave. This does not apply strictly to very narrow waves, such as those going through a pinhole or through a very narrow slit.

The second fact is that light is slowed down a little after entering a transparent substance that is denser than the one it has been traveling in. That is, light travels more slowly in air than in a vacuum, more slowly in water than in air, and more slowly in glass than in water.

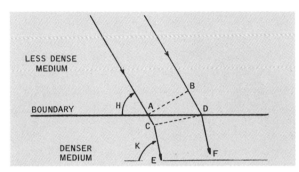

Refraction.
Passage of light from a less dense medium to a denser one.

With these facts and a simple diagram, we now can see why light changes direction. The points A and D are on the boundary of two transparent substances, with the denser the lower in the diagram. The ray reaches A and B at the same instant. The part at A enters the denser medium. By the time the part at B reaches D, the other part has traveled only the shorter distance AC. As a result, the front of the ray is now in the direction CD and it must now travel in the new direction shown by CE and DF. If our eye is at E, we see the object as if it were out along the line EC, instead of its original lower direction. For this reason we always see the stars higher than they actually are, their light being bent by the earth's atmosphere. Of course, if the light were coming perpendicularly to the boundary surface, there would be no change in direction.

Let us now consider a similar ray coming in more nearly parallel to the boundary surface, such as the ray from a star only a few degrees above the horizon. BD is much longer than it was before for a ray of the same width. As a result, the part from A is slowed down for a longer time before the slowing starts for the part from B. This changes the direction of the

wave front more than in the first case. In other words, there is more refraction, or bending, of the ray.

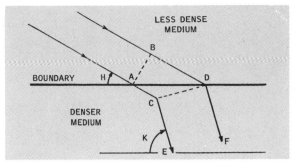

Refraction.
Similar to preceding diagram but with light entering more nearly parallel to boundary between media.

The angle of bending is quite large when we go from air to glass or to quartz, etc., but it is rather small from a vacuum to air. In the two diagrams, H is the angle that the light ray from a star makes with the true horizon before the light enters our air. K is the angle at which the star seems to us, who are immersed in the air, to be above the horizon. If K is zero—that is, if a star appears to be on the true horizon—the ray under average conditions is bent 33' 51". This is more than the diameter of the sun; therefore, when the lower limb, or edge, of the sun appears to be on the horizon, the sun actually is entirely below it. If a star appears to be half a degree above the true horizon, the refraction is only 28' 11". As a result of this, the upper limb of the sun appears to be raised less than the lower, making the sun near the horizon seem to be flattened into an oval shape. Contrary to the belief of many people who have not studied the matter, refraction makes the sun and moon appear to be smaller when they are near the horizon than they actually are. The illusion of an increase in size comes from the fact that when they are near the horizon we realize that they are far away and mentally adjust our impressions. When a star is 45° above the horizon, its refraction is only about 57", roughly 1/35 as much as when it was rising. The diagrams oversimplify the refraction of light into our atmosphere, for the air gets denser and denser as we get down to the surface. However, the simplification does not at all change the qualitative description given.

Cause of the Spectrum, or Band of Color. One more fact must be considered about refraction if

Sun setting behind Santa Monica Mountains. Greater refraction of lower limb of sun gives oval appearance to disk. Notching is caused by common irregularities of refraction near horizon. (Paul Roques, at Griffith Observatory.)

posed on the blue from a lower one and blend to make the colors produce ordinary white. However, at the highest point of the sun, there should be a slight bluish edging, and at the lowest point a slight reddish edging. These cannot be observed when the sun is above the horizon for two reasons: first, the rest of the sun is far too bright by contrast; and, secondly, with the small amount of refraction, there is very little edging to observe. In addition to this, our air usually absorbs nearly all the green and blue light from objects near the horizon, leaving very little blue to form the upper edging. If the air is unusually clear and if our horizon is a nice, straight line, as we find it over a calm ocean, we often can observe a flash of color after all or nearly all of the sun, except the edging itself, has disappeared. It will seldom last more than a second or so. Almost always so little blue gets through our air near the horizon that the color we get from the edging is green. This phenomenon is therefore usually called the *green flash* or the *green ray*.

Each of the chemical elements when glowing and in gaseous form emits radiation different from each other element. Also, the radiation is changed by temperature, pressure, passing through gases on its way to us, by presence of electrical and magnetic fields, and by the relative velocity of the star, or other object, toward or away from us. By passing the light through the instrument we call the spectroscope (described in Chapter 40) we can disentangle these differences and often can determine the peculiar condition of the star or nebula that produced them.

we are to understand the explanation of the breaking up of light into colors. If two wave lengths of light enter the air as a single ray, the shorter one is slowed down more than the longer one. As a result, it is bent more than the longer one. If we would examine the image of a star near the horizon very carefully, we would see that its light is broken up into a band of colors with the blue highest and the red lowest. For objects having a visible disk, as our sun has, each point actually has this color effect, but the red from a higher particle must be super-

CHAPTER **61** MEASURING ASTRONOMICAL DISTANCES

PART I—IN THE SOLAR SYSTEM

One of the outstanding features of astronomy is the immensity of the distances involved—for example, 35 million miles to Mars at its closest point, 93 million miles to the sun, and 25 trillion miles to the nearest star. How these are measured is a mystery to most people, even to those who know some astronomy. In fact, many seem to believe that astronomers are merely guessing at the

distances. It is not easy to explain all the methods used for measuring astronomical distances, but an attempt will be made here to present the simplest ones as clearly as possible.

The first measurement of the size of the earth, based upon accurate reasoning, was made in Egypt by Eratosthenes about 250 B.C. At Syene (the modern Aswan) he observed that there was no shadow in the bottom of a well at noon at the time of the summer solstice, when the sun is farthest north of the equator. This place happens to lie very nearly on the Tropic of Cancer, so that the midday sun on the longest day is almost exactly overhead.

At Alexandria on the same day in June, but not necessarily in the same year, Eratosthenes found that the noonday sun was 1/50 of a circle (or 7°.2) south of the zenith, and he concluded correctly that the difference of their latitudes is 1/50 of a circle (see Fig. 1.). Since Alexandria is almost directly north of Syene, the circumference of the earth must be 50 times the linear distance between these two places. His estimate of that distance was 5,000 stadia, giving a result of 250,000 stadia for the earth's circumference. Unfortunately there were several kinds of stadia, and we are not sure which one Eratosthenes used. If he used the stadium that was 1/8 of a mile, his measurement was about 20 percent too large. But according to one authority, the distance was probably measured in paces by specially trained men, and the stadium used was 517 feet, or about 1/10 of a mile. This would give a result within about one percent of the true polar circumference of 24,860 miles.

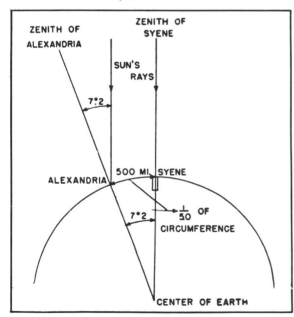

Fig. I. How Eratosthenes measured the earth.

He found the sun 7°.2 from the zenith of Alexandria when it was overhead at Syene. The line from the center of the earth through Alexandria makes equal angles with the parallel rays of the sun. So the angle at the earth's center made by the lines to the two stations is also 7°.2, which is $^{1}/_{50}$ of a circle. If the linear distance between the places is 500 miles, the earth's circumference is 50 times this, or 25,000 miles.

From this we see that there are two kinds of measurement necessary to determine the size of the earth. One is to find the angular distance between two selected stations (several hundred miles apart) in degrees of the earth's circumference. The other is to find their linear distance apart in miles or kilometers. Since the time of Eratosthenes, great improvements have been made in the accuracy of such measurements. An instrument called the zenith telescope will give the latitude of a point with a probable error of about 1/10 of a second of arc, which corresponds to a distance of 10 feet.

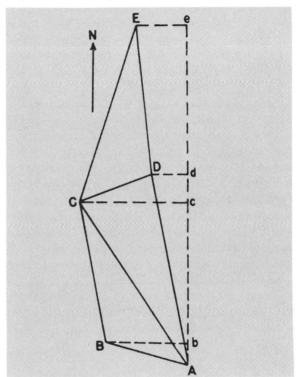

Fig. 2. Triangulation.

Finding the north-south distance between A and E by using one measured line AB and a chain of triangles.

To find the linear distance between two places several hundred miles apart, it is not practicable to measure the entire distance by means of a tape line. Instead, a process called triangulation is used. It depends upon the fact that, given the length of one side of a triangle and the values of two of its angles, one can compute by trigonometry the lengths of the two other sides. In Fig. 2, suppose that the north-south distance between A and E is desired. A base line AB of moderate length, usually from 10 to 20 miles, is carefully laid off on leveled ground. Its ends are marked in stone or concrete and its length is accurately measured with steel tapes. So many precautions are taken that the error does not exceed half an inch in a measured length of 10 miles.

Points C and D are chosen so that each is visible from at least two other points of the series. For example, in the figure it is assumed that point C is visible from points A and B, while point D is visible from points C and A. A surveyor's transit is set up over the point B, and its telescope is directed first to A and then to C. The difference of the readings of the horizontal circle is the angle ABC. The angle BAC is similarly measured with the transit set over the point A. Having the base AB and its two adjacent angles of the triangle ABC, the surveyor can compute the distances AC and BC. The distance AC then serves as a base for solving the triangle ACD, whose side CD, in turn, becomes the base of triangle CDE. The process can be continued through a chain of many triangles.

The north-south distance between A and E is the line Ae. This is the sum of the projections Ab, bc, cd, and de of the lines AB, BC, CD, and DE upon the meridian. These projections can be computed when the direction of any line has been observed with the aid of Polaris, the North Star. The north-south distance Ae in miles divided by the number of degrees in the difference of latitude between A and E gives the length of a degree. This number

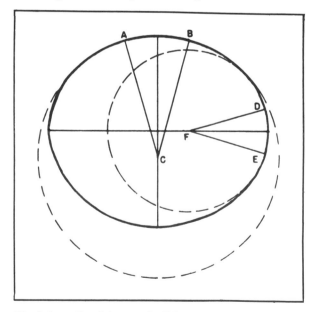

Fig. 3. Lengths of degrees in different latitudes.

The radius of curvature is longest at the pole and shortest at the equator.

multiplied by 360 is the earth's circumference, providing the earth is a sphere. However, it turns out that the earth is not a perfect sphere.

As a result, a degree of latitude varies in length from 68.7 statute miles at the equator to 69.4 at the poles, its average length being 69. The higher the latitude the greater is the length of each astronomical degree. This is because the earth's surface is flatter near the poles, as illustrated in the accompanying figure. If such minor irregularities as mountains and valleys are neglected the earth's surface is at every point perpendicular to the direction of gravity. Therefore, vertical lines near the pole, such as AC and BC, converge to point C beyond the center of the earth. Verticals near the equator, such as DF and EF, meet at point F, which lies between the surface and the earth's center. The angles ACB and DFE have been made equal. Since AB is part of the large circle centered at C and DE is part of the small circle centered at F, AB is longer than DE.

The shape of the earth has been exaggerated in Fig. 3. The actual difference between the longest and shortest diameters of the earth is only 27 miles. The equatorial diameter is 7,927 miles and the polar diameter is 7,900 miles. The earth bulges at the equator because it is rotating. All parts of the spinning earth experience a centrifugal tendency to move away from its axis. The effect on a body on the earth's surface is partly to lift it so that its weight is diminished and partly to slide it toward the equator. A long time ago the sliding effect of the earth's rotation operated to make the equatorial bulge. We can think of the material being moved toward the equator until the upward slope was steep enough to offset the sliding effect. If the earth should stop rotating (a very unlikely eventuality, to say the least), the oceans would rush to the polar regions, since the water at the equator is more than 13 miles farther from the earth's center than the water at the poles.

Now that we have seen how the earth is measured, we take our first step out into space to find the distance of the moon. The simplest method to explain is the method involving two widely separated observatories on the same meridian. In Fig. 4, let A represent Stockholm about 60° north of the equator and B the Cape of Good Hope about 34° south of the equator. When the moon crosses the meridian, simultaneous measurements of the moon's angular distance from the zenith are made at each station. A sharp, starlike mountain peak on the moon is chosen to make the results accurate. Thus the angles 1 and 2 are determined.

In the triangle ABC, the angle ACB at the center of the earth is known from the difference in latitudes of A and B and in this case is equal to about 94°. AC and BC, the sides of the triangle, are radii of the earth and are known from measurements of the size and shape of the earth. Since two sides and the included angle are known, the third side AB can be computed.

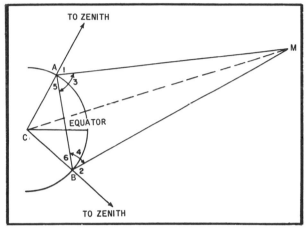

Fig. 4. Finding the moon's distance.

Of course, the distance to the moon cannot be shown correctly to scale in the diagram. Let M represent the moon's position. In the triangle MAB, angle 3 is found by subtracting the sum of angles 1 and 5 from 180°. Similarly angle 4 is found by subtracting the sum of angles 2 and 6 from 180°. The triangle MAB can then be completely solved, since the side AB and two adjacent angles are known. After the distances MA and MB from a point on the moon's surface to points on the earth's surface are found, the distance from the center of the moon to the center of the earth can be easily calculated. The real size of the moon can be found from its angular diameter and its distance.

The mean distance from the center of the earth to the center of the moon is 238,857 miles. However, as a result of the elliptical orbit of the moon and the effects of the attraction of the sun, this distance varies from 221,463 miles to 252,710 miles. These figures are probably correct within a mile or two.

Perhaps some day an observatory will be set up on the moon, and then the line from the earth to the moon could be used as a base for measuring the distance to a planet. However, we do not need to

wait for that, because we can still use the earth as a base line for reaching out to the planets. The method is similar to that used for the moon, but the results are not as accurate because the distances are so much greater. The method does not work for a direct determination of the sun's distance for several reasons. The sun has no fixed point on which to focus in making measurements, the heat of the sun disturbs the measuring instruments, and the great distance of the sun makes a very small angle to be measured. However, as will be explained later, as soon as we know the distance from the earth to any planet, we can calculate the distances to all the other planets and to the sun.

The first determination of a planetary distance that was approximately correct was made in 1672 when Mars was in opposition and closest to the earth. The French Academy of Sciences sent Jean Richer to Cayenne, French Guiana, South America, to observe Mars. Meanwhile, G. D. Cassini made observations of it in Paris. From these the distance of Mars was found, and then the distance of the sun was calculated as 87,000,000 miles. This result, though inaccurate, was better than any previously obtained.

This method has a disadvantage in that it requires two observers working with different instruments at a distance from each other. A single observer can obtain better results by using the principle, first used by Hipparchus for the moon, of sitting still and allowing the rotating earth to carry him, so to speak, from one side of it to the other. For the sake of simplicity, we shall imagine that the orbital motion of Mars has stopped temporarily, that the planet is on the celestial equator, and that the observer is on the earth's equator. When the observer is at A, in the accompanying Fig. 5, Mars is rising and appears projected against the background of stars at position C. After the earth has rotated for 12 hours and carried the observer to B, Mars is setting and appears at D. The apparent angular displacement of Mars can be accurately measured from the known positions of the stars. This is equal to the angle CMD which also equals AMB. Combining this angle with AB, the earth's diameter, we find the distance from the earth to Mars.

It is not possible to obtain the ideal conditions just described, but the variations do not affect the principles involved. Mars will not stop in its orbit for even 12 hours, and it may not be on the

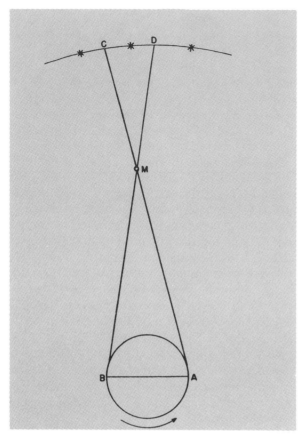

Fig. 5. The apparent shift of Mars.
How Mars appears to shift against the background of stars as seen by an observer who is carried in 12 hours from A to B by the rotation of the earth.

celestial equator. Also, the observer may not be on the equator. These things complicate the calculations, but do not affect the accuracy of the result.

This method was very successfully used by David Gill in 1877. He went to Ascension Island, which is 8° south of the equator and was used during World War II as a stopping place for airplanes flying between South America and Africa. With the aid of Mrs. Gill, he made 350 sets of measurements of Mars during nearly five months. This work was done along what is now called Mars Bay on an isolated beach covered with a fine volcanic dust and infested with flies and mosquitoes. The results were excellent, giving a distance from the earth to the sun of almost exactly 93,000,000 miles, which is the accepted figure today.

When Mars is closest to the earth, it is about 35,000,000 miles away. Venus comes within a distance of about 26,000,000 miles, but at that time it appears very close to the sun and it is not possible to see stars near it to determine its position. On rare occasions Venus passes directly across the disk of the sun, producing what is called a *transit*. This is not the same as an eclipse of the sun by the moon because Venus is 100 times farther away than the moon and appears as a small black spot on the sun's disk. Observations from different parts of the world of the instant when Venus is in contact with the limb of the sun could be used to determine the distance of the sun.

In 1761 and 1769 two transits of Venus were observed in different parts of the world. The results were not as good as expected because of the difficulty of determining the exact instant when Venus was in contact with the sun.

The best determination of the scale of distances in our solar system was made from photographic observations of the asteroid Eros in the winter of 1930-31. There are thousands of asteroids, and these miniature planets, like stars, appear in a telescope as points of light. This makes their positions easier to measure than that of a planet, which appears as a disk. Also, there are some asteroids that come closer to the earth than any planet. But with the exception of Eros, the closely approaching asteroids are so small and faint that it is doubtful that any will ever be used for finding the scale of our solar system.

The orbits of Eros, Venus, the earth, and Mars are shown in Fig. 6. The orbit of Eros is quite eccentric, part of it lying outside the orbit of Mars and part lying inside. The least possible distance between Eros and the earth is about 14,000,000 miles. In 1931 the minimum distance was about 16,000,000 miles. It will not be as close again until 1975. Of such importance was this 1931 approach that a comprehensive program was laid out to observe it. The positions of nearly 6,000 stars were measured photographically to serve as reference points for determining the changing positions of Eros. Fourteen observatories in nine countries cooperated in this preliminary work.

The actual observations of Eros were made from October, 1930, to May, 1931. Altogether 2,847 photographic plates were taken with thirty different telescopes at twenty-four observatories in fourteen countries. Most of these plates contained several images of Eros. Each plate had to be measured to determine the position of Eros. All this information was sent to Dr. H. Spencer Jones, Astronomer Royal at the Greenwich Observatory, who worked for ten years before obtaining the

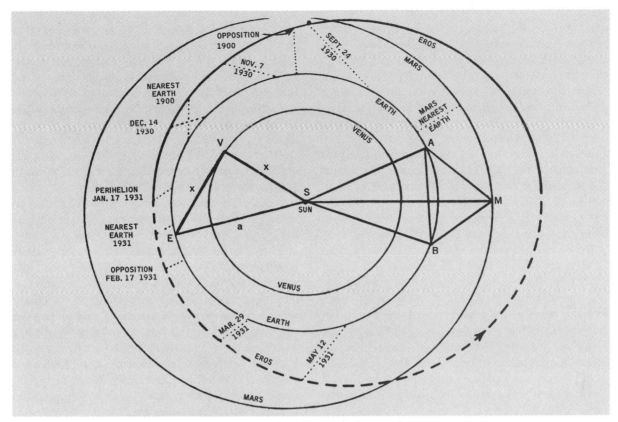

Fig. 6. Orbits of Venus, earth, Eros, and Mars.

final results. From the distance separating Eros and the earth, a figure of 93,000,000 miles was found for the distance from the earth to the sun.

We must now explain how the sun's distance can be calculated, when we know the distance of some other member of the solar system, such as Mars, Venus, or Eros. As early as the time of Copernicus, the distances of the planets from the sun were approximately known in terms of the earth's distance from the sun as a unit. Assuming the orbits to be circles, which is not quite true, let us take the case of Venus shown in the accompanying diagram. When Venus is at its greatest angular distance from the sun as seen from the earth, the line EV from the earth to Venus is tangent to the orbit of Venus and perpendicular to the line VS, the radius of the orbit of Venus. The angle VES can be measured and is about 47°. With two angles in the triangle known, the relative lengths of the sides can be calculated.

It is necessary to use trigonometry to get the correct answers, but in this case we can get approximate values by using geometry and calling the angle at the earth 45°. Then since the angle at

Venus is 90°, the angle at the sun is 45°. The two sides of the right triangle, EV and VS, are equal. Let us call each of them x and the distance from the earth to the sun a. Since the sum of the squares of the sides of a right triangle is equal to the square of the hypotenuse, $x^2 + x^2 = a^2$. Then $2x^2 = a^2$, $x^2 = 0.5a^2$, and x equals the square root of 0.5 times a, or 0.7a. That is, the distance from Venus to the sun is 0.7 the distance from the earth to the sun.

The case of a planet farther from the sun than the earth can be illustrated by the method used by Johannes Kepler in his thorough investigation of the orbit of Mars. In the same diagram suppose that Mars is at M when the earth is at A. The angle SAM is measured. Then Mars goes once around the sun and gets back to M at the end of 687 days, which is the length of its year. The earth, going in the same counterclockwise direction, will make nearly two trips around the sun in the same time and will be at B. The difference between two earth years ($730\frac{1}{2}$ days) and the 687 days is $43\frac{1}{2}$ days. The angle ASB can be calculated, since it is the angle the earth describes in $43\frac{1}{2}$ days. Angle SBM can be measured. SA and SB are radii of the earth's

327

orbit, each of which we shall assume is equal to the astronomical unit, the earth's mean distance from the sun. Then the triangle ABS can be solved for the third side AB and for the other two angles. We can then proceed to solve the triangle ABM. Angle MAB is the difference between the observed angle MAS and the computed angle BAS. Similarly, angle MBA is the difference between the observed angle MBS and the computed angle ABS. With the side AB and the two adjacent angles known, we can find AM and BM, the distances of Mars from the earth in terms of the astronomical unit. Finally, we can get SM, the distance from the sun to Mars, from either of the triangles, SAM or SBM, having been given two sides and the included angle.

From many such pairs of observations separated by the interval of 687 days, Kepler was able to find the distance of Mars from the sun at many points and to determine the size and shape of its orbit. For this purpose he used Tycho Brahe's observations, which extended over a period of twenty years. Thus we see how Kepler could make quite an accurate map of the solar system. The scale of the map is known as soon as any distance represented on it is measured, such as the distance from the earth to Mars or Eros.

It is reassuring to know that the distance from the earth to the sun can be checked by several independent methods. One of the most interesting and ingenious of these is a result of Roemer's original method for the determination of the velocity of light, discussed in Chapter 31.

Now that we know the velocity of light with great accuracy from other methods, we can reverse

Roemer's process which used the period of eclipse on Jupiter as a constant. Modern observations of the eclipses of Jupiter's satellites show that the time required for light to travel across the earth's orbit is less than that found by Roemer, being about 16⅔ minutes, or 1,000 seconds. Multiplying 186,000 miles per second, the speed of light in round numbers, by 1,000 seconds, we get 186,000,000 miles as the diameter of the earth's orbit. One half of that is 93,000,000 miles, the earth's distance from the sun. This agrees with the result obtained by triangulation.

Another method of checking this figure is by means of what is known as the aberration of light. As the earth revolves around the sun, all stars at right angles to the direction of its motion seem to be displaced from their true positions by an angle of 20″.5. The cause of this is the same as that which makes vertically falling rain strike one's face if he is moving rapidly and makes raindrops trace a slanting path on the windows of a moving train. The motion of the observer makes the vertically falling drops seem to come from a point above and a little in advance of him. This is illustrated in the model which is shown here. Turning a crank causes a group of small balls, representing light waves from a star, to move down vertically. An open piece of tubing, representing a telescope, is mounted obliquely on a movable frame, which is carried horizontally across the apparatus to the left as the balls move downward. It is obvious that the top of the tube must be tipped forward to the left in order to make the balls go down the center of the tube and not hit its sides. So the telescope on the moving

Model for demonstrating aberration of light.

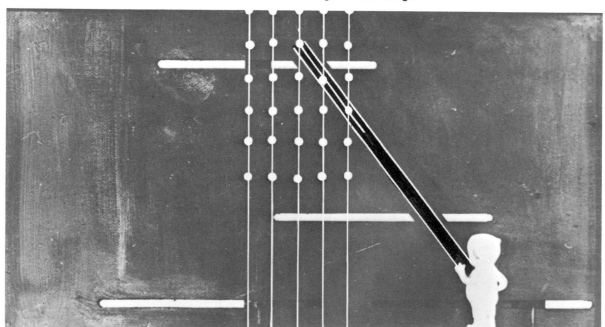

earth must be pointed slightly in advance of the true position of the star in order to make the rays of light go down the tube of the telescope and reach the eye.

If the earth were fixed, there would be no aberration of light. If the earth moved uniformly in a straight line, we would not be aware of the existence of aberration, because the displacement would always be the same. Observations show that the star's displacement changes direction, always keeping ahead of us, so that the star seems to describe a small orbit in one year. The shape of this apparent orbit is circular for a star whose direction is perpendicular to the plane of the earth's orbit. A star in that plane appears to oscillate in a straight line. Stars in other positions appear to describe ellipses.

The maximum aberration of a star is about 20".5. This angle depends on the velocity of the earth in its orbit and the velocity of light. For those who have had trigonometry, it should be pointed out that this angle is slightly less than 1/10,000 of a radian (206,265"). Therefore, we know that the earth's speed is slightly less than 1/10,000 of the speed of light and is equal to 18.5 miles per second. Multiplying this speed by the number of seconds in a year (60x60x24x365.25), we get the circumference of the earth's orbit, which can be regarded as circular without any appreciable error in this case. Dividing this circumference by 2x3.1416, we find the radius of the orbit, or the earth's mean distance from the sun, to be again 93,000,000 miles.

The earth's speed in its orbit can also be found from the spectra of stars. As the earth alternately approaches and recedes from a star, the lines in the stellar spectrum shift back and forth. To understand this, let us imagine that we are driving an auto rapidly around a race track. Some distance away a bell is ringing. We notice that the pitch of the bell is higher when we are approaching the bell than when we are moving away from it. This results from the wavelike nature of sound. Our ears receive more waves per second when we are approaching the bell than when we are receding from it. If we had the proper instruments, we could make a rough calculation of our speed from the change of the pitch of the bell.

Light also is of a wavelight nature, but we do not see changes of spectra of objects moving on the earth, because their speeds are so small compared with the velocity of light. However, when the earth is approaching a star, the waves of starlight arrive with a little higher frequency and with a little shorter wave length than when the earth is not getting any closer to the star. This results in placing any given line of the star's spectrum a little toward the blue end from its normal position. Blue light has a higher frequency and shorter wave length than red light. Similarly when the earth is receding from the star, the light has a lower frequency and greater wave length than when the earth is stationary with respect to the star. Such a line would be shifted toward the red end of the spectrum. This phenomenon is known as the Doppler effect.

The amount of the displacement of a line in the spectrum depends on the relative speed of the earth and the star. Each star has a velocity of its own which is superposed on that of the earth, but the two effects can be easily disentangled. For example, a certain star in the plane of the earth's orbit shows a relative velocity of approach of $30\frac{1}{2}$ miles per second. Six months later there is a relative velocity of recession of $6\frac{1}{2}$ miles per second. The difference of 37 miles per second is due to the earth's velocity of $18\frac{1}{2}$ miles per second. The star is approaching the sun with a speed of 12 miles per second. The earth's speed of $18\frac{1}{2}$ miles per second agrees with the value found from the aberration of light. The same result comes from two entirely independent methods.

It seems rather remarkable that by observing the stars we can get the distance to the sun. But we have seen how these observations give us the earth's speed. From that we get the size of the orbit. Then we calculate the radius to find the sun's mean distance. How great the orbit is can perhaps be better appreciated when we realize that while the earth is traveling $18\frac{1}{2}$ miles in one second, its curved path differs from a straight line by only one-ninth of an inch out of the $18\frac{1}{2}$ miles.

PART 2—BEYOND THE SOLAR SYSTEM

Today we have become so accustomed to hear of the motion of the earth around the sun that we do not realize how difficult it was for this idea to be accepted. The first known clear statement that the earth revolves around the sun was made by the Greek astronomer Aristarchus of Samos, who lived in the third century B.C. He was led to this conclusion by his measures of the distances and sizes of the sun and moon. Even his crude methods told him that the sun must be many times larger than the earth. He could not believe that such a huge sun could go around so much smaller an earth.

An argument against the motion of the earth was that the stars should appear to change their relative positions as the earth moves from one side of the sun to the other. No such apparent shifts of the stars were observed. Aristarchus gave the correct explanation that the stars are so far away that these shifts could not then be detected. Aristarchus was nearly 2,000 years ahead of his time. The authority of Aristotle and many others stood against him. In the second century A.D. the Alexandrian astronomer Ptolemy set forth arguments to prove that the earth could not move in space. These enabled man to ridicule and resist the truth for a long time. Not until Copernicus in the sixteenth century resurrected the ideas of Aristarchus and amplified them a good deal was man put back on the right road.

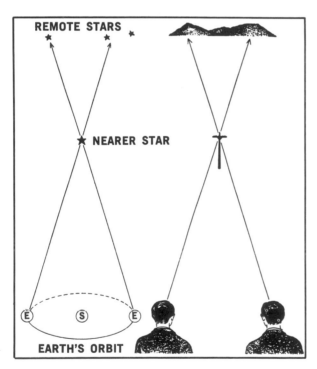

Fig. I. Diagram showing the principle used in measuring the distance of a star.

The apparent shift of the nearby star with respect to the more remote ones, due to the earth's motion, is used to determine the star's distance. This is similar to the apparent shift of the telephone pole against the mountains, due to the observer's motion.

Copernicus could not prove the earth's motion, but that has since been demonstrated by several phenomena too delicate for observation without a telescope. Two of them were described in Part 1 of this chapter. These are the aberration of starlight and the annual shift of the lines in the spectra of the stars. A third proof is the apparent annual shift of the nearer stars with respect to those more distant. This was finally detected in 1838 and is illustrated in Fig. 1. This diagram reminds us of others used to explain how distances are measured from the earth to objects in the solar system. In those diagrams the base line is confined to the earth itself. But as soon as the diameter of the earth's orbit is measured, it can be used as a base line to reach out to the stars.

This new base line is nearly 25,000 times longer than the longest base line on the earth, but it is not nearly as large as we should like to have it. More than two centuries elapsed after the invention of the telescope before instruments were made with sufficient accuracy to measure the very small angles involved. Success was finally attained by three different observers almost at the same time in 1838. The three friendly rivals were F. W. Bessel, a German; F. G. W. Struve, a Russian; and T. Henderson, an Englishman. Bessel observed the star 61 Cygni with an instrument called a heliometer. Struve observed Vega with a micrometer, and Henderson at the Cape of Good Hope in South Africa observed Alpha Centauri with the meridian circle. Thus three different observers in three different countries measured three different stars with three different kinds of instruments.

Bessel was awarded the medal of the Royal Astronomical Society and his two rivals received honorable mention. This award was based not only on the slightly earlier publication of Bessel's results but on their convincing character, which left no doubt that a reliable measurement had been made of the apparent shift of a star caused by the earth's motion around the sun.

The apparent displacement of an object as seen from two different points is called parallax. A star's parallax is defined with reference to the earth and the sun. In Fig. 1, it is one-half the angle made by two lines from the star to the earth at the opposite ends of its orbit. Thus it is the angle subtended at the star by the 93,000,000-mile line from the earth to the sun.

The relation between parallax and distance is

shown in Fig. 2. A part of a circle is drawn, centered on the star, with radius equal to the distance from the star to the sun. This circle will pass through the sun. Starting at the sun, go along the arc until the distance traversed along the arc is equal to the distance of the star from the sun. Let

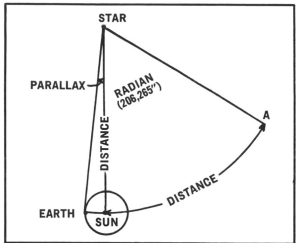

Fig. 2. Relation between parallax and distance of star.

The distance from the star to the sun is used as a radius of a circle and is laid off as an arc around the circumference from the sun to A. The angle subtended by this arc is called a radian and equals 206,265".

the final point that is reached be A. The angle subtended by the resulting section of the circle is called a radian. Now there are 2 times 3.1416 (2π) radians in a circle of 360°, so that the angular equivalent of one radian is equal to 360° divided by 2π. This is equal to about 57°.3. Since there are 3,600" in one degree, there are 206,265" in one radian. Now, the parallax of the star is the angle at the star subtended by the radius of the earth's orbit. Thus, if the parallax were 1", the arc sun-earth in Fig. 2 would be 1/206,265 as long as the arc sun-A. The distance sun-A, which is equal to the distance sun-star, would therefore be 206,265 times as long as the radius of the earth's orbit, or 206,265 astronomical units. If the parallax were one-tenth as large, the distance would be ten times greater. The distance in astronomical units is equal to 206,265 divided by the parallax.

The nearest star to the sun is the double star Alpha Centauri, whose parallax is 0".761. Dividing this into 206,265, we find that the distance is 271,000 astronomical units. Multiplying 93,000,000 miles (one astronomical unit) by 271,000, we get the distance of the nearest star as about 25,000,000,000,-000 (25 trillion) miles.

Since the distance of a star expressed in miles, or even in astronomical units, is an inconveniently large number, larger units are used. The best known of these is the light-year, the distance light travels in one year. It is equal to the speed of light (186,000 miles a second) multiplied by the number of seconds in a year. The result in round numbers is six trillion (6,000,000,000,000) miles. Therefore, the distance of the nearest star is about four light-years.

To appreciate the size of a light-year, let us notice the following relations. A light-year is 63,310 astronomical units. In one mile there are 63,360 inches. Therefore, on a map on which the earth is one inch from the sun, a light-year would be represented almost exactly by one mile. The nearest star would appear on such a map about four miles from the sun, whereas Pluto, the farthest planet, would be only 40 inches away. This is a convenient scale to remember; if one inch equals one astronomical unit, then one mile equals one light-year.

Venus, the nearest planet to the earth, comes within about 25 million miles of us. Thus the nearest star, 25 trillion miles away, is one million times more distant than the nearest planet. This distance of 25 trillion miles was vividly described by John A. Brashear, the famous maker of lenses and mirrors for telescopes. Mounted in the eyepieces of certain optical instruments are cross hairs, which serve as a reticle to define the line of sight with accuracy. Spider web is usually employed for this purpose. It is not the ordinary web, but the very fine web wrapped around the cocoon. Some of this web was weighed on a delicate balance. It was then found that one pound of this spider web would be 25,000 miles long, enough to go around the earth. Brashear then calculated that the amount of spider web necessary to reach from the earth to the nearest star would be 500,000 tons. To transport it would require a freight train 150 miles long, to say nothing of the spiders needed to spin it.

Another unit of distance is the parsec, the distance at which a star would have a parallax of one second of arc. It is equal to 3.26 light-years. No star is as close as one parsec, but the unit is convenient for astronomers to use. The distance of a star in parsecs is equal to one divided by the parallax. For example, a star having a parallax of 1/10 of a second of arc is 10 parsecs away. This equals 32.6 light-years.

This distance of 10 parsecs is the standard distance used for comparing the luminosities of stars. The apparent brightness of a star depends upon both its intrinsic brightness and its distance. The apparent brightness is expressed by the apparent magnitude and the intrinsic brightness by the absolute magnitude. The latter is the apparent magnitude a star would have at the standard distance of ten parsecs.

CHAPTER 62 SOME FAMOUS TELESCOPES

The increase in size and in quality of the telescopes used by astronomers is an excellent example of the rapidity with which recent generations have improved new developments from their crude beginnings. The 200-inch Hale reflector at Palomar has now been in successful use for many years. Yet it is only about three and a half centuries since Galileo pointed his first tiny spyglass at the sky. With it he discovered four moons of Jupiter and the phases of Venus, observed spots on the sun, and saw there was something peculiar about Saturn. When he turned his crude glass toward the clouds of the Milky Way, he found them largely composed of many thousands of stars, which no one had dreamed existed.

These discoveries were made in 1609 and 1610. Galileo's telescope had no mounting. He held it or rested it against a wall. He had no way of eliminating the bad effects of dispersion of color, which are created by a simple lens. Later astronomers, trying to overcome this difficulty, made their telescopes longer and longer. Some of these early telescopes were far longer than any in use today and were mounted by ingenious pulley arrangements. It took almost unbelievable skill for a man to observe objects through such instruments.

The next great advance in telescopes was that made by Sir Isaac Newton, who invented a telescope that used a concave mirror instead of a convex lens. It reflected light to form the necessary image. This type of telescope usually has a very short focal length and therefore makes observing far easier than with refractors. Today the majority of our great telescopes are improvements on this Newtonian invention.

Somewhat later John Dollond, an English optician, found he could combine pieces of flint and of crown glass in one lens that would have less color dispersion than previous lenses. Dollond's invention was followed soon by great improvements in the manufacture of glass, which made available pieces of far superior optical quality.

The next great advances were the building of larger and more nearly perfect telescopes, both of the lens (or refracting) type and of the mirror (or reflecting) type. In the making of better refracting telescopes the names of Joseph von Fraunhofer and of Alvan Clark stand out most significantly. Clark, and his sons, made the gigantic objectives for the 36-inch refractor at the Lick Observatory and the 40-inch refractor at Yerkes Observatory. William Herschel, the musician who began work as an amateur astronomer, gave the development of large reflecting telescopes its greatest impetus.

For a long time reflecting telescopes were handicapped by the fact that the mirrors had to be made of a metallic alloy called a speculum metal, which was capable of taking a very high polish. After a time the polished surface tarnished and it was necessary to have the surface refigured by someone who had the skill of the original maker. As a result, the refracting telescope, in general, led the way.

Shortly after the middle of the nineteenth century a method was invented of depositing a coating of silver on a glass mirror. When this became tarnished, it was necessary merely to dissolve the coating with nitric acid, clean the mirror thoroughly, and deposit on it a new coating. Any astronomer could learn to do this accurately and quickly. This invention tipped the scale toward the reflecting telescope, for its advantages over the refractor were numerous.

Even with Dollond's invention there still was much difficulty from the unwanted color fringe around the images of objects. It was necessary still to provide a very long focus for a refractor. The length of the telescope made mountings very expensive, and the large glass disks were also expensive. Furthermore, it was difficult to make large disks of glass with the nearly perfect optical quality required. With the largest refractors, such as the 40-inch at Yerkes, the absorption of light in passing through the objective lens began to be serious, and the slight sagging of the lens as the

Mount Wilson Observatory. Principal buildings are (left to right): Snow horizontal solar telescope, 60-foot solar tower telescope, 150-foot solar tower telescope, 60-inch telescope dome, 100-inch telescope dome.

The 100-inch "Hooker" reflecting telescope at Mount Wilson. For many years this was largest telescope in world. (Mount Wilson and Palomar Observatories.)

Moonlight view of the 200-inch telescope on Palomar Mountain. Mount Wilson and Palomar Observatories operate as one. (Mount Wilson and Palomar Observatories.)

DATA ON THE 200-INCH TELESCOPE

Mirror

Diameter	200 inches
Thickness at edges	24 inches
Thickness at center	$20\frac{1}{4}$ inches
Weight when cast	$19\frac{3}{4}$ tons
Weight of glass removed	$5\frac{1}{4}$ tons
Final weight	$14\frac{1}{2}$ tons
Weight of abrasives used	31 tons
Diameter of central hole	40 inches
Accuracy of shape	1/500,000 inch

Telescope

Diameter of tube	22 feet
Length of tube	55 feet
Diameter of horseshoe	46 feet
Diameter of each of 2 yoke girders	$10\frac{1}{2}$ feet
Diameter of each of 2 worm wheels	$14\frac{1}{2}$ feet
Angular error from tooth to tooth of worm wheels—less than one second of arc	
Power for moving telescope by fast motion to east or west—2 h.p. motor	

Power for moving telescope to west at celestial rate	1/12 horsepower motor
Weight of mounting	500 tons
Length of wiring	over 400 miles

Dome

Diameter	137 feet
Height above ground	135 feet (about 12 stories)
Height above observing floor	110 feet
Weight of revolving part	1000 tons
Width of shutter-opening	30 feet
Weight of each shutter	50 tons

Observatory

Height above sea level	5600 feet
Distance from San Diego	66 miles
Distance from Pasadena	125 miles

Cost

$6,500,000 given by Rockefeller Foundation for whole Palomar project.

The 200-inch Hale telescope on Palomar Mountain. This is the largest telescope in the world

Observer's cage at upper end of 200-inch telescope. An astronomer, working at the prime focus, rides in this cage while making his observations.

Above, the parabolic mirror of the 200-inch telescope at Mount Palomar before it was coated with aluminium. The 40-inch central plug was removed before the mirror was placed in position. Below, the 48-inch Schmidt telescope, used to photograph wide fields of the sky. Through it, all the sky observable from Palomar has been photographed in both blue and red light. For differences recorded by this method, see photographs on page 273. (Ted Watterson/ Mount Wilson and Palomar Observatories.)

telescope was pointed in different directions showed noticeable distortions.

None of these difficulties applied to the reflectors. The light was reflected from the front surface and therefore it did not matter whether the glass behind it was of an excellent quality. Since there was no color aberration at all, the telescope could be made of far shorter focal length. There was no absorption from the glass, and the largest mirrors lost no larger a proportion of light than did the very smallest. Only one surface needed to be shaped and polished, and even an amateur could learn quickly to do a rather excellent job.

The improvements considered so far have concerned the optical qualities of the instruments. The improvements of mountings were almost as important. Herschel's giant reflectors had no automatic method of moving them to show the stars in their daily paths across the sky. Rather early in the nineteenth century clocks were introduced to move telescopes in such a way as to compensate for the rotation of the earth. These made it possible to take full advantage of photographic plates in making observations and, by exposing for hours on a single object, to observe bodies far too faint for any visual view. In the twentieth century the development of electric drives resulted in mountings superior to the earlier clock-driven ones. Engineering developments of these two centuries greatly reduced the friction of the moving parts and largely avoided the bending of tubes and mountings that had made trouble earlier. The latest telescopes at Palomar, Lick, and other observatories carry these engineering improvements to a degree scarcely dreamed of a generation ago.

In addition to the telescopes we have described above, the largest telescopes currently in operation are the 236-inch telescope at the Crimea Observatory in the Soviet Union, a 158-inch reflector at the Kitt Peak National Observatory in Arizona, and a twin reflector at the Southern Station of the U.S. National Observatory in Chile.

The dramatic advances that have been made in astronomy during the past few years have put an increasing premium on our ability to reach to fainter and fainter objects, pushing existing telescopes to their limits. Strenuous efforts are made to improve their efficiency, through the adoption of better optical coatings, increasingly sensitive detectors, and computer control. Just so much can be done along these lines, however, and it has be-

come increasingly evident that larger telescopes will be required if further progress is to be made. Unfortunately, a mere scaling-up of existing telescopes becomes prohibitively expensive and is probably technologically impossible. For instance, it has been estimated that a telescope with five times the aperture of the 200-inch at Palomar would cost $2 billion and take 50 years to complete. Thus, the next generation of telescopes will be quite different, and draw heavily on recent advances in computer technology and lasers. While the many different designs that have been proposed may seem quite exotic, most are currently technologically feasible. All of them have one feature in common: They envision the replacement of the traditional one-piece mirror with either a segmented mirror or several mirrors bringing the starlight to a common focus.

An already operational harbinger of things to come is the multiple mirror telescope on Mount Hopkins in Arizona. It consists of six telescopes, each of 70-inch aperture, in a hexagonal array on the same mounting. A complicated optical system brings the light from the six telescopes to a common focus, while laser beams are used to ensure that the telescopes maintain their alignment as the telescope moves under computer control. The $7.5 million price tag of this instrument is roughly ⅓ the cost of a conventional telescope of similar light-gathering power.

Above, the 120-inch reflecting telescope at Lick Observatory, completed in 1959. A number of the pictures in this book were made with this instrument. Note observer's cage at upper end. Below, the 158-inch reflector at Kitt Peak National Observatory, put into operation in late 1972. A breakthrough in telescope design, its extreme compactness is due to the combination into one assembly of the declination axis and the large "horseshoe" bearing. (Kitt Peak National Observatory.)

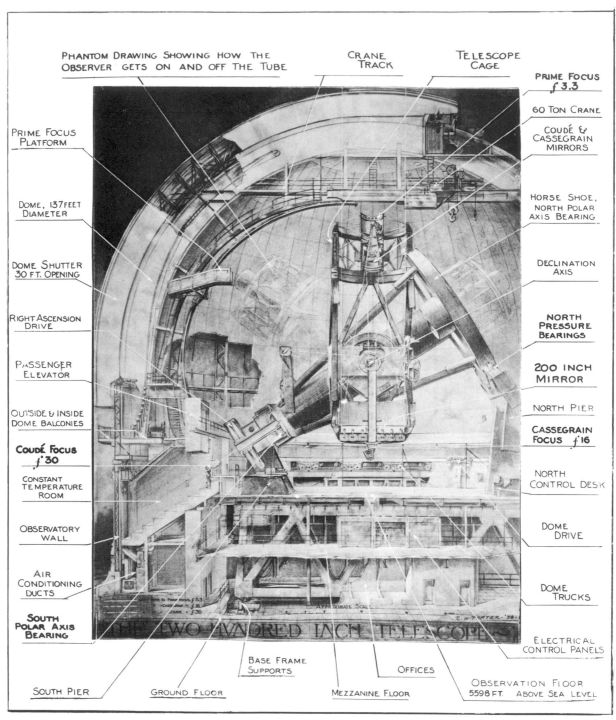

PHANTOM DRAWING SHOWING HOW THE OBSERVER GETS ON AND OFF THE TUBE

CRANE TRACK

TELESCOPE CAGE

PRIME FOCUS *f* 3.3

60 TON CRANE

COUDÉ & CASSEGRAIN MIRRORS

PRIME FOCUS PLATFORM

DOME, 137 FEET DIAMETER

DOME SHUTTER 30 FT. OPENING

RIGHT ASCENSION DRIVE

PASSENGER ELEVATOR

OUTSIDE & INSIDE DOME BALCONIES

COUDÉ FOCUS *f* 30

CONSTANT TEMPERATURE ROOM

OBSERVATORY WALL

AIR CONDITIONING DUCTS

SOUTH POLAR AXIS BEARING

HORSE SHOE, NORTH POLAR AXIS BEARING

DECLINATION AXIS

NORTH PRESSURE BEARINGS

200 INCH MIRROR

NORTH PIER

CASSEGRAIN FOCUS *f* 16

NORTH CONTROL DESK

DOME DRIVE

DOME TRUCKS

ELECTRICAL CONTROL PANELS

OBSERVATION FLOOR 5598 FT. ABOVE SEA LEVEL

SOUTH PIER

GROUND FLOOR

BASE FRAME SUPPORTS

MEZZANINE FLOOR

OFFICES

THE TWO HUNDRED INCH TELESCOPE

The 200-inch telescope. This drawing from the set made by Mr. Porter shows the 200-inch telescope, cut away along a vertical plane through the polar axis, and with explanatory notes on the margin giving essential features of the instrument. The tube weighs about 140 tons, the entire instrument 500 tons, and the rotating dome weighs 1,000 tons.

338

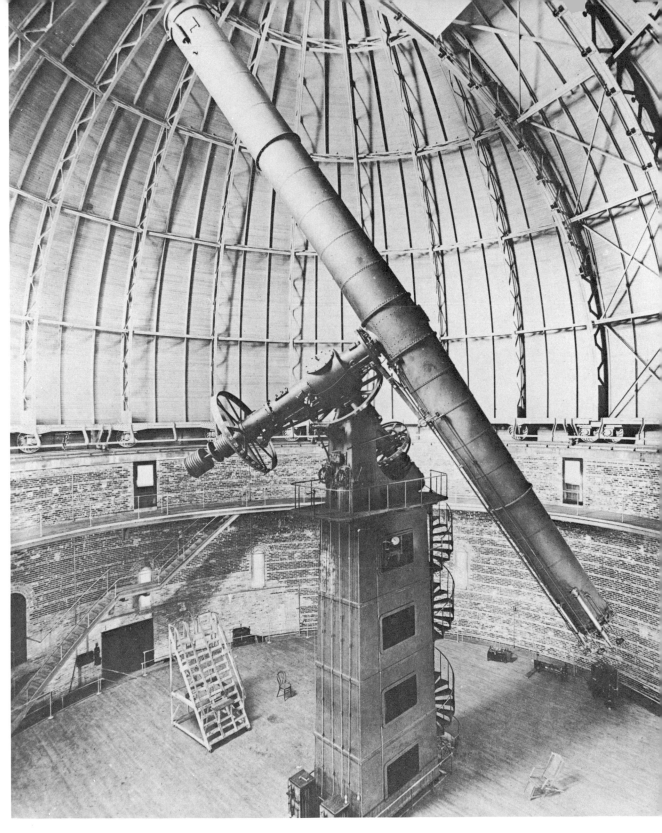

The 40-inch refracting telescope at Yerkes Observatory of University of Chicago, located at Williams Bay, Wisc. This is largest refracting telescope in the world. Second largest is at Lick Observatory. For many reasons, no more very large refracting telescopes are likely to be built.

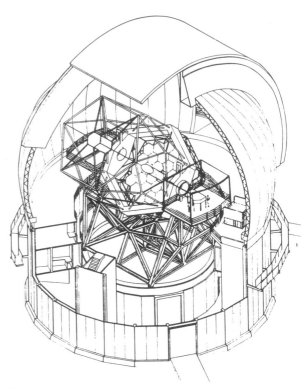

The Ten Meter Telescope and its housing, at the University of California. It is planned to locate this telescope at Mauna Kea Observatory, Hawaii. (University of California.)

Telescopes of dramatically larger light-gathering power are on the drawing board. A study group at the Kitt Peak National Observatory considered several versions of a telescope equivalent in power to a single dish 1,000 inches in diameter. One design envisions a fully steerable dish resembling a radio telescope, with a mirror made of more than a thousand "petals." Another consists of rings of identical telescopes, each mounted and steered independently, and all feeding into a common focus. There may be a single ring of six 400-inch telescopes, or sixteen 250-inch reflectors, or even a four-ring complex of one hundred and eight 100-inch telescopes!

The University of California is studying the feasibility of building a 400-inch mirror made up of hexagonal segments arranged honeycomb fashion. It would have a central mirror surrounded by 54 separate, independently controlled hexagonal units arranged in three rings. It is to be a general-purpose telescope, useful in the infrared as well as the visible, and so will have to be located in a dry, high altitude site.

Large as these telescopes are, they are dwarfed by the enormous antenna required by the radio

astronomers, not only because the radio signals from space are so very weak but also because large apertures are required to see any detail in the sources of this long-wavelength radio radiation. For several years, a radio telescope with a dish 1,000 feet across has been in operation at Arecibo in Puerto Rico. The dish nestles in a natural, cup-shaped valley and reflects radio waves onto a receiver suspended high overhead between three towers. An even more dramatic radio telescope is located 52 miles west of Socorro, New Mexico, on a Pleistocene lake bed known as the Plains of San Augustin. It is actually not one telescope but a collection of 27—each a monster 82 feet in diameter—arranged in a Y-shaped array. The "short" arm of the Y extends northward for 11.8 miles, while the other two arms are 13 miles long. The 27 telescopes can be moved and arranged along the arms in a wide variety of configurations. The radio signals from the telescopes are brought together to a central control building where they are processed by computer. The assemblage becomes, in effect, an enormous interferometer, so that the fineness of detail that can be seen in the results equals that achievable by a single dish with a radius large

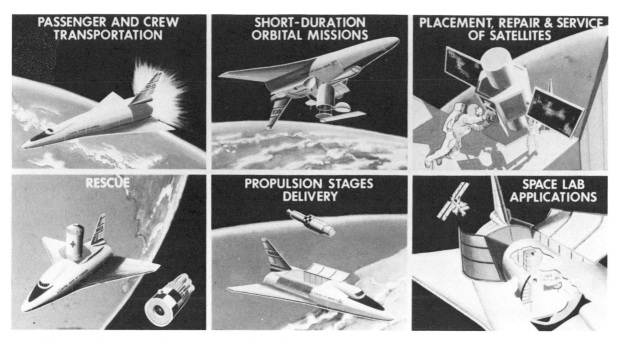

An artist's conception showing the variety of missions which a Space Plane can perform. This reusable vehicle provides an economical means of performing industrial and scientific experiments, defense missions, the placement of weather and communications satellites and servicing or orbiting space stations. Unmanned satellites also can be inspected and serviced in space or be retrieved for return to earth. (NASA.)

An artist's conception depicting activity at a possible manned, modularized space station in earth orbit. The modules would house various equipment functions and activities of the space station and could be carried to earth by a Space Shuttle Orbiter. (NASA.)

enough to enclose the entire assemblage. Thus, when extended to the ends of the three tracks, the telescope (appropriately called the "Very Large Array") becomes equivalent to a single dish nearly 30 miles across! Radio maps can be made with ten times finer detail than can be seen with the largest ground-based optical telescope.

Now that the space shuttle is a proven success, observational astronomers are going into space in a big way. While small telescopes have been launched into orbit for several years, these have all been specialized instruments designed for a specific purpose, such as studying X-rays from stars. If all goes according to schedule, in 1983 the space shuttle will be used to help launch a 96-inch telescope into orbit. It will be, effectively, an orbiting observatory. The telescope will be able to focus light onto any one of six different instruments. There will be two cameras capable of photographing stars as faint as the 29th magnitude. Free of

The Space Telescope with an aperture of 2.4 meters (8 feet) would enable scientists to gaze deep into space—seven times further than has now been done, possible to the outer edges of the universe. The hinged light shield (shown at right end) would be closed during launch; the two solar panels would provide electrical power for environmental systems, on-board computers and instruments. (NASA.)

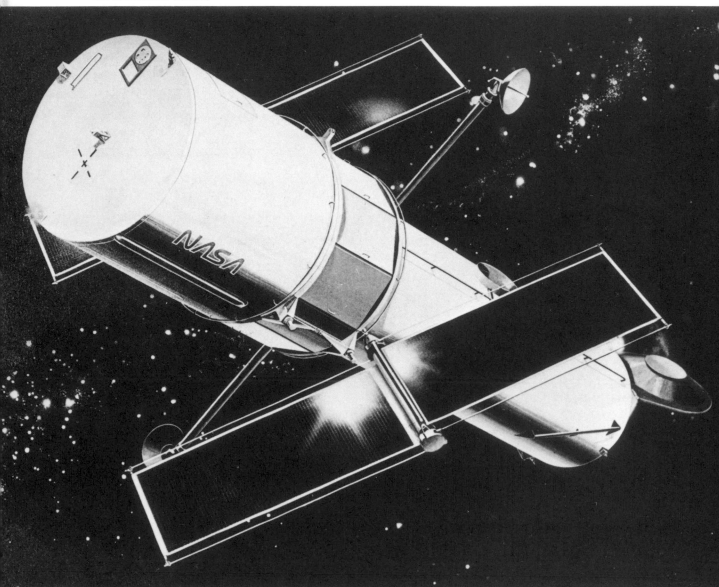

The Multiple Mirror Telescope is located on the 2606-meter summit of Mount Hopkins about 60 kilometers south of Tucson, Arizona. (Smithsonian Institution and University of Arizona.)

the blurring effect of the turbulent earth's atmosphere, the images of stars will be ten times sharper than is typically the case on the earth, making it possible to recognize individual stars in much more distant galaxies than is presently the case. There will be two spectrographs fitted with Digicon detectors that will, for instance, be able to produce the spectrum of a blue B0 star of about 17th magnitude in an exposure time of 15 minutes. There will be a high-speed photometer, consisting of a single photoelectric cell and a variety of filters, that will be able to detect rapid fluctuations in the brightnesses of close binary systems. Finally, the precise pointing capability of the telescope will make it a valuable astrometer, able to pinpoint the locations of stars with ten times the precision attainable on the ground. This will make possible much better measures of the separations of stars in close binary systems and more precise parallaxes, which are the annual shifts in the locations of stars that result from the orbital motion of the earth around the sun.

The space shuttle will be used not only to put the space telescope into orbit, but also to make it possible for technicians to visit the instrument for routine maintenance or repair, and to update the instrumentation as additional needs become apparent. It is expected that the lifetime of the space telescope will be at least 15 years. By the end of that time even more ambitious space observatories will undoubtedly be on the drawing boards.

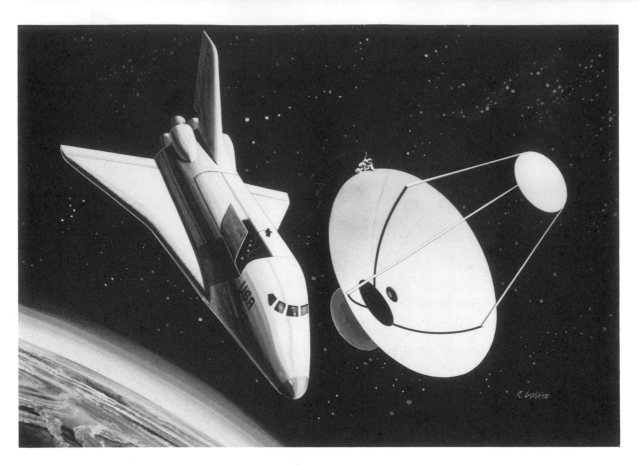

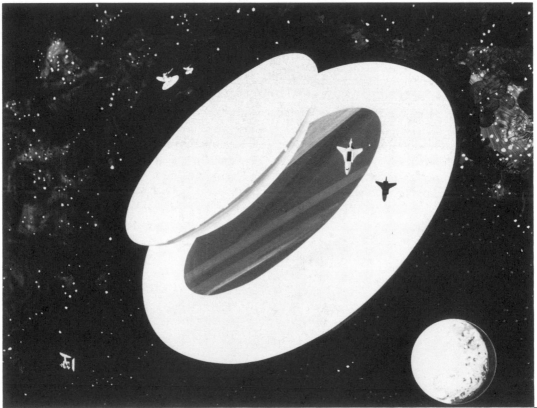

Space SETI Systems. Artist's conceptions of (*above*) a small (30-meter) space SETI (Search for Extraterrestrial Intelligence) system, to be erected with a single Shuttle flight and (*below*) an intermediate (300-meter) system antenna showing two feeds, a relay satellite, RFI (radio frequency and interference) shield and a shuttle type vehicle. (NASA.)

If there is one trait that characterizes the human race and differentiates it from the animals, it is the seemingly irresistible urge to explore new frontiers; to "see what is on the other side of the mountain." This urge has spread mankind into all corners of the world, frequently requiring that he adapt himself to an extremely hostile environment. As soon as one frontier is conquered, we embark on the task of conquering the next. We have reached into the air and are invading the depths of the oceans. In this same spirit we are now making our first tentative steps into space.

Man has long dreamed of reaching out beyond the confines of his earthbound environment. Lacking the technological capability, for generations he had to relegate this dream to the confines of space fiction. Imaginative writers like Jules Verne titillated imaginations with all sorts of captivating possibilities "out there in space." As soon as our civilization had achieved the technological sophistication that was required, it was inevitable, given our ingrained urge to surmount yet another frontier, that we go and see for ourselves "what there is" in space.

The progress we have already made is extremely impressive. We have demonstrated our ability to survive in the harsh lunar environment. Astronauts have been able to adapt to a weightless state and to perform useful tasks over periods of weeks and months in the Russian Soyuz and American Skylab satellites. The demonstrated success of the space shuttle provides a capability of putting significantly large masses into space at still large, but not intolerable, cost. Each shuttle flight can lift up to 65,000 pounds into space, and such flights will become increasingly frequent. Current plans call for 12 flights per year by 1984 and 40 flights per year by 1990.

Quite understandably, these triumphs have spawned literally hundreds of proposals for expansion. A spirit of adventure will undoubtedly bring many of these suggestions to fruition. However, unless it is found that space provides us with resources essential to our continued existence and development as a human race, it is probable that future activities will be largely limited to satisfying our scientific curiosity and demonstrating our technological prowess.

So, the extent to which the human race rushes into space in large numbers will depend on the advantages to be gained. What are these advantages?

The environment of space differs from what we have on the surface of the earth in three ways. First, there is weightlessness, then, an availability of abundant solar heat, and, thirdly, an absence of an atmosphere. Each of these three circumstances offers intriguing opportunities for exploitation.

The sun will inevitably be a progressively more important contributor to the energy we need on the earth to keep our explosively developing technology running, as our supply of fossil fuels runs out, as it eventually must. For the U.S., it has a more immediate urgency in the light of the $50–100 billion we expend for OPEC oil each year, justifying an expenditure of comparable magnitude in space if it would make a major contribution to the country's energy independence. Several proposals have been advanced to do just that.

Only a small fraction of the sunlight falling on the earth penetrates the atmosphere and reaches the surface, so the collection of this energy above the atmosphere becomes extremely attractive. It is proposed that large arrays of solar cells be assembled in space. An area of 5 square kilometers would collect 1000 megawatts of energy, as much as a typical nuclear power station. Placed in a geosynchronous orbit, the array would transmit its power at microwave frequencies to a fixed antenna on the earth. It has been calculated that the maximum economically feasible size of such an antenna would cover some tens of square kilometers,

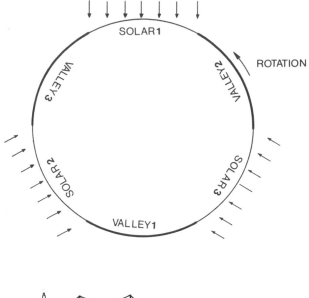

NATURAL SUNLIGHT
FROM PLANAR MIRRORS

SOLAR 1

VALLEY 3

VALLEY 2

ROTATION

SOLAR 2

SOLAR 3

VALLEY 1

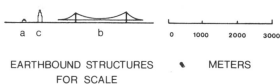

a c   b

0   1000   2000   3000

EARTHBOUND STRUCTURES   METERS
FOR SCALE

End view of space community. The cylinder is four miles in diameter, with a length of sixteen miles and an environment that has normal air, blue sky, natural sunshine, a day/night cycle, seasons and controlled weather. The valleys have lakes, rivers, grass, trees, animals, bird and up to 100,000 people.

producing 5000 to 10,000 megawatts, with a transmitting antenna about a kilometer in diameter. All this is well within our current technological capability. Quite aside from the cost of such a venture, however, there are a number of environmental concerns that would have to be addressed. Problems might arise with the transmission of microwaves through the ionosphere, and one must investigate potential adverse health effects.

Already, some 200 commercial companies have reserved space on future space shuttle flights to conduct experiments that take advantage of either the zero-gravity or the vacuum of space. Purer crystals can be grown for microelectronic applications, clearer glass for fiber optics, or stronger alloys from metals that refuse to mix in a gravitational field. By 1982 McDonnell Douglas will be conducting pharmaceutical experiments, and the company plans to be conducting full-scale drug manufacture by 1986.

To aid in the development of these plans, NASA has on the drawing boards a space station for the 1990s that will serve as a staging point between shuttle flights. It will consist of two living modules connected together, along with communication antennae and solar panels, and manned by a crew of eight. The station will be assembled in space; it has been estimated that six shuttle flights will be required to transport all of the various parts. At an altitude of 200 to 300 miles above the earth, one of its first functions will be to act as a base from which manned orbital transfer vehicles could service the increasing number of communication satellites in orbit. One of the early tasks of the station would be the assembly of a large communication satellite with an antenna 300 feet across. It would be so powerful that it could handle up to 250,000 calls from earthbound wrist radios, relay hundreds of TV channels, and even deliver mail electronically.

Toward the waning years of this century, the advantages to be gained by carrying out manufacturing and scientific activities in space will become so apparent that pressure will become intense to place even larger space stations into orbit. However, even if the comparatively cheap space shuttle is used to transport all of the required materials from the surface of the earth, the total cost would rapidly become prohibitive. This is because so much effort is required to overcome the earth's gravity. Accordingly, if large space stations are to be built, the materials will have to come from space itself—from the moon or from asteroids. Less than four percent as much booster energy is required to lift material from the surface of the moon. Almost all of the elements required are already known to be present on the moon from analysis of the moon rocks returned by the lunar astronauts. Most of the rocks are what are called anorthrosites, high in abundance of aluminum and calcium. The basaltic lavas in the maria are rich in iron. There are localized areas rich in what has been dubbed KREEP, an acronym for potassium (K), rare earth elements (REE), and phosphorus (P). There are further trace elements, including radioactive potassium, thorium, and uranium. The bombardment of the lunar surface by the solar wind over billions of years should have enhanced the abundance of the cosmically abundant carbon and nitrogen. Rich mineral deposits such as are found on the earth are

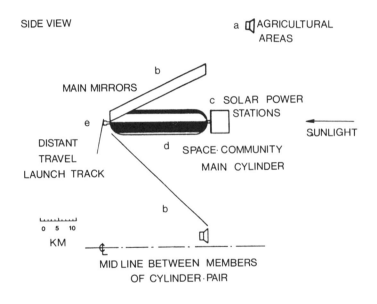

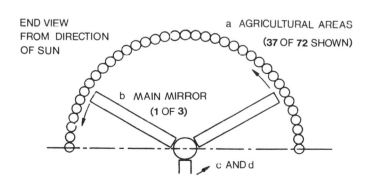

Side and end view of a space community as a complete ecosystem.

347

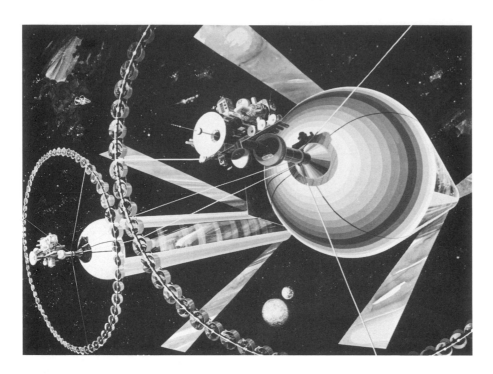

A futuristic space colony and its interior. (NASA.)

not expected, but there may be pleasant surprises. Only hydrogen is a doubtful constituent of the lunar material, and, of course, there is no water; these would have to come from the earth, or possibly water could be produced in space from hydrogen and oxygen.

Another promising source of material is provided by asteroids. Carbonaceous chondrites in many meteorites (believed to be representative of asteroids) contain bound water, among many other useful materials, and iron meteorites are largely alloys of iron, nickel, and cobalt. About 50 asteroids are known to be in near-earth orbit, and there are undoubtedly hundreds if not thousands of small undetected flying mountains that come close enough to be useful.

What might the next century have in store for us? Men of vision, such as Gerard O'Neill of Princeton University, believe that we will see the

establishment of increasingly large space stations capable of providing self-sustaining environments for large colonies. O'Neill has designed a series of such stations of increasing size and complexity that we would be quite capable of building in space using technology already available in the 1970s. His design for the first to be built envisions a station made up of two cylinders connected together and rotating in opposite directions with a period of 20 seconds. Each cylinder would be over half a mile long and six to seven hundred feet in diameter. The surface of each cylinder would be divided into six longitudinal zones, three of them transparent. The remainder would provide the land areas within which the inhabitants would live, grow crops, and carry out industrial activities. Large mirrors would reflect sunlight into the cylinders through the transparent zones. About 95 percent of the required material would come from the

Lunar Arecibo. Artist's concept of an array of three Arecibo type spherical antennas constructed within natural craters on the far side of the moon.

moon, largely in the form of aluminum and glass. The rest would have to come from the earth, largely in the form of liquid hydrogen for the conversion into water. According to O'Neill, a station of this size could provide a self-sustaining environment for as many as 10,000 inhabitants. Light industry could be carried out on the inside of the land surfaces under conditions of normal gravity, while heavy industry would be concentrated along the axes of the cylinders where they would have zero gravity. Living in the first station may be a bit spartan, though O'Neill envisions that it would be provided with swimming pools, shops, libraries, and movie theatres. The estimated cost would be about the same as the cost of the Apollo project.

O'Neill proposes that a major task of the first space colony should be the construction of the next, larger space station that might be ten times as big and provide a home for 100,000 inhabitants. Eventually such stations would comprise living cylinders 15 to 20 miles in length, and provide some hundreds of square miles of land area. Crops would be grown outside the cylinders in appendage satellite containers.

While this all may be technologically feasible, one is left wondering whether sufficiently large numbers of inhabitants would be willing to spend their entire lives in space. It is conceivable that by that time conditions on the earth will have become so crowded that the human race will have no choice. The amenities will have to be very attractive. One also wonders what sort of social and political organizations will emerge in these isolated communities.

Is all this impossibly visionary? Perhaps. But we must not forget that, a mere hundred years ago, nobody would have predicted that we would casually fly across the oceans in this century, or sit in our living rooms in front of television sets watching a man walking on the moon.

CHAPTER **64** **LIFE IN THE UNIVERSE**

At intervals of a little more than two years, the orbits of Mars and of the earth cause these two planets to approach to a distance that, in unusually favorable years, may be as little as 35,000,000 miles. Halfway between these approaches the planets swing apart until they are separated by nearly 250,000,000 miles. The approaches of the planet Mars always bring to the layman, and to some extent to the professional astronomer, the question of life in other parts of our universe. There is only a small phase of this interesting subject upon which the astronomer can speak definitely as an astronomer. He can tell us that life of any sort is very improbable on any of our sun's planets except Mars. He can tell us that his observations indicate the possibility but not the certainty of vegetable life on that little world. There is absolutely no evidence pro or con regarding the presence of animal life, though the observations being accumulated by Mariner IX and other satellites are beginning to suggest strongly that the existence of life in some exotic form is possible, yet highly unlikely. We do know that we ourselves could not live for a single second if exposed on the Martian surface.

It will be impossible for astronomers ever to build telescopes that, used on the earth, can show any planets that may move around other stars. Perhaps it seems foolish to make such an absolute statement in view of the many times that later discoveries have shown the truth to be contrary to earlier views. For example, an astronomer of the last generation stated very definitely that airplanes would never fly around the world and that it was foolish to consider that they ever could drop dynamite on fortifications or on cities. In the case of the telescope of the future, however, the situation is somewhat different. The nearest of all stars is about 270,000 times as far away as our sun. Let us suppose it to have a planet as large as Jupiter. It is possible for us to conceive a telescope big enough to show such a planet, although the telescope's construction would be beyond present engineering skill. However, even if it could be built, a telescope of this size would increase the light of the star's image so tremendously that the

350

planet would be lost in the increased glare. A far larger telescope would be needed under such conditions to make the planet visible. This larger telescope, in turn, would further increase the glare of the star's image, and a vicious circle would be started. If we are ever to observe planets that may move around other stars, it must be by one of two means: manned space travel or an observational instrument entirely different from the telescope. It is impossible for us to deny either of these eventualities, but let us consider their probability.

Light traveling 186,000 miles per second takes $4\frac{1}{3}$ years to come to us from the nearest star. Let us suppose that some space ship of the future may attain a speed 1/10 that of light. It would require more than 40 years for it to make the journey to our nearest neighbor. The great majority of stars visible to the naked eye are many times as far away as the nearest one. The average star of our own system is thousands of times as far away. One cannot say positively that space ships will not visit the stars, but it does seem very improbable. It would be necessary for generations of men to journey their whole lives upon the vessel.

With respect to the invention of some other instrument than the telescope to reveal such a planet, we can say merely that at present we cannot conceive of the form of the device. We do not believe that such an invention ever will be made, but we cannot deny the possibility. Since the observation of planets other than those of our own solar system is either an impossibility or something that belongs to the distant future, we now can merely examine the stars themselves and from them speculate concerning the question of whether life exists elsewhere in the universe.

Our sun is an ordinary star in a galactic system of perhaps 100,000 million stars. This system is flattened and is rotating rather irregularly about its center. The stars are distributed in spiral form just as we find in the case of many other systems of stars with the same characteristic. As we examine the light from these stars, we find that they differ a great deal from each other. Many give thousands of times as much light as our sun, and many give thousands of times less light. On the other hand, a great many stars are very much like the sun. It is quite reasonable to guess that planets may move around those stars that resemble our sun, much as they do in our own solar system. Unfortunately we do not know for certain the method of forma-

tion of our planets. It may be, as many believe, that they have formed directly from the sun through the normal following out of the laws of physics. If this be true we can be quite certain that stars similar to our sun will in general have planetary systems. But some astronomers believe our system to be a sort of freak, one caused by the very close approach of a star to our sun in the distant past. If this second hypothesis is true, only a very small fraction of the stars are likely to have planetary systems unless our galaxy of stars was at one time much more closely crowded together than it is today. If so crowded, such close approaches would have been much more common than under present conditions. There are a good many indications that our system of stars at one time did occupy a much smaller volume than it does now. If this hypothesis of a close approach of two stars were the true story of the origin of planets, we would expect some systems to be formed for other stars, but would expect them to be much less common than under the first supposition. Nevertheless, among 100,000 million stars the probabilities would seem very great that the absolute number of planetary systems would be quite large even if the percentage of stars forming them should be very small. Without any certainty on the subject, we feel it probable, then, that other worlds do exist within our own system of stars.

Our modern telescopes with their photographic attachments make it possible for us to observe many other galaxies of stars that resemble the one that we have been discussing. It has been estimated that within reach of the 100-inch telescope at Mount Wilson there are in the neighborhood of 100 million such systems, each one with thousands of millions of stars. Even if the formation of planets were such a rare event that only one star in a billion could have a planetary system, there would be something like 10 billion planetary systems within this observable part of our universe. One would guess that within the present observable universe we would have anywhere from 10 billion planetary systems up to a million times that many, or even more.

The question of life on such planets becomes, then, one that involves the cause of life. If life is merely a chemical process, the betting on the existence of life in the depths of the universe would be greatly in favor of it. If life is something entirely different from that, there seems no practicable way of calculating even roughly the odds in the betting.

The observed probability of life on one planet of our system other than the earth seems to increase very much the probability of life in distant places. If there is life on the planet Mars it follows quite definitely that our system was not created just so the earth uniquely would have life. Of course in such speculation one would have to consider the slight possibility that life on the earth and on Mars had a common origin. Under such a condition it could still have been created uniquely for the earth.

If intelligent life exists in other solar systems there are certain a priori considerations to which we can attach much credence in guessing which are the most likely places to find it. We would not expect it to exist on a planet moving about a highly variable star. Also, multiple stars are ruled out, unless the stars are very distant from one another, because a planet could not, in general, exist for long in a stable orbit, due to the gravitational field of multiple stars. We can discard the giant blue stars because their short lifetimes have been entirely insufficient for the slow process of biological evolution. With less certainty we can omit the giant red stars because of the suspected rather short interval since they came into their present conditions.

Present-day information tells us that the fainter stars of the main sequence (see page 207) are the most stable. Our sun is one of these, and we expect little change in our sun for billions of years to come, and probably two billion years ago it was not much different from what we observe. The dwarf red stars of this sequence would have too small a planetary volume within which water normally would be in liquid form for them to be probable sources. Some also are subject to sudden flares that would destroy life. It seems probable, therefore, that if civilizations exist, they are on planets whose primary stars are yellow or orange stars of the main sequence. Our sun is a yellow star.

If intelligent races exist, with scientific and technical cultures, some can be expected to be less advanced than we are, others much more advanced. We can expect some of the more advanced races to have developed, among other things, much more sophisticated devices of radio astronomy. They may well be able to send out very powerful and very tight radio beams accurately directed in space. They should be able to receive much weaker radio signals than we can today.

A civilization such as this undoubtedly would wish to establish contact with other cultures of the cosmos. This thought has led to the development of two "long shot" observational programs of the type that may lead to a contact with such a culture *if* one exists in our part of the galaxy. The scientific value of a contact would be great enough to warrant a moderate effort even though the odds may be thousands to one against success.

The first program was carried out at the National Radio Astronomy Observatory in West Virginia in 1960 (Project Ozma, after the princess of the Land of Oz). They monitored radio radiation from two of the closest, yellow, main sequence stars, and looked for any abnormal bursts of energy that could come from an unnatural source. Some 200 hours of observing time were devoted to the search, but without success. Later, in 1973, a more systematic search, dubbed Ozma II, recorded the radiation from more than 600 likely nearby stars. Using computers, they scanned their records for telltale signals. Again, none were found, but it is safe to say that every radio astronomer, scanning his results, has in the back of his mind the hope that he will see an unnatural single signal or series of signals. Imagine, then, the excitement of the Cambridge radio astronomers when in 1967 they observed regular pulses of radiation with a eriod of 1⅓ seconds from the direction of the constellation Vulpecula. Very understandably, they called the source an LGM, for "little green men." However, when three additional pulsating sources were found, it became obvious that they were not signals from extraterrestrial intelligent life after all, but rather, the first of the very interesting and important pulsars.

The probability of success of a long, systematic search for detectable signals from an extraterrestrial civilization would, of course, be directly proportional to the number of likely sources that could be reached. This means studying the vicinities of more distant, fainter stars, requiring larger, more sensitive receivers. What is undoubtedly the ultimate in receiver area is envisioned in a proposal dubbed Project Cyclops. It calls for an enormous array of over 1,000 individually mounted radio telescopes, each 100 meters in diameter, covering an area 6 miles in diameter. The array would be so sensitive that it could detect radiation equal to the strength of the earth's leakage signals from a distance of 100 light years. Even if no ex-

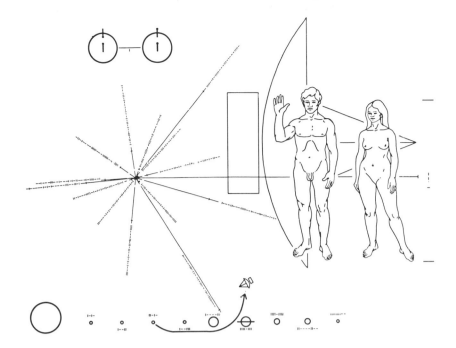

Pioneer X spacecraft, which will orbit out of the solar system after passing Jupiter, bears the above message, showing the size of humans in relation to the spacecraft, the orbit of the spacecraft through the solar system, and the position of the solar system in relation to 14 pulsars and the center of the galaxy. (Carl and Linda Sagan and Frank Drake.)

traterrestrial civilizations were ever found, such an array would be extremely valuable for a wide variety of astronomical studies, but the $5 billion price tag for the array probably means that it will be a while before it is built.

The Pioneer X spacecraft that was launched late in 1972 was the first spacecraft to leave the solar system. In about two years it reached the vicinity of Jupiter and underwent an acceleration in Jupiter's gravitational field, which swung the satellite out of the solar system with an eventual constant velocity of 11½ kilometers per second. After a voyage of possibly millions of years there is the extremely remote possibility that it may pass close enough to an inhabited planetary system to attract attention and be captured. Small as this possibility is, it was considered worthwhile to compile a message from earth for the edification of this unknown civilization. This message is reproduced here. A thin 6 × 9-inch aluminum plate, mounted on the exterior of the spacecraft, has etched on it a message deemed to be decipherable by any advanced technological society. Toward the right two human figures are drawn in front of an outline of the spacecraft to establish scale. On the left the sun's location is represented relative to fourteen pulsars and the direction to the center of the galaxy. The present periods of the pulsars are indicated in octal form along the fourteen lines. Since pulsars are all slowing down at known rates, this diagram should not only fix our location but

establish the time of launch. Along the bottom is a schematic representation of our solar system and the interplanetary orbit of the spacecraft. In the upper left a schematic representation of the two hyperfine states of hydrogen should hopefully establish the unit of time on which the message is based.

The foregoing summarizes almost all that an astronomer can say about life in the universe when he speaks purely as an astronomer. Any further speculation is not astronomy but simply man's guesses concerning the origin of the universe. It may be interesting to consider the two principal forms of guessing that man has engaged in ever since he first began to speculate about anything else than his daily food, his living quarters, and the material effects of the people and lower animals around him.

The speculative thinkers have been divided into two great groups: first, those who believe that originally matter existed and that this matter gradually evolved to give us a universe of our present form. This group believes that some of this matter organized itself purely by chemical and physical means in such a way as to become aware of its existence, as in the case of the lower animals and ourselves. The other group believes that originally there was a thinker and that in one manner or another through his thoughts our universe has been constructed. This latter group is divided into many subgroups, some maintaining that the

353

Artist's concept of ground level view of Cyclops system antennas, showing the central control and processing building. (NASA.)

external universe is a reality, and others thinking that is is all an illusion of the mind. There appears to be no way to decide which group may be the nearer right. Our difficulties start with some sort of an innate quality of the human mind that causes most people, when they try to speculate on beginnings, to utter the statement, "I do not see how anything can be at all," and following such a feeling to experience, "But I am, even though I cannot understand how it is possible for me to be. Therefore, something exists." This contradiction has been experienced by nearly every one of us at times throughout his life, as well as by the greatest philosophers of the ages.

The guess that the universe formed itself purely by chemical and physical processes is called *materialism*. The guess that it came from thoughts or ideas is called *idealism*. Some philosophers have believed in a sort of dualism, but their number is small. Most of the philosophers of all generations and beliefs have accepted some form of idealism. Most of the so-called practical, hardheaded men have believed in materialism. In general, we all act

in most of our everyday contacts as if materialism were true, although the great majority of us give at least lip service to idealism. If materialism is true, there would seem to be no doubt at all of the existence of life in countless places. If some form of idealism is true, the probability would depend on what that form of idealism may be and on the desires of the original thinker. It may be well to consider for a moment why there has been such a strong preponderance of thinkers who have favored idealism.

Instinctively, as stated above, we tend to a feeling that it would be impossible for anything to exist of itself without something having formed it. Such an original existence is equally repugnant to our mind—whether we consider it as inorganic matter or whether we consider it as an uncreated thinker. The fact of our own existence, however, forces us, despite our dislike, to accept the fact that something exists. The philosophers have experienced this just the same as the rest of us do everyday. Pondering it the philosophers have realized that the thing they experience is a thinker. Therefore,

354

Artist's rendering of the Cyclops antenna array, based on a drawing produced by a computer programmed with the system specifications. Each parabolic "dish" is 100 meters (nearly 330 feet) in diameter. The antennas are separated by a sufficient distance (about three times their diameter) in order that they will not be in each other's shadow when they are set to receive a signal coming in at a limiting angle of 20° above the horizon. (NASA.)

generally it has seemed more reasonable to them to accept as the original existence something that resembles their only experience—that is, a thinker. The materialist has his hypothesis complicated by the necessity of assuming that out of matter eventually evolved the awareness that is himself. Except for this one great advantage of idealism, idealism and materialism are both equally satisfactory and unsatisfactory.

Summarizing the points that have been discussed, we find countless stars so similar to the sun that from a materialistic point of view there surely must be inhabited planets. From the idealistic point of view there may or may not be such life, though the observed probability of at least vegetable life on Mars increases very much the probability of life in other places.

Having carried our speculation so far, we naturally are impelled to go still farther. We do not know how long man has existed on the earth. It is possible that it has been but a very few thousand years, though none of our scientists would accept such a short interval. It seems much more probable that we have existed for less than 2,000,000 years as man and that our ancestors existed for hundreds of millions of years before that. When we consider the stars it seems probable that such stars as our sun have existed for billions of years and will continue to give out heat for billions more. As good a guess as we can make is that our sun will make this earth a comfortable place for our descendants to inhabit for a dozen billion years to come.

Assuming such a future lifetime for our descendants, it becomes rather interesting to compare the lifetime of the human race to that of a man who will live to be 100 years old. If we consider that the first being who truly could be called a man lived 2,000,-000 years ago, this man who represents our race is today less than two weeks old. We of the twentieth century are in the very dawn of man's existence. We cannot expect the human race to be civilized or intelligent today any more than we could expect a two-week-old baby to write the thesis of a doctor of philosophy. We would guess that as the millions of years pass, our race will change as much as that baby will change when he grows to maturity. In all

355

this we find a message of hope. Most of us can endure discomfort for ourselves if we believe our children will receive the things we desire for them. The discomfort that the past and present generations have brought upon us today and that may well persist for a dozen generations more is, in the analogy, merely a pain experienced during a few seconds. It will be rather unimportant to the race if it should take us a million years to become civilized. Even then our race will be still at the very dawn of its existence.

If, on other worlds—perhaps in other systems of stars—there are planets and life, it would seem quite probable that in many cases hundreds of millions of years would have passed since a race began and that old races would exist that might well have attained the calm philosophical viewpoint that aged men so often find. Other races would exist in the full vigor of their civilization. Still others, like us, would be dreaming of the future, and on yet other planets the very beginnings of life would be spawned.

We marvel at the changes that the last century has brought to us—changes with which we have not been able to keep pace emotionally, so that we have, as H. G. Wells has said, an "age of confusion." With our little knowledge of today, many of us guess that soon we will have found everything there is to learn, not realizing that we are merely children playing on the sand and calling to each other to look at the pretty pebbles that are our discoveries.

In a book published in 1884, S. P. Langley ended his discussion with a parable:

"I have read somewhere a story about a race of ephemeral insects who live but an hour. To those who are born in the early morning the sunrise is the time of youth. They die of old age while its beams are yet gathering force, and only their descendants live on to midday; while it is another race which sees the sun decline, from that which saw it rise. Imagine the sun about to set, and the whole nation of mites gathered under the shadow of some mushroom (to them ancient as the sun itself) to hear what their wisest philosopher has to say of the gloomy prospect. If I remember aright, he first told them that, incredible as it might seem, there was not only a time in the world's youth when the mushroom itself was young, but that the sun in those early ages was in the eastern, not in the western, sky. Since then, he explained, the eyes of scientific ephemera had followed it, and established by induction from vast experience the great 'Law of Nature,' that it moved only westward; and he showed that since it was now nearing the western horizon, science herself pointed to the conclusion that it was about to disappear forever, together with the great race of ephemera for whom it was created.

"What his hearers thought of this discourse I do not remember, but I have heard that the sun rose again the next morning."

Glossary

Aberration of light: the apparent displacement of an astronomical object in the direction in which the observer is moving. In a somewhat analogous situation, when a person runs in the rain, the rain drops appear to come chiefly from the direction in which he is running. For an observer on the earth, nearly all aberration of light is caused by the velocity of the earth in its orbit around the sun. The maximum displacement is almost 20½ seconds of arc. As seen from the earth, a star seems to move, annually, in an ellipse, the center of which is the true position of the star.

Aerolite: a stony meteorite, in contrast to a metallic meteorite composed principally of iron and nickel (see *siderite*).

Airglow: a faint glow observed at night from the highest layers in the earth's atmosphere. It is produced when atoms or molecules in these layers gradually release energy stored up from sunlight during the previous day.

Albedo: the fraction of light which a body reflects of that which falls on it. The average albedo for the surface of the moon is 0.07; for cloud-covered Venus, 0.76; for the earth, about 0.40.

Altitude: the angular height of an object above the true horizon.

Amplitude: usually either the total variation in the velocity of a spectroscopic binary star as observed from the earth, or the variation in magnitude of a variable star.

Analemma: commonly a figure-eight-shaped table placed on a globe of the earth to show the variations between mean solar and sundial (apparent) time and the changes in the sun's declination throughout the year.

Angular diameter: the angle subtended by the actual diameter of an object as seen by an observer. Angular diameter varies inversely with distance; at twice the distance the angular diameter is half its former value. The angular diameter of the moon varies during the month from 33′ 30″ when closest to us to 29′ 21″ when farthest.

Aphelion: the point in its orbit where a planet or a comet is farthest from the sun. See *perihelion*.

Apogee: the point in its orbit where the moon or an artificial satellite is farthest from the earth. See *perigee*.

Asteroid: a synonym for *minor planet:* any body that is smaller than a regular planet but large enough to be observed. Sizes of the nearly 2,000 known asteroids vary in diameter from about a mile to about 485 miles (for Ceres).

Astronomical unit: the mean distance between the centers of the earth and sun. It is close to 93,000,000 miles.

Atmosphere (of earth): the gaseous envelope that surrounds the earth. It includes the molecules of water vapor, but not the dust particles. At the surface of the earth its average pressure is approximately 15 pounds per square inch.

Aurora: light emitted by the thin upper atmosphere of the earth at heights roughly between 50 and 600 miles. *Aurora borealis* is a synonym for northern lights, and *aurora australis* for southern lights. Auroras are caused principally by charged hydrogen particles and by electrons from the sun.

Azimuth: the angular distance measured eastward from the north point of the horizon to the point on the horizon directly below an object (*i.e.*, at the foot of the object's vertical circle). Engineers commonly measure azimuth either east or west from the north point.

Binary: a physical double star; that is, two stars that are close together and move in ellipses around their mutual center of mass. *Optical double stars* merely look like double stars because they are almost on the same line from the earth. One star may be far closer to the earth than to the other star. If the stars of a binary are far enough apart that they can be separated by a telescope, the two are called a *visual binary*. If that is impossible and their nature can be determined only by the Doppler variations in their spectrograms, they are called a *spectroscopic binary*. A binary in which one star, as seen from earth, during its orbital motion about the other star passes, at least partially, between us and the companion is called an *eclipsing binary*.

Black body: A body that absorbs all the radiation that falls on it and reflects none. The absorbed radiation heats

357

the body to its *black body temperature*, which determines how the resultant emitted radiation is distributed among the various wave-lengths, or colors, according to the *black body radiation law*. Thus, if the temperature is 1000°C, it will emit mainly at short wavelengths, and will glow red, while if the temperature is 50,000°C, it will emit mainly at short wavelengths, and appear bright blue. Strictly, no object is a black body, but the moon, sun, planets, and stars may approximate black bodies, and it is convenient to consider them as such in order to determine their temperatures by their general radiation.

Black hole: The theoretical final fate of a star with more than twice the mass of the sun. Its diameter is so small and its density and surface gravity so high that light cannot escape.

Celestial equator: the great circle on the celestial sphere cut by the projected plane of the earth's equator. It passes through the east and west points of the horizon for an observer at any place on the earth (except at the poles, where there is no east or west). At the earth's equator, it passes through the observer's zenith.

Celestial latitude: The angular distance of an object from the plane of the ecliptic measured in degrees towards the north or south ecliptic poles. Does not correspond to terrestrial latitude.

Celestial longitude: The angular distance one must go along the ecliptic eastward from the vernal equinox to the point on the ecliptic crossed by a great circle from the ecliptic pole through the object.

Celestial meridian: the great circle that passes through the zenith, the nadir, and the north and south points of the horizon. It and all other circles that pass through the zenith and nadir are called vertical circles.

Celestial poles: A line connecting the North Celestial Pole to the South Celestial Pole passes perpendicularly through the Celestial Equator.

Chromatic aberration: the failure of a lens made from a single piece of glass to bring all wave lengths of light to focus at the same distance. This produces a colored halo around the image. The fault can be partially corrected by making the lens from two pieces of glass of different densities. Such a lens is called *achromatic*.

Chromosphere: the "layer" of the sun directly outside the photosphere. It is reddish in color, chiefly because of hydrogen emission. The chromosphere is only a few thousand miles thick, but prominences may rise from it for hundreds of thousands of miles.

Cluster: a physically connected group of stars. They vary in size from what might be called multiple stars to globular clusters with hundreds of thousands of stars. Clusters such as the Pleiades that are located within our galaxy, near its central plane, are called *galactic clusters* or *open clusters*.

Coma: (a) the blur of light around the nucleus of a comet; the coma and the nucleus constitute the comet's head. (b) One of the defects in the image formed at the focus of a telescope.

Conjunction: the position in which one body of the solar system is most nearly in line with another, as seen from the earth. For example, the moon is in conjunction with the sun at the instant of the new moon. The planets Mercury and Venus are in *inferior conjunction* when they pass between the earth and the sun. They are in *superior conjunction* when they pass on the far side of the sun.

Constellation: all the stars in a certain limited region of the sky, no matter what their distance from us. We have inherited the concept of constellations from mythology, and it is convenient to use.

Corona: the outermost part of the sun. Except by use of a complicated apparatus, the coronagraph, it can be observed only during the moments of a total solar eclipse. It has been observed to extend a dozen million miles out from the photosphere. Its form changes markedly during the sunspot cycle.

Crater: in lunar studies, any depression of the moon's surface, other than a valley. The depression may have been caused by chance or by impact of a meteorite or an asteroid; or it may be a volcanic crater or a *caldera* (sunken area). The term *endocrater* is coming into use to denote any crater that has resulted from internal causes. *Ectocrater* has been suggested for any crater thought to have been formed by external causes. However, *impact crater* covers this group quite well, is self-explanatory, and has been in use for generations.

Day, apparent solar: the interval of time as measured by the sundial between successive passages of the sun across the meridian.

Day, lunar solar: the time interval equal to the synodic month. It varies slightly.

Day, lunar sidereal: the time interval equal to the sidereal month.

Day, mean solar: the average length of the apparent solar day throughout the year—24 hours. Astronomers use a longer, technical definition, but this short one is sufficient for almost all purposes. The mean solar day is, of course, the one used in everyday living.

Day, sidereal: the interval of time between the passage of a star across the observer's meridian and the next passage of the same star. It is approximately 3^m 56^s shorter than the mean solar day, and is the true period of rotation of the earth on its axis.

Declination: the angular distance measured, northward or southward, along the star's hour circle from the celestial equator to the star. On the celestial sphere, declination corresponds almost perfectly with latitude on the surface of the earth. It is not related to celestial latitude.

Doppler effect: the apparent change in frequency of waves when their source and an observer are approaching or receding from one another. If a train whistles when approaching us, the note sounds higher than it actually is; in other words, we receive more sound waves per second than the train actually emits. For receding sounds, exactly the reverse is true. This is the Doppler effect for sound. The same sort of effect holds true for light, but can be determined only by means of careful, quantitative measurements with a spectrograph. If a star is moving away from us, the lines of its spectrum all are shifted toward the red. If it is approaching us, they are shifted toward the blue end of the spectrum. Doubling the speed doubles the shift. For slow speeds, such as a dozen miles per second, the shifts are very small. By measuring the shifts of the lines of its spectrum, we can tell at how many miles per second a celestial object is approaching or receding, no matter how far away it may be.

Dwarf: any of the stars on the main sequence line of a modern Russell-Hertzsprung diagram. In this diagram (see pages 285–289), the farther a star is toward the right, the cooler it is; and the greater the total radiation from it per second, the higher it is in the diagram. The main sequence line, along which most of the stars are found, is a conspicuous diagonal line beginning near the top left. The stars above that line are called *giants;* those below it are *white dwarfs.* Dwarfs are average stars on the main sequence line. Those near the left end of the diagram are far more luminous than the average. Today the term *main sequence stars* is being used more and more.

Eccentricity: the mathematical constant that describes the shape of an elliptical orbit. A circle may be thought of as an ellipse with an eccentricity of zero; a parabola as one with an eccentricity of one; a hyperbola as one with an eccentricity greater than one.

Eclipse: (a) the passage of one body into the shadow of another. In a *lunar eclipse* the full moon passes into the shadow of the earth. If it enters completely, there is a *total lunar eclipse;* if not completely, a *partial lunar eclipse.* (b) If the new moon completely hides the sun as seen from any point of the earth, there is a *total solar eclipse.* For this to occur, the moon must be at a point in its orbit where its angular diameter is greater than that of the sun. If conditions are right for a total solar eclipse, except that the moon appears smaller than the sun, there is a thin ring of the sun seen around the moon. This is an *annular eclipse.* If the new moon does not completely block the sun from view for any point on the earth, but does hide part of the sun, there is a *partial eclipse.* Often total and annular solar eclipses are called *central eclipses.* The *eclipse year* is the interval of time the sun takes to move around the ecliptic from one of the moon's nodes to the same one again. Because the nodes of the moon's orbit move backward around the ecliptic, the eclipse year is shorter than the tropical year (see *Year*). It is about 346.62 days long and is important in eclipse predictions.

Ecliptic: The great circle cut in the celestial sphere by the extension of the plane of the earth's orbit. The ecliptic is inclined to the celestial equator at an angle of about 23° 27'. The *ecliptic poles* connect a line perpendicular to the ecliptic plane; they are thus 23° 27' from the *celestial poles.*

Elongation: the angular distance that an object is east or west of the sun, as measured around the ecliptic. For objects farther than the earth is from the sun, the elongation may have any value up to 180°. For Mercury and Venus, which are closer than we to the sun, there is a *greatest elongation.* This varies between 18° and 28° for Mercury, depending upon that planet's position in its quite eccentric orbit. For Venus, which moves in an almost circular orbit, the greatest elongation is close to 47°.

Equinox: either of the two points where the center of the sun crosses the celestial equator. The crossing point from south to north is the *vernal eqinox;* from north to south, the *autumnal equinox.* Another way to express the same fact is that the equi-noxes are the points where the ecliptic cuts the celestial equator. Writers often loosely substitute the times at which these events occur for the points. Spring begins at the instant when the sun passes the vernal equinox, and fall commences when it passes the opposite point.

Escape velocity: the outward velocity needed for an object to escape from a celestial body. For example, a rocket needs a velocity of about 25,000 miles per hour to escape the earth—that is, to keep from falling back to the earth. Although "popular" accounts often speak of a body escaping the earth's gravitation, this is not strictly true, because the earth's gravitational field pervades *all* space.

Exploding star: see *nova.*

Faculae: see *flocculi.*

Flare: a region, usually near a sunspot group, that suddenly brightens tremendously. When seen at the limb of the sun, flares become a type of prominence. The spectrograph shows hydrogen to be their most common element.

Flare stars: stars that in a minute or so double or more than double their brightness, then slowly revert to their usual brightness. It takes an hour or so to complete the cycle. Probably the phenomena are extreme cases of the flares that occur on our sun. All of the observed flare stars are dwarf, reddish ones, which are far less luminous than our sun. Even such flares as occur on our sun would, therefore, have a much greater percentage effect on the brightness of these stars.

Flocculi: clouds usually of very hot calcium gas found most often near sunspots. Some are of tremendous size. The photosphere is so bright that they can be observed only by special equipment, except when near the limb of the sun. There the darkening of the limb renders them visible to ordinary observation, and they are given the special name of *faculae.* Hydrogen flocculi are also observable.

Frequency: usually the number of waves per second of a given length of light emitted or received. (See *Doppler effect* for the reason why the number of waves emitted and received often differ.) The highest frequencies visible to the human eye are those of violet light—about 750,000,000,000,000 waves per second. The deepest red has a little more than half this frequency. Frequency is the characteristic that distinguishes light waves from radio waves, X-rays, and other electromagnetic waves.

Galactic latitude: Analogous to celestial latitude, but using the galactic plane (mean plane of the Milky Way) as the reference plane.

Galactic longitude: Analogous to celestial longitude, but measured along the galactic plane (mean plane of the Milky Way) to the east from the direction to the center of the galaxy.

Galaxy: any of the great communities of stars that occupy space. Our galaxy includes all the stars that belong to the Milky Way, plus the more than a hundred globular clusters that surround it. Our galaxy is estimated to have something like 100,000,000,000 or more stars and to have a diameter of about 100,000 light-years. (One light-year is 5,880,000,000,000 miles.) The nearest galaxies to our own, about 180,000 light-years away, are a pair of much smaller ones, called the Magellanic Clouds. At a little more than 2,000,000 light-years distance and just visible to the naked eye is the nearest other major galaxy, the Andromeda Galaxy. Galaxies are classified into three main types: *spiral galaxies*, *irregular galaxies*, and *elliptical galaxies*.

Giant: A star that is above and to the right of the main sequence in the Russell-Herzsprung diagram. See *dwarf*. A giant is a super-luminous star. Since its surface temperature is low, its surface area must be very great. Some giants have diameters up to 400 times the diameter of the sun.

Granules: Small whitish spots crowded together on the solar photosphere. Up to a thousand miles in diameter, they change form within a few minutes. Slightly hotter than the surrounding solar gases, they are the tops of rising columns of gas in the solar atmosphere.

Hour circle: any great circle that passes through the poles of the celestial sphere. Hour circles are perpendicular to the celestial equator. Their principal use is in the measurement of the positions of objects in terms of right ascension and declination.

Inclination: (a) the angle between two planes or, sometimes, the complement of this angle. For example: the inclination of the plane of the moon's orbit to the plane of the ecliptic is 5° 08′. The inclination of the orbit plane of the inner satellite of Mars to the Martian equator is 0° 57′. (b) Sometimes, when one of the planes is the equator of a body, the context makes it more convenient to tell the same fact as the inclination of the *axis of rotation* toward the second plane. For example: the obliquity of the ecliptic usually is defined as the angle of inclination of the earth's axis of rotation toward the plane of the ecliptic. It is 23° 27′ at the present time.

Insolation: the total amount of radiative energy from the sun received at any given place during a specified interval of time. It depends on the intensity of radiation and the length of time that it is received. For example: at the earth's poles, because of the low altitude of the sun, the intensity of the radiation received always is far less than it is in tropical regions at noon. However, the length of the polar day may more than compensate. Around June 21 at the north pole and December 21 at the south pole, the daily insolation is greater than it ever is at the equator.

Ionosphere: a moderately high layer of the earth's atmosphere, so named because it is the lowest layer for which ionization is a general characteristic. The ionosphere has great practical importance in our everyday lives because it is the principal agent in the reflection of radio waves back to the earth. Without such reflection, we could not have long-distance radio reception.

Libration: usually the real or apparent oscillation of the moon that allows us to see some of the hidden side; sometimes the oscillation of other celestial bodies.

Light-year: the distance light travels in one year— 5,880,000,000,000 miles. The light-year is the most commonly used unit for expressing the distances of bodies beyond the solar system.

Limb: the edge of the sun, moon, or a planet as seen from the earth. Probably the definition will soon be extended to read, ". . . seen by an observer, no matter where he may be."

Luminosity: the total amount of radiation emitted by a body. It does not include reflected radiation. Usually the term is applied to stars and commonly is measured in terms of the sun's luminosity as a unit. See *Magnitude*.

Magnetopause: The sharply defined boundary between the earth's magnetosphere and the flow of charged particles in the solar wind.

Magnetosphere: the extreme upper regions in the earth's atmosphere extending out to at least 40,000 miles above the surface and occupied by belts of charged particles trapped in the magnetic field of the earth.

Magnitude: the apparent brightness of a star. It is indicated by a number: the higher the number, the dimmer the star. *Absolute magnitude*, which measures intrinsic brightness, is the apparent magnitude a star would have if seen from a distance of 10 parsecs. Absolute visual magnitudes (which differ from photographic magnitudes) range from about -7 for the most luminous stars to $+19$ for the least luminous dwarfs. The sun is $+4.85$. By definition, a difference of 5 in magnitude equals a difference of 100 in *luminosity*.

Mare (plural **maria**): any of the large dark areas on the moon or Mars. Because the earliest observers believed such an area to be a body of water, they named it "mare," which is Latin for "sea." Apparently lunar maria are tremendous plains of lava with a thin covering of dust.

Meteor: the flaming light observed during the passage of a solid body, traveling at high speed, through the earth's atmosphere from outer space. The friction between the thin upper air and the body's surface raises the temperature of each so high that they become luminescent. Above roughly 100 miles the air is too thin for us to observe this phenomenon. Meteors enter our atmosphere with speeds between less than 10 and more than 40 miles per second.

Meteorite: a meteor that has landed as a solid on earth. The solid particle in the average meteor faintly visible to the naked eye compares in size with a grain of sand. Normally a meteorite is entirely evaporated by the heat of friction during its passage through the atmosphere and settles to earth as a fine dust. But sometimes considerably larger bodies land as solids on the earth's surface in a solid state. Some astronomers use the term meteorite also to designate bodies before they reach our air. Others speak of such a body, when it is moving about the sun before reaching the earth, as a *meteoroid*. Probably the principal difference between a meteorite and an asteroid is size. A meteoroid (meteorite) in space is too small to be observed by means available to us, whereas an asteroid can be observed.

Milky Way: the broad band of faint light that forms a great circle on the celestial sphere and that marks the central plane of our galaxy. The light is, principally, that of thousands of millions of stars, each of which is too faint to be seen individually by the naked eye.

Month, anomalistic: the interval required for the moon to pass from apogee (or perigee) to the same point again. It is important in distinguishing between predictions of total and annular solar eclipses. Its length is 27.55455 days, slightly longer than the sidereal month, because of a slow forward motion of the apogee.

360

Month, calendar: one of twelve unequal intervals into which the civil year is divided.

Month, nodical: the interval required for the moon to pass around its orbit from one node to the same node again. It is essential in eclipse predictions because the new, or full, moon must be rather close to a node for any kind of eclipse to occur. Its length of 27.21222 days is slightly less than the sidereal month because of a regression of the nodes around the ecliptic.

Month, sidereal: the interval required for the moon to pass around its orbit until it again reaches the same place with respect to the stars, as seen from the earth. This interval amounts to 27.32166 days.

Month, synodic: the interval from new moon to new moon. It averages 29.53059 days.

Motion: the process of changing place or position. *Linear motion* is the rate (e.g. number of kilometers per second) an object moves with respect to another object chosen as origin. *Angular motion* is the angle on the celestial sphere passed over in some arbitrary unit of time—as, for example, 13° per day eastward around the path of the moon's orbit. *Direct motion* is the eastward direction in which the earth moves around the sun. All planets have direct motion around the sun. However, some satellites move in *retrograde* orbits (from east to west) around their primaries, and some comets move in retrograde orbits around the sun. Because we observe them from the moving earth, the planets and comets *apparently* take up a *retrograde* motion among the stars when the earth is between them and the sun. When Mercury and Venus are near inferior conjunction they appear to *retrograde*, as seen from the earth. *Proper motion* is the slow change of a star's position against the stellar background because the star is moving in space. For most stars this motion is only a fraction of a second of arc per century. However, for some exceptional stars, the proper motion may be many times larger. For example, "Barnard's star" has a proper motion of 10.3 seconds of arc per year.

Nadir: the point on the celestial sphere through which a plumb line would pass if extended downward; it is exactly opposite the *zenith*. Since no two places on the earth have the same plumb line, each location has a different nadir and zenith.

Nebula: a cloud of gas or dust or a mixture of the two. Nebulae in the general vicinity of an extremely hot star become luminous and emit light of their own, most of it from hydrogen. Nebulae in the vicinity of cooler giant stars reflect some starlight. Such nebulae are called *bright nebulae*, in contrast to *dark nebulae*, which are not near any star and can be observed only because they hide the stars beyond them. The "dark rift" of the Milky Way is the most conspicuous dark nebula. Originally astronomers used the word nebula, which is Latin for mist or cloud, for any such appearing object in the sky. However, large telescopes resolved many of these "nebulae" into distant clusters of stars, and the spectroscope showed that others, not so resolved, must be either clusters or galaxies of stars. The last such use of the term was in speaking of certain galaxies as *spiral nebulae*.

Neutron star: The highly-collapsed, extreme-density remnant of a supernova explosion. A mile or two in diameter, its mass is nearly twice the mass of the sun. A rapid axial spin produces a pulsed emission observed as a *pulsar*.

Node: one of the two opposite points where great circles intersect each other. In astronomy, usually one circle is the ecliptic, and the other is the circle traced on the celestial sphere by the orbit of the moon, a planet, a comet, or an artificial satellite. The *ascending node* is the point where the object passes the ecliptic from south to north; the *descending node* is the reverse point. Often the Greek capital letter *omega*, Ω, is used to denote the ascending node, and the same letter, printed upside down, is used for the descending node.

Nova: Latin for a "new" star made conspicuous by the sudden explosion of a previously unobservable or inconspicuous star. For a short time it may give off more intense radiation than any other star, but within months or years it fades and becomes inconspicuous again. Most novae are of the *ordinary nova* type with less than, perhaps, a hundred-thousand-fold increase in light. A *supernova* may emit more light than all the rest of the stars in its galaxy. The most famous nova was Tycho's star, which he observed in 1572 and which became as bright as the planet Venus, observable in full daylight. The brightest during this century has been Nova Aquilae, in 1918, which rivaled Sirius, the brightest star. It seems fairly certain that the explosion is atomic, and it is credible that the white dwarfs have resulted from such explosions. Certain extreme, irregular, variable stars have been called *recurrent novae*. T Coronae Borealis is the most famous of these. Probably a star would require a mass at least a third greater than that of our sun ever to become a true nova.

Nutation: see **precession**.

Oblateness: the amount of flattening that occurs at the poles of a rotating body. For gigantic Jupiter, which rotates more rapidly than does any other planet, the oblateness is observable to the eye. In the case of the earth, the equatorial diameter is about 27 miles more than the polar diameter.

Occultation: the shutting off of light from one body by the interposition of another. Usually this involves the passages of the moon (or, rarely, a planet) between the earth and a star. Strictly speaking, the passage of the moon between the earth and the sun is an occultation, but we call it an eclipse. Occultation also is used for the hiding of a planet by the moon or of any object by a planet.

Opposition: a position halfway around the celestial sphere from the body of reference. For example, the full moon is in opposition to the sun. A superior planet, when nearest to us, misses being in opposition only because of the eccentricities of its orbit and that of the earth. Conjunction is the antonym of opposition.

Parallax: the apparent shift in position of a celestial body against the background of stars, caused by a shift in position of the observer. From the amount of shift, the distance of the celestial body can be calculated. The parallax of some asteroid passing close to the earth is measured from the ends of the earth's diameter, and, using Kepler's harmonic law, the distances of all other bodies that revolve around the sun can be computed. This is *planetary parallax*. We choose an asteroid for two reasons. First, some asteroids pass much closer to the earth than any other planets, which permits us to obtain a more accurate value, percentagewise, of the parallax. Secondly, because an asteroid appears merely as a point of light in the telescope we can measure its position more accurately. As we observe the parallax from the earth, the maximum value is found when it is on the horizon. This maximum value is the *horizontal parallax*. Radar measurements of distance now supplant the direct measurements which we always have used. For the closer stars we measure from the ends of the diameter of the earth's orbit rather than the diameter of the earth, and proceed as before. This is *stellar parallax*. Except for a few thousand of the nearest stars, we must compare stars with these closest ones by various theoretical means to obtain distances.

Parsec: the distance of an object that has a stellar parallax of one second. One parsec equals 3.259 light-years, or 19,160,000,000,000 miles. Both parsec and light-year are used in expressing stellar distances.

Perigee: the point in its orbit where the moon or an artificial satellite is nearest earth. See *apogee*.

Perihelion: the point at which a body moving around the sun is closest to the center of the sun. See *aphelion*.

Period: the time required for a body to move once around its orbit. The period as computed from gravitational theory is the *sidereal period*, except for negligible complications. The *synodic period* is the time required for a body to come again to the same position with respect to the sun, as seen from the earth. For example, Jupiter has a sidereal period of 11.86223 years, but a synodic period that averages only 399 days. This is the time required between successive passages of opposition to the sun as seen from the earth. See *month*.

Perturbation: usually, the departure of a planet or comet from elliptical motion around the sun, because of the attraction of other planets. The term however is sometimes used to denote various types of disturbances.

Phase (lunar or planetary): the ratio of lighted to unlighted surface, as viewed from the earth. It may be measured as the length of time since the "new moon" phase, or as the fraction of the surface illuminated. For planets such as Mars that are farther than the earth from the sun, there can be no "new moon" phase. Consequently for them phase is measured as the fraction of surface illuminated. Even for Mercury and Venus this is the better choice.

Photosphere: the lowest level of the sun that is visible to direct observation. It is called, loosely, the surface of the sun, and is the level from which we receive most solar radiation. The observed temperature of the photosphere varies from the center of the disk to the limb. At the center we actually can see through to a lower level and therefore measure a higher temperature. At the limb we are looking through a greater thickness of the cooler gases above the photosphere and cannot see as far into the sun.

Plains, lunar: flat areas on the moon. The largest of these are the maria (*see Mare*). Smaller but similar are the *mountain-walled plains* such as Ptolemaeus and Grimaldi. Mountain-walled plains are areas that are flatter than the terrain around them and that appear to have sunken from it, leaving a boundary wall. They are a type of crater. A *ringed plain* is a crater that exhibits evidence of strong initial explosions, whether impact or endoforce (internal) in nature. Ringed plains have external

as well as internal boundary walls, in contrast to mountain-walled plains, which may completely lack the external wall.

Poles: on the celestial sphere, the points cut by the extension of any chosen line. When the line is the axis of the earth, it cuts the sphere in the north and south *celestial poles*. If the line is perpendicular to the plane of the ecliptic, the poles are the *poles of the ecliptic*. If the line is perpendicular to the main plane of the Milky Way the points are the *galactic poles*. All three of these poles are much-used in astronomy. Also, a line through the center of any body cuts its surface in poles. The axis of the earth's rotation thus gives us the everyday poles of the earth. The axis of the earth's magnetic field cuts the surface in the north and south *magnetic poles*.

Precession: the cone-shaped motion traced in space by the axis of a rotating body when a force tends to tip the direction of the axis. The "wobbling" motion of a toy top is an example. The earth undergoes precession because the gravitational force of the sun tends to pull the plane of the earth's equator into the ecliptic, and the moon and even the planets exert smaller, similar forces. The greater the precessional force, the more rapid is the precession. The precessional period of the earth is approximately 25,800 years. Precession has no effect on the obliquity of the ecliptic or on the latitude of a place on the earth as is so commonly believed. It does, however, interchange the seasons and causes the north celestial pole to move in a 47° circle around the north pole of the ecliptic. Variations of the precessional force do cause slight variations in the obliquity of the ecliptic. These are called *nutations*.

Prominence: a great volume of incandescent gas, mainly hydrogen, on the sun. When observations are made by use of light from hydrogen, prominences show up as dark-appearing filaments, sometimes very long, against the bright photospheric background. When these filaments extend beyond the solar limb, they can be observed by appropriate instruments as bright, flamelike, red, outward extensions of the chromosphere. They have been divided into several sub-classifications among which *active* and *eruptive* prominences are the most spectacular.

Pulsar: A star producing regular sharp pulses of emission with periods ranging from 0.03 seconds to 4 seconds. It is probably a neutron star with a very strong magnetic field interacting with its surroundings to produce a highly-directional pulse, similar to a rotating light-house beacon.

Pulsation: usually, the variations of emission of radiation by variable stars; often referred to as *stellar pulsation*. The regularity of the pulsations divides such stars into regular, irregular, and semi-regular variables. A relationship between the periods of regular variables and their luminosities has enabled us to determine stellar distances that are too great to be measured directly by parallax.

Quasars: faint starlike objects with strong radio emission. If shifts of spectral lines to longer wavelengths are interpreted as Doppler shifts, they are receding with velocities up to 9/10 the velocity of light and are the most remote objects known.

Radial velocity: the speed of an object toward or away from the observer. It is measured directly by means of the *Doppler effect*. The experimental use of radar to measure radial velocity of the closest objects has been a recent development.

Radiant: the point at which the paths of meteors, projected backward, seem to meet. The meteors of a shower penetrate our atmosphere in nearly parallel paths. These apparent paths appear to meet at a distant point, just as parallel railroad rails seem to do as we look down the track.

Ray: a beam of light; also, one of the bright streaks that can be observed on the moon, especially near the full phase. Rays are related to lunar craters that appear to brighten under a high-altitude sun. Some astronomers believe that rays are caused by the impact of material expelled from the craters. Others believe that rays follow geological faults that have been either caused by or enhanced by terrific explosions in the craters and that the rays are the result of local exhalations from the faults.

Reversing layer: the part of the sun's atmosphere that produces the dark lines in the sun's spectrum. The light from a luminous gas appears in a spectroscope as a series of bright lines, the spectrum. The lines correspond to certain wave lengths of light, which are determined by the elements (or compounds) composing the gas. If the light passes through the same gas that emitted it, the gas absorbs the same wave lengths emitted. If the absorption is sufficient, as in the sun, the spectrum becomes a series of dark lines against the bright background of continuous radiation emitted by the sun's interior. It was once believed that a definite layer of the solar atmosphere produced the dark lines, but we now know that they are caused by the gases above the photosphere, principally by those of the chromosphere, but slightly by those of the corona. Nevertheless, the term *reversing layer* persists. See *spectrum.*

Right ascension: a coordinate that with *declination* forms the most commonly used system of celestial coordinates, the equatorial system. Nearly all modern stellar catalogs and planetary ephemerides are expressed in these coordinates. Right ascension is analogous to longitude on the earth, but not to celestial longitude. It is defined as the angular distance, measured eastward around the celestial equator from the vernal equinox to the foot of the hour circle that passes through the object. It may be expressed in either degrees of arc or in hours.

Rills: narrow lunar valleys that may be hundreds of miles long. They are found in all sorts of terrain, but never near the centers of maria despite being common near their shores.

Saros: an interval used by ancient astronomers for empirical prediction of eclipses. After one saros—about 19 years—we can expect repetition of an eclipse, although not in the same place on the earth.

Satellite: strictly, any body that moves under the control of another one. Actually, the planets are satellites of the sun, but the term usually is reserved for bodies moving around a planet.

Sea: see **mare.**

Siderite: a meteorite composed of a nickel-iron alloy. Compare with *aerolite* and *siderolite.*

Siderolite: a meteorite that is partly *siderite* and partly *aerolite.*

Solar constant: the amount of solar radiation received per second by an area one centimeter square at the distance of the earth from the sun. The constant equals 0.328 calories per square centimeter per second. (A *calorie* is the amount of energy required to raise the temperature of one gram of water one degree centigrade.) From a knowledge of the constant and of the area of a sphere with radius equal to the earth's distance from the sun, we can compute the amount of radiation emitted by the sun each second. Sometimes the solar constant is expressed in another unit called the *erg.* Actually the "constant" is not a constant, but undergoes slight fluctuations.

Solar system: the sun and all other bodies close enough to have their motions controlled by the sun's gravitational attraction. All such permanent bodies move in ellipses around the sun, although in the case of comets the orbital eccentricities may be large enough for the orbits to be almost parabolic.

Solar wind: charged particles ejected from the sun and flowing outward throughout the entire solar system. The particles typically travel at speeds between 200 and 400 miles per second.

Solstice: either of the two points on the ecliptic which are midway between the equinoxes. The sun passes the summer solstice at the instant summer begins for the northern hemisphere, about June 21. The name came from the fact that at the beginnings of summer and of winter the declination of the sun is changing very slowly. Before the invention of clocks the eastward motion was difficult to determine. Therefore, the ancients believed that the "sun stood still."

Spectral class: one of the groups into which the spectra of stars are divided according to their differences in composition. These depend principally on the temperature of the star's photosphere and to some extent on pressure and the chemical content of the stellar gases. The chief classes are designated *O, B, A, F, G, K, M, N.* In this list, each class is cooler than the preceding one. The peculiar order of the letters and the omission of some are the result of our ignorance of *spectral classes* late in the last century.

Spectrum: the colored pattern produced when light is passed through a spectroscope. The wave lengths are placed in order from the shortest at one end (the violet end) to the longest at the other (the red end). The dispersion into the separate wave lengths (colors) is produced either by a prism or by a "grating." In either case each bit of light is the image of a very narrow slit through which the light has entered. If all wave lengths are present the band is a continuously changing one from red to violet, as it is for light that comes from a luminous solid, from a liquid, or from a gas under too high a pressure, or for a gas of which the temperature is great enough to highly ionize it. This is the condition at and below the photospheres of stars. See *reversing layer.*

Spherical aberration: the failure of a spherical mirror to bring to a common focus point all the light from any one point of the light source. This defect is corrected by substitution of a paraboloid mirror for the spherical one.

Sunspot: a phenomenon of the sun, characterized by a dark area, the *umbra,* with a sharp boundary that separates it from a not-so-dark area, the *penumbra,* which in turn is separated by an equally sharp boundary from the bright photosphere. A sunspot is darker than the photosphere because its temperature is lower. Sunspots vary in number from a few in some years to many in others, their number fluctuating in a variable cycle that averages about eleven years. Some are visible to the naked eye.

Supernova: see **nova.**

Terminator: the boundary circle between day and night on the moon or a planet.

Transit: (a) the passage of the center of an object across the celestial meridian. *Upper transit* is a crossing made above the pole, and *lower transit* is the one made a half revolution later below the pole. Both transits are visible for circumpolar objects; indeed that fact may be used as a definition of such objects. (b) The passage of an object between the observer and a second object of much greater angular diameter. Examples: passages of Mercury and of Venus across the face of the sun; passage of a planet's satellite across the face of the planet.

Umbra: see **sunspot.**

Variable star: any star whose light varies or appears to vary in intensity. Apparent variation occurs in an *eclipsing variable.* Stars whose period of variation is almost constant are called *regular variables.* Others are *irregular variables* and *semi-regular variables.* Our sun is a slightly variable star. See *nova* and *flare stars.*

Wave: a disturbance traveling in a solid, liquid, or gaseous medium. Waves are of two general types: *longitudinal* and *transverse.* Sound is a longitudinal wave. The particles of the air or other medium oscillate backward and forward in the direction in which the sound is traveling. Waves in water are transverse waves. The particles of the medium (water) move up and down at the surface, but do not move in the direction the wave is traveling. If one generates a series of ripples in a pond and then watches a tiny stick which is floating on the surface, the truth of this statement becomes apparent at once. *Electromagnetic waves* are electromagnetic fields that move through space at 186,000 miles per second. They include all radiation from the longest radio waves, through the infrared, visible light, ultraviolet, X, and finally gamma rays, which are shortest of all.

Wave length: the distance traveled per second by a sound wave or electromagnetic wave, divided by the number of waves emitted per second by the wave source.

Year: the time interval required for the earth to move completely around the sun. The completion of the trip can be considered from several reference points, each of which denotes a different type of year useful for certain problems. (a) *Sidereal year:* the time required for one revolution with respect to the stars— about $365\frac{1}{4}$ mean solar days. It is the year obtained when we calculate the orbit of the earth. (b) *Tropical year:* the year of the seasons; the interval of time required by the earth to pass completely around the ecliptic, beginning and ending with the vernal equinox. This is the year used in everyday life. Because precession causes a small regression of the vernal equinox, the tropical year is shorter than the sidereal year by about $20^m\ 24^s$. (c) *Anomalistic year:* the time interval required for successive passages of the earth through perihelion. Since perihelion moves slightly forward in the ecliptic, this year is longer than the sidereal year by $4^m\ 43^s$. It is useful in theoretical astronomy. (d) *Eclipse year:* see *eclipse.*

Zenith: the point where a plumb line would cut the celestial sphere if extended upward. See *nadir.*

Zodiac: a band that extends 8° on each side of the ecliptic. Of the regular planets, only Pluto ever is seen outside it. The term "zodiac" has no use in modern astronomy.

Index